Bioenergetics Primer for Exercise Science

Primers in Exercise Science Series

Bioenergetics Primer for Exercise Science

Jie Kang, PhD

College of New Jersey

Human Kinetics

Library of Congress Cataloging-in-Publication Data

Kang, Jie.
  Bioenergetics primer for exercise science / Jie Kang.
    p. cm. -- (Primers in exercise science series)
  Includes bibliographical references and index.
    ISBN-13: 978-0-7360-6241-1 (soft cover)
    ISBN-10: 0-7360-6241-6 (soft cover)
 1. Energy metabolism. 2. Bioenergetics. I. Title.
  QP176.K36 2008
  612.3'9--dc22

                        2007023568

ISBN-10: 0-7360-6241-6
ISBN-13: 978-0-7360-6241-1

Acquisitions Editor: Michael S. Bahrke, PhD; **Developmental Editor:** Christine M. Drews; **Assistant Editors:** Laura Koritz, Katherine Maurer, and Melissa McCasky; **Copyeditor:** Joyce Sexton; **Proofreader:** Red Inc.; **Indexer:** Betty Frizzell; **Permission Manager:** Carly Breeding; **Graphic Designer:** Fred Starbird; **Graphic Artist:** Carol Smallwood; **Cover Designer:** Keith Blomberg; **Photo Asset Manager:** Laura Fitch; **Art Manager:** Kelly Hendren; **Associate Art Manager:** Alan L. Wilborn; **Illustrator:** Roberta Polfus; **Printer:** Total Printing Systems

Printed in the United States of America.   10   9   8   7   6

The paper in this book is certified under a sustainable forestry program.

Human Kinetics
Web site: www.HumanKinetics.com

United States: Human Kinetics
P.O. Box 5076
Champaign, IL 61825-5076
800-747-4457
e-mail: info@hkusa.com

Canada: Human Kinetics
475 Devonshire Road, Unit 100
Windsor, ON N8Y 2L5
800-465-7301 (in Canada only)
e-mail: info@hkcanada.com

Europe: Human Kinetics
107 Bradford Road
Stanningley
Leeds LS28 6AT, United Kingdom
+44 (0)113 255 5665
e-mail: hk@hkeurope.com

Australia: Human Kinetics
57A Price Avenue
Lower Mitcham, South Australia 5062
08 8372 0999
e-mail: info@hkaustralia.com

New Zealand: Human Kinetics
P.O. Box 80
Mitcham Shopping Centre, South Australia 5062
0800 222 062
e-mail: info@hknewzealand.com

E3612

To my wife, Julie, for her patience, understanding, encouragement, and love. To my children, Jason and Justin, for keeping my life balanced and giving me such joy. To my parents for their belief, direction, support, and love.

Contents

Preface   xi

Acknowledgments   xiii

Part I   Essentials of Bioenergetics 1

CHAPTER 1   Energy and Energy Metabolism 3

Energy.. 3
Energy Consumption 5
Energy Transformation 8
Energy Transformation in Sport and Physical Activity 14

CHAPTER 2   Metabolism of Macronutrients During Exercise 19

Carbohydrate 19
Lipid 24
Protein and Amino Acids 27

CHAPTER 3   Regulation of Energy Metabolism.......... 33

Overview of a Biological Control System................. 33
Neural and Hormonal Control Systems 35
Regulation of Substrate Metabolism During Exercise 38

Part II   Application of Bioenergetics in Physical Activity............47

CHAPTER 4   Measurement of Energy Metabolism 49

Laboratory Approaches 49
Field-Based Techniques 54
Subjective Measures............................. 60

CHAPTER 5   Energy Cost of Physical Activities and Sports 65

Principles of Energy Utilization During Exercise 65

Energy Cost During Various Exercises 69

Metabolic Calculation 75

CHAPTER **6** **Exercise Strategies for Enhancing
Energy Utilization 79**

Physical Activity and Energy Balance 79

Enhancing Energy Expenditure Through Exercise
and Physical Activity........................ 80

Exercise Intensity and Fat Utilization 83

Other Exercise Strategies 85

Limitations of Exercise Alone in Weight Management 87

Part III Bioenergetics in Special Cases . . 91

CHAPTER **7** **Metabolic Adaptations to Exercise Training . . 93**

Cellular Adaptations to Aerobic Training 93

Changes in Fuel Utilization 95

Responses of Oxygen Uptake and Endurance Performance . . 98

Adaptations to Anaerobic and Resistance Training 101

CHAPTER **8** **Influence of Gender and Age on Metabolism 105**

Gender Differences in Substrate Metabolism 105

Pregnancy 109

Exercise Metabolism in Elderly People 111

Energy Metabolism in Children and Adolescents 113

CHAPTER **9** **Energy Metabolism in People With Obesity
and Diabetes 117**

General Description of the Diseases 117

Insulin Resistance 120

Alterations in Metabolism During Exercise in Obesity
and Diabetes 123

Role of Exercise in Improving Insulin Sensitivity 125

Part IV Influences on Bioenergetics Apart From Physical Activity 129

CHAPTER 10 **Resting Metabolic Rate** 131

General Description of Resting Metabolic Rate 131

Measurement of Resting Metabolic Rate 132

Factors Influencing Resting Metabolic Rate 134

Influence of Exercise on Resting Metabolic Rate 136

Role of Resting Metabolic Rate in the Pathogenesis of Obesity 139

CHAPTER 11 **Thermic Effect of Food** 143

Thermic Effect of Food and Its Measurement 143

Influence of Protein, Carbohydrate, and Fat Consumption . . 144

Other Factors Influencing Thermic Effect of Food 146

Interaction Between Physical Activity and Thermic Effect of Food 148

Thermic Effect of Food and Obesity 149

CHAPTER 12 **Selected Pharmacologic and Nutritional Substances** 153

Sibutramine 153

Leptin 155

Ephedrine 156

Caffeine 159

Appendix: Answers to Review Questions   163

Glossary   173

References   181

Index   201

About the Author   209

Preface

Bioenergetics is one of the fastest-growing disciplines in the area of exercise science. It concerns issues related to energy storage and transformation, and in this regard ties directly to our ability to perform physical activity. Preserving energy or having efficient energy transfer is a gateway to the success of many athletic performances. On the other hand, in our contemporary world, filled as it is with conveniences, maximizing energy expenditure has been advocated as a top priority for each of us. In light of the fact that energy is needed to power muscular activity but that excessive energy storage can lead to obesity and other metabolic disorders, it is important to understand what energy transformation is all about, how energy expenditure can be assessed and quantified, and what strategies either can augment energy provision to enhance performance or can maximize energy utilization to facilitate weight loss or maintain optimal body weight.

This book was written in recognition of the fact that energy storage and transfer are essential to health, fitness, and sport performance. The book provides a contemporary review of a broad spectrum of materials related to energy metabolism. It brings together diverse issues that are of theoretical interest and of practical importance. As reflected in the table of contents, the book not only covers the physiological basis of bioenergetics but also provides practical information that can be readily used, such as guidelines for measuring energy expenditure and fuel utilization, choosing the right exercise program to maximize energy expenditure, working with energy balance equations, and selecting an appropriate nutritional supplement to facilitate energy expenditure. The 12 chapters are organized into four parts, each of which has a distinct focus. Part I conveys basic information that readers need to know before pursuing more advanced topics in bioenergetics; part II centers on ideas for applying information pertaining to bioenergetics; part III characterizes the uniqueness and specificity of energy transformation; and part IV emphasizes overweight and obesity and strategies for maximizing energy expenditure.

This book is unique in many respects. It encompasses all aspects of current information regarding energy metabolism. It combines theory with practice and serves as a "one-stop shop" for those who are interested in building their knowledge base and at the same time want to know how they may apply what they learn in their daily practice. To make learning more effective, each chapter includes an introduction and a summary, key points, and study questions. Also included in the book are a list of abbreviations (see inside front cover), a glossary of important terms used in the chapters, and an appendix that provides answers to the study questions in each chapter. A particularly novel aspect of the book is its inclusion of chapters on energy metabolism in special populations such as women, children, and elderly persons as well as in individuals with metabolic disorders.

Energy is what makes us function, but energy in excess can be detrimental. It is perhaps this dilemma that makes bioenergetics an exciting area to study. It is hoped that this volume will provide readers with a wide array of perspectives on energy and energy transformation and perhaps also entice some readers to continue their academic pursuits in this growing branch of science.

This book is the second volume in Human Kinetics' Primers in Exercise Science Series, which provides students and professionals alike with a nonintimidating basic understanding of the science behind each topic in the series, and where appropriate, of how that science is applied. These books, written by leading researchers and teachers in their respective areas of expertise, present, in an easy-to-understand manner, essential concepts in dynamic, complex areas of scientific knowledge. The books in the series are ideal for researchers and professionals who need to obtain background in an unfamiliar scientific area or as an accessible basic reference for those who will be returning to the material often.

Acknowledgments

This book would not be possible without the contributions of the dedicated team of professionals at Human Kinetics. In particular, I would like to thank Mike Bahrke for his support and advice during my early pursuit of this writing idea, Chris Drews for her thoughtful and constructive feedback in developing the manuscript, and Joyce Sexton for her editorial expertise in bringing this edition to completion. Special appreciation must also go to those who have influenced my career and taught me the excitement of bioenergetics and exercise metabolism: William McArdle, Mike Toner, Robert Robertson, Fredric Goss, James Hagberg, and David Kelley. Finally, I would like to express my gratitude to the many colleagues that I am fortunate to have worked and collaborated with: Alan Utter, Jay Hoffman, Edward Chaloupka, Nicolas Ratamess, and Avery Faigenbaum. Our regular interaction and exchange of ideas has been a tremendous inspiration for me in completing this writing project.

I

ESSENTIALS
OF BIOENERGETICS

All forms of human movement are energetic events. We need energy in order to carry out various physiological functions that are essential to our life. Foods are the source of the energy we use each day, but energy stored in foods needs to be transformed into the type that is usable within the biological system. The rate at which energy transformation occurs can vary greatly depending on the demands placed on the body. However, because such variation is under biological control, it is unlikely that we over- or underproduce energy.

Part I serves as a foundation for the entire volume. It conveys basic but important information needed by anyone who plans to pursue more advanced topics in bioenergetics. The three chapters are arranged in an order that should facilitate understanding of the materials. Chapter 1 provides an overview of various energy fuels and metabolic pathways. Chapter 2 demonstrates more specifically how metabolic pathways operate during exercise of varying intensity and duration. Chapter 3 explains in greater depth how the metabolic pathways are intrinsically regulated.

1

ENERGY AND ENERGY METABOLISM

This chapter begins with the concept of energy and then discusses the three energy-containing nutrients—carbohydrate, fat, and protein—with particular reference to the bioenergetic roles they play in support of cellular function and human movement. The chapter also deals with the principle of energy transformation and the various metabolic pathways by which energy stored in the energy-containing nutrients is converted to biologically usable energy in daily life and for sports and physical activities.

ENERGY

All forms of physical activity in daily life and athletics are energetic events. Therefore it is imperative to understand what energy is and how the body acquires, converts, stores, and utilizes energy. **Energy** is defined as the ability to produce change and is measured by the amount of work performed during a given change. Unlike the physical properties of matter, energy cannot be defined in concrete terms of size, shape, or mass. The presence of energy is revealed only when change occurs. Energy is neither created nor destroyed. It exists in many forms that can be converted from one to another. For example, the energy in flowing water can be converted into energy for a lightbulb. In the body, energy is first obtained from energy-containing nutrients in food and, in most circumstances, then converted to potential energy stored in the body tissues. Via cellular respiration, this potential energy is then converted to the high-energy compound **adenosine triphosphate** (ATP), as well as heat. The energy in ATP is used for a variety of biological work, including muscle

contraction, synthesis of molecules, and transportation of substances. That energy is neither created nor destroyed during any physical or chemical process is one of the most important axioms of science, the **law of the conservation of energy.** With reference to the human body, this law dictates that the body does not produce, consume, or use up energy; it merely transforms energy from one state to another.

Units of Energy

Within a biological context, energy is measured in **joules** (J) or **kilojoules** (kJ), which are units of work, or in **calories** (cal) or **kilocalories** (kcal or Cal), which are units of heat. The joule is named for the British scientist Sir Prescott Joule (1818-1889). A kilojoule is the amount of work required to move a 1-kilogram object a distance of 1 meter under the force of gravity. In Europe and most parts of Asia, the joule or kilojoule is the standard measure of energy in food and the body. In the United States and Canada, however, the calorie or kilocalorie is the measure most commonly used. In theory, a kilocalorie is the amount of heat required to raise the temperature of 1 kilogram of water by 1 degree Celsius. Any measure in kilocalories or kilojoules is 1000 times greater than that in calories or joules, respectively. To convert calories to joules or kilocalories to kilojoules, the value in calories or kilocalories needs to be multiplied by 4.186; that is, 1 cal = 4.186 J or 1 kcal = 4.186 kJ.

Potential and Kinetic Energy

In exercise science, the form of energy that powers muscle contraction is often considered **mechanical**

energy, and activities such as walking, running, swimming, jumping, and throwing require the production of mechanical energy. This form of energy is possessed by an object due to its motion or its position or internal structure. Mechanical energy can be either **kinetic energy** (energy of motion) or **potential energy** (energy of position). For example, a book on a shelf has stored potential energy. By stretching a rubber band, you give it potential energy. Kinetic energy, on the other hand, is illustrated when individuals perform physical activity. Think of a gymnast on the balance beam; the movements and flips that the gymnast does display kinetic energy. When you are running, walking, or jumping, your body is exhibiting kinetic energy. Both forms of energy can exist at the same time, but one form often changes to the other. For example, the water at the top of a waterfall has stored potential energy. Once the water leaves the top of the waterfall, the potential energy is changed to kinetic energy. Within a biological system, such a transfer of energy can be exemplified as the release of the energy stored in energy-containing nutrients through **catabolism,** a process in which more complex substances are broken down into simpler ones. In this case, the released potential energy is transformed into kinetic energy of motion. On the other hand, **biosynthesis** may be viewed as a reverse process in which energy in one substance is transferred to other substances so that their potential energy increases.

Biologically Usable Energy

In a living cell, ATP is the most important carrier of the energy needed for the performance of many complex functions. This energy-containing compound stores potential energy extracted from food and can yield such energy to power various biological activities via **hydrolysis,** a process in which a compound is split into other compounds by reacting with water. Adenosine triphosphate is the only form of chemical energy that is convertible into other forms of energy used by living cells. As such, ATP is often regarded as an energy currency. Fats and carbohydrate are the main storage forms of energy in the body. However, energy derived from the oxidation of these two fuels is not released fast enough to meet the energy demand of activities that are short and explosive. The liberation of energy from food is a rather complex process that is well controlled by enzymes and that takes place within the watery medium of the cell. But because of the production of ATP, the slowness of energy transformation from foods is not a concern. Adenosine triphosphate may be viewed as a temporary reservoir that functions to provide instant energy to the cells whenever it is needed.

Adenosine triphosphate has a three-part structure: (1) an adenine portion, (2) a ribose portion, and (3) three linked phosphates (figure 1.1). The formation of ATP, which occurs when **adenosine diphosphate** (ADP) and inorganic phosphate (P_i) combine, requires a considerable amount of energy. A portion of this energy is then stored as potential energy in the chemical bond that links P_i to the adenosine group. During hydrolysis, **adenosine triphosphatase** (ATPase) catalyzes the reaction when ATP joins with water. In the degradation of 1 mol of ATP, the outermost phosphate bond splits and liberates approximately 7.3 kcal of free energy that is available for work. This results in the production of ADP and P_i. In some cases, additional energy is released when another phosphate splits from ADP, and this results in the production of **adenosine monophosphate** (AMP). The energy liberated during the breakdown of ATP transfers directly to other energy-requiring molecules. In muscle, for instance, the energy is used to power the myosin cross-bridge, causing the muscle fiber to shorten. The splitting of an ATP molecule takes place immediately before muscle contraction begins and does not require oxygen.

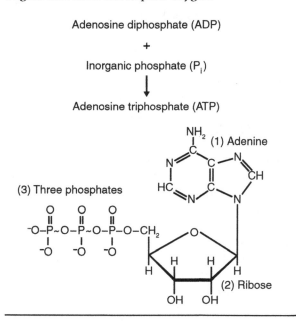

Figure 1.1 An adenosine triphosphate (ATP) molecule. The symbol "~" represents energy stored in the phosphate bond.

Reprinted from S.K. Powers and E.T. Howley, 2004, *Exercise physiology: Theory and applications to fitness and performance,* 5th ed. (New York, NY: McGraw-Hill), 31, by permission of The McGraw-Hill Companies.

The body can store only a very limited amount of ATP; most activities are powered by ATP produced mainly through the oxidation of carbohydrate and fat. An example of a time when the body relies on its stored ATP is during breath holding, a short sprint, or lifting. Chemical processes in which ATP is formed from other energy fuels are discussed in detail later in this chapter.

> ### ▶ K E Y   P O I N T ◀
>
> Adenosine triphosphate serves as the body's energy currency, although its quantity is very limited. The free energy yielded by the splitting of the phosphate bond of ATP powers all forms of biological work. In most activities, ATP is generated instantly from the degradation of carbohydrate and fat.

ENERGY CONSUMPTION

The energy needed to fuel the body comes from the food we eat as well as from the energy already stored in the body. Carbohydrate, fat, and protein are the three energy-containing nutrients that people consume regularly. Upon entering the body, these macronutrients undergo a series of hydrolytic reactions including the digestion of starches and **disaccharides** to **monosaccharides,** protein to amino acids, and lipids to glycerol and fatty acids. These simpler forms of the macronutrients are then absorbed and assimilated via the **hepatic-portal vein,** which routes blood from the capillary beds of the gastrointestinal tract into the liver. While some of these molecules are used to meet immediate energy needs of the body, others are stored as potential energy in more complex forms, such as **glycogen** in muscle and liver and **triglycerides** in muscle and adipose tissue. The amount of energy taken in depends on the total amount of food consumed and the nutrient composition of the food.

Measurement of Energy Content of Foods

The energy content of food can be measured through the use of a bomb calorimeter, which consists of a sealed steel chamber surrounded by a jacket of water (figure 1.2). An amount of food of known weight is placed in the chamber under high oxygen

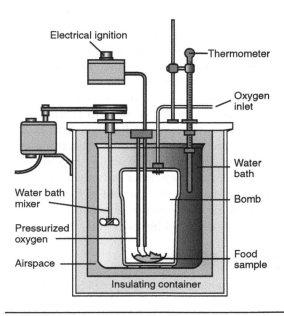

Figure 1.2 A bomb calorimeter. When dried food undergoes combustion inside the chamber of a bomb calorimeter, the rise in temperature of the surrounding water can be used to determine the energy content of the food.

Reprinted, by permission, from A. Jeukendrup and M. Gleeson, 2004, *Sports nutrition: An introduction to energy production and performance* (Champaign, IL: Human Kinetics), 63.

pressure, and an electrical current is used to start the reaction. As the food undergoes combustion, heat is produced and transferred through the metal wall of the chamber and heats the surrounding water. The increase in water temperature can be used to calculate the amount of energy in the food on the basis that 1 kcal is the amount of heat needed to increase the temperature of 1 kg of water by 1 °C. For example, if the water volume surrounding the chamber is 5 L and the temperature of the water rises by 2 °C, then the amount of energy contained in the food is 5 × 2 = 10 kcal (or 10 × 4.186 = 41.86 kJ). If the mass of the food that underwent combustion was 5 g, then the energy density of the food was 10 / 5 = 2 kcal/g.

This method determines quite accurately the total energy content in foods. However, it is not without drawbacks (Smolin and Grosvenor 2003). The technique is expensive to use and provides no information as to the composition of carbohydrate, fat, and protein in the food that underwent combustion. Because the body cannot completely digest, absorb, and utilize all of the energy in a food, caloric values from this technique are often slightly higher than the amount of energy the body can actually obtain from the food. This pertains particularly to proteins, because the body cannot oxidize the nitrogen component of **amino acids,** the building blocks of a protein. Consequently, nitrogen atoms combine with hydrogen to form **urea** to be excreted via the kidneys. Because energy is stored in the hydrogen bond, such a loss of hydrogen results in a reduction in the energy of an amino acid that is available for use (McArdle et al. 2005). Quantitatively, the energy the body can actually obtain from 1 g of protein consumed is about 4.6 kcal on average, rather than 5.65 kcal as measured by the bomb calorimeter. This is a loss of approximately 20% of the potential energy stored in a protein molecule. As neither carbohydrate nor fat contains any nitrogen, the amount of fuel the body acquires from each of these two nutrients is similar to the amount determined by the bomb calorimeter.

Digestive Efficiency

How much energy stored in foods can become available to the body is also affected by the efficiency of the digestive process. Digestive efficiency, often defined as the coefficient of digestibility, is the percentage of ingested food that is digested and absorbed to serve the body's metabolic needs (Jeukendrup and Gleeson 2004). A coefficient of digestibility of 50% means that only half of the energy consumed was ultimately absorbed. As this digestive parameter provides information about how much energy from the food consumed is actually available inside the body, it has become a major factor guiding the design of dietary programs for weight loss or maintenance. The coefficient of digestibility in both lipids and carbohydrates is relatively higher than in proteins, reaching 90% or more. However, carbohydrate products containing dietary fiber have lower digestibility. Therefore, consuming carbohydrate rich in fiber helps in reducing the amount of energy available to the body. According to early data published in a U.S. Department of Agriculture handbook (Merrill and Watt 1973), for instance, the coefficient of digestibility of wheat bran carbohydrate is only 56%, suggesting that the body will obtain only a little over half of the energy stored in this food. Protein has a greater range of coefficients of digestibility (i.e., 80-97%). The reason is that a protein molecule can vary in terms of its constituent amino acids or its food source. In general, the coefficient of digestibility is lower in plant protein than in protein from animal sources.

Table 1.1 shows coefficients of digestibility, combustion heats, and net energy values for nutrients in various food groups. As shown in the table, the average coefficients of digestibility for protein, lipids, and carbohydrate are 92%, 95%, and 97%, respectively.

For lipids and carbohydrate, the net energy values are identical to the product of the coefficient of digestibility and heat of combustion. However, for protein, the net energy value is much lower than the product of the coefficient of digestibility and heat of combustion (i.e., 4.05 vs. 5.20 kcal/g). This difference is explained by the fact that some of the energy stored in amino acids is lost due to the production of urea, which incorporates the hydrogen bond.

Atwater General Factors

Conveniently, average net energy values can be rounded to simple whole numbers, often referred to as Atwater general factors. The Atwater factors for carbohydrate, lipid, and protein, respectively, are 4, 9, and 4 (1 g of carbohydrate = 4 kcal; 1 g of lipid = 9 kcal; and 1 g of protein = 4 kcal). The Atwater factors are named for Wibur Olin Atwater (1844-1907), an American chemist. They provide a viable and fairly accurate means of estimating net energy consumption. They can be used to determine the caloric content of any portion of food or of an entire meal from the food's composition and weight. As a result of the application of the Atwater general factors, virtually all food items on the market today are labeled with an overall and nutrient-specific energy content. Table 1.2 illustrates how these factors are used to calculate the caloric values of chocolate chip ice cream.

Bodily Energy Stores

Energy is stored in the body primarily as fat in the form of triglycerides, though a much smaller amount is also stored as glycogen in the muscle and liver. The body must have a steady supply of energy, and some of it comes from glucose, the simplest form of carbohydrate. As we eat, energy is supplied by the food. Between meals, the breakdown of stored glycogen and fat helps in meeting energy needs. If no food is eaten for more than several hours, the body must shift the way it uses energy to ensure that **glucose** continues to be available. It accomplishes this by increasing the use of stored fat and by mobilizing liver glycogen. The maintenance of blood glucose is of particular importance to the survival and functioning of the central nervous system. As shown in table 1.3, glycogen stores are limited; for an 80 kg (176 lb) person the body contains approximately 500 g glycogen, which in theory could be depleted within several hours of strenuous exercise. When glycogen stores decrease significantly, protein degradation,

Table 1.1 Digestibility, Heat of Combustion, and Net Energy Values of Protein, Lipid, and Carbohydrate

Food group	Digestibility (%)	Heat of combustion (kcal/g)	Net energy (kcal/g)
Protein			
Meat and fish	97	5.65	4.27
Eggs	97	5.75	4.37
Dairy products	97	5.65	4.27
Cereals	85	5.80	3.87
Legumes	78	5.70	3.47
Vegetables	83	5.00	3.11
Fruits	85	5.20	3.36
Overall average	92	5.65	4.05
Lipid			
Meat and eggs	95	9.50	9.03
Dairy products	95	9.25	8.79
Vegetable foods	90	9.30	8.37
Overall average	95	9.40	8.93
Carbohydrate			
Cereals	98	3.90	3.82
Legumes	97	4.20	4.07
Vegetables	95	4.20	3.99
Fruits	90	4.00	3.60
Sugars	98	3.95	3.87
Animal foods	98	3.90	3.80
Overall average	97	4.15	4.03

Adapted from A.L. Merrill and B.K. Watt, 1973, *Energy value of foods: Basis and derivation*, Agriculture Handbook No. 74 (Washington, DC: United States Department of Agriculture).

Available: http://www.nal.usda.gov/fnic/foodcomp/Data/Classics/ah74.pdf

Table 1.2 Method for Calculating Caloric Value of Food From Its Composition of Macronutrients

Ice cream (100 g or 3.5 oz)	Composition (%)	Weight (g)	Atwater factor	Calories (kcal)
Protein	3	3	4	12
Lipid	18	18	9	162
Carbohydrate	23	23	4	92
Water	56	56	0	0
			Total calories	266
			% kcal from lipids	162 / 266 = 61%

Note: Calories = weight (g) × Atwater factors (kcal/g).

Adapted from W.D. McArdle, F.I. Katch and V.L. Katch, 2005, *Sports and exercise nutrition*, 2nd ed. (Baltimore, MD: Lippincott, Williams & Wilkins).

Table 1.3 Availability of Energy Substrates in the Human Body

Substrate	Weight (g)	Energy (kcal)
Carbohydrate		
Muscle glycogen	400	1600
Liver glycogen	100	400
Plasma glucose	3	12
Total	503	2012
Lipids		
Adipose tissue	12,000	108,000
Intramuscular triglycerides	300	2700
Plasma triglycerides	4	36
Plasma fatty acids	0.4	3.6
Total	12,304	110,740

Note: These values were estimated based on an average 80 kg (176 lb) man with 15% body fat.

Adapted from W.D. McArdle, F.I. Katch and V.L. Katch, 2005, *Sports and exercise nutrition*, 2nd ed. (Baltimore, MD: Lippincott, Williams & Wilkins), and A.J. Vander, J. Sherman and D.S. Luciano, 1985, *Human physiology: The Mechanisms of Body Function*, 4th ed. (New York, NY: McGraw-Hill).

which produces amino acids, increases. Some amino acids are converted to glucose, while others are directly metabolized for energy. Protein is not stored as an energy fuel in the body. It serves mainly as a structural component of muscle tissue as well as many other organs. Thus, the breakdown of protein to produce glucose and hence energy can result in the loss of muscle and other lean tissues.

Of the three energy-containing nutrients, the fat molecule carries the largest quantities of energy per unit weight. The reason is the greater quantity of hydrogen in the lipid molecule. In a well-nourished individual at rest, catabolism of lipids provides more than 50% of the total energy requirement (Vander et al. 2001). Although most cells store small amounts of fat in their cytosol, most of the body's fat is stored in specialized cells known as **adipocytes,** which function to synthesize and store triglycerides during periods of food intake. As shown in table 1.3, the potential energy stored in fat molecules for an 80 kg (176 lb) individual equals 110,740 kcal. Given an energy expenditure of 100 kcal per mile, this amount of energy could fuel a run of over 1100 miles (1770 km). This contrasts sharply with the limited 2000 kcal of stored carbohydrate, which could fuel only a 20-mile (32 km) run. During prolonged energy restriction, substantial amounts of fat are used to provide energy. However, when the supply of glucose is limited, as during starvation or in the diabetic state, **fatty acids** cannot be completely oxidized and chemical **ketones** are produced. Ketones are by-products produced mainly in the mitochondrial matrix of liver cells when carbohydrate is so scarce

that energy must be obtained from the breakdown of fatty acids. Ketones can be used as an energy source by many tissues. In sustained starvation, even the brain adapts to meet some of its energy needs by utilizing ketones (Powers and Howley 2001).

> ### ►K E Y   P O I N T◄
>
> Carbohydrate, fat, and protein are the three energy-containing nutrients consumed daily. Compared to fat, the amount of carbohydrate stored as glycogen is rather limited. However, carbohydrate is a preferable source of energy. Protein contains energy but contributes little to energy metabolism. Carbohydrate and protein each provide about 4 kcal of energy per gram, compared with about 9 kcal per gram for fat.

ENERGY TRANSFORMATION

Energy transformation occurs in both living and nonliving systems and is essential to life. As mentioned earlier, the transformation of energy from one form to another follows the law of the conservation of energy. This law states that energy is neither created nor destroyed, but instead is transformed from one state to another without being used up. For example, in photosynthesis, solar energy is harnessed by plants, which take in carbon, hydrogen, oxygen, and nitrogen from their environment and manufacture carbohydrate, fat, or protein. In the body, via cellular respiration, the energy possessed by macronutrients is changed into chemical energy, which is then stored within energy substrates or converted to mechanical and heat energy. The body stores energy in a variety of chemical compounds, including ATP, **phosphocreatine** (PCr), glycogen, and triglycerides. As an energy currency, ATP can be readily used to meet immediate energy needs. However, this high-energy compound is stored in a limited quantity. In fact, the body stores only 80 to 100 g of ATP at any one time (McArdle et al. 2005). This provides energy that can sustain maximal exercise, such as a 60 yd (55 m) sprint, the high and long jump, base running, and football play, for only several seconds. Consequently, in most sporting events and daily physical activities, ATP is replenished continuously through a series of chemical reactions involving energy transformation. Three distinct energy systems have been identified as playing a role in replenishing ATP: the ATP-PCr system, the glycolytic system, and the oxidative system.

> ▶ **K E Y   P O I N T** ◀
>
> The potential energy stored in nutrients is captured through three energy-yielding systems: (1) the ATP-PCr system, (2) the glycolytic system, and (3) the oxidative system. The operation of these systems is of essence to the continual supply of ATP in support of various biological functions.

ATP-PCr System (Phosphagen System)

The ATP-PCr system is also known as the phosphagen system because both ATP and PCr contain phosphates. This system, which serves as the immediate source of energy for regenerating ATP, has three components. First, there is ATP itself. This high-energy compound, stored in the muscles, rapidly releases energy upon the arrival of an electrical impulse. Adenosine triphosphate is degraded to ADP by the enzyme ATPase. Because the reaction involves combination with H_2O, the splitting of ATP is often referred to as hydrolysis. This process can be illustrated as follows:

$$ATP \xrightarrow{\text{ATPase}} ADP + P_i + energy$$

The second player in the system is PCr, another high-energy compound that exists in five to six times greater concentrations in muscle than does ATP (Brooks et al. 2005). Unlike what happens with ATP,

energy released by the breakdown of PCr is not used directly to accomplish cellular work. Instead, PCr provides a reserve of phosphate energy that is used to regenerate ATP as a result of muscle contraction and to prevent ATP depletion. In this process, ADP combines with P_i to become ATP using the bonding energy stored in PCr. This reaction is catalyzed by the enzyme **creatine kinase.** The process can be illustrated as follows:

$$PCr + ADP \xrightarrow{\text{creatine kinase}} ATP + Cr$$

The third component of the system involves ADP and the action of the enzyme **adenylate kinase** (or myokinase with reference to muscle). This enzyme catalyzes the production of one ATP (and one AMP) from two ADPs:

$$ADP + ADP \xrightarrow{\text{adenylate kinase}} ATP + AMP$$

The three components of this immediate energy system and the respective kinase enzymes are all water soluble. As such, they exist throughout the aqueous part of the cell and in close proximity to the contractile elements of the muscle outside of the **mitochondria.** They can be immediately available to support muscle contraction. With some ATP resynthesized from PCr, this system is able to fuel all-out exercise for approximately 5 to 10 s, as in a 100 m sprint. It is frequently observed that during the last few seconds of a 100 m race, runners tend to slow down. If maximal effort continues beyond 10 s, or if more moderate exercise continues for longer periods, ATP replenishment requires energy sources in addition to PCr.

Glycolytic System (Glycolysis)

The energy used by the glycolytic system to replenish ATP needed by cells is limited to the energy stored in carbohydrate molecules such as glucose or glycogen. This system is also referred to as **glycolysis.** Glycolysis involves a cascade of chemical reactions, each catalyzed and regulated by a specific enzyme. As shown in figure 1.3, glycolysis produces pyruvic acid. The production of pyruvic acid occurs regardless of whether or not oxygen is available. However, the availability of oxygen determines the fate of pyruvic acid. When oxygen is lacking, pyruvic acid is converted to lactic acid. Glycolysis is also referred to as the Meyerhof pathway in honor of the German biochemist Otto Fritz Meyerhof (1884-1951), who

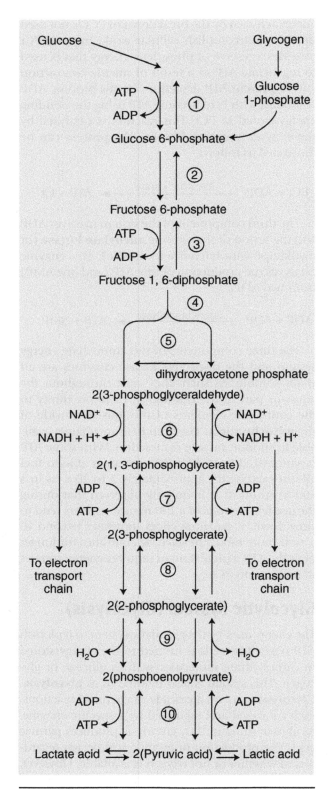

Figure 1.3 Summary of the glycolytic pathway in which glucose or glycogen is degraded into pyruvic acid.

Reprinted, by permission, from J.H. Wilmore and D.L. Costill, 2004, *Physiology of sport and exercise*, 3rd ed. (Champaign, IL: Human Kinetics), 125.

was awarded the Nobel Prize in Medicine in 1922 for his discovery of the pathway. Glycolysis can be summarized as follows:

$$\text{Glucose} \rightarrow 2\ \text{ATP} + 2\ \text{lactate}^- + 2\ \text{H}^+$$

Glycolysis requires 12 enzymatic reactions for the breakdown of glycogen to lactic acid (1 for glycogen to become glucose, 10 for glucose to become pyruvic acid, and 1 for pyruvic acid to become lactic acid). Because of its more complex chemical pathway, this system is relatively slower in generating ATP as compared to the ATP-PCr system. This energy-yielding pathway is similar to the ATP-PCr system in that both systems form ATP in the absence of oxygen and function in the watery medium of the cell outside the mitochondria.

The oxygen-independent glycolytic system works predominantly within skeletal muscle tissue. This is especially the case in muscles consisting primarily of fast-twitch (i.e., type IIb) fibers. This type of muscle fiber contains considerable amounts of glycolytic enzymes. In muscle, glycogen is usually first broken down into glucose molecules via a process called **glycogenolysis.** These individual glucose molecules are then able to enter the glycolytic pathway. The glycolytic pathway also allows the entry of glucose derived from liver glycogenolysis and transported via the circulation. At the onset of glycolysis, ATP is used for the conversion of glucose to glucose-6-phosphate, a compound that is needed in order for the pathway to proceed. This reaction is followed by another energy-requiring reaction in which fructose 6-phosphate is converted to fructose 1,6-diphosphate. During the later reactions in glycolysis, the energy released from glucose intermediates stimulates the direct transfer of a phosphate bond to ADP. This results in the production of up to four ATPs. Because two ATPs are lost in the initial steps of **phosphorylation** using ATP, for each glucose molecule entering the pathway the system generates a net gain of two ATPs. The process in which energy is transferred from an energy substrate to ADP via a phosphate bond that does not require oxygen is called substrate-level phosphorylation.

The glycolytic system is rather inefficient in terms of how much of the energy stored in a glucose molecule can result in ATP resynthesis. In fact, the amount of ATP produced from anaerobic glycolysis is only 5% of the amount that a glucose molecule is capable of generating. In addition, this pathway is associated with the production of lactic acid, which may be involved in the onset of fatigue. This by-product of glycolysis can release hydrogen ions that increase acidity within the muscle cell, thereby disturbing the normal internal environment necessary for maintaining muscle contraction as well as other physiological functions. The glycolytic system, however, has the advantage of replenishing ATP more rapidly than the oxidative pathway. With this system, most cells are able to withstand very short periods of low oxygen by using anaerobic glycolysis. Consequently, the glycolytic system plays a major role in fueling sporting events in which energy production is near maximal for 30 to 120 s, such as 200 to 800 m runs. There are special cases in which glycolysis supplies most, and in some cases all, of the ATP that a cell needs to survive and function. For example, red blood cells contain the enzymes for glycolysis but have no mitochondria; all their ATP production occurs by glycolysis. Also, as mentioned earlier, fast-twitch muscle fibers contain considerable amounts of glycolytic enzymes but have few mitochondria. During intense exercise, these muscle fibers rely mainly on ATP derived from glycolysis.

Oxidative Pathway

Most of the energy used daily comes from the oxidation of carbohydrate, lipid, and, in rare cases, protein consumed in the diet. This aerobic production of energy occurs within mitochondria, often called the "powerhouses of the cell." Mitochondria are found scattered throughout the cytoplasm. As shown in figure 1.4, mitochondria are oval-shaped bodies surrounded by two membranes, and their internal space or matrix contains numerous enzymes that are capable of catalyzing oxidative energy transformation. As mentioned earlier, the glycolytic system captures only a very small portion of the energy stored in a glucose molecule. However, the oxidative system makes it possible for the remaining energy to be extracted from the glucose molecule. This is accomplished via the conversion of pyruvate into **acetyl-CoA** (acetylcoenzyme A) rather than lactic acid, which is possible when oxygen is sufficient. Acetyl-CoA can then enter the citric acid cycle, also known as the Krebs cycle in honor of biochemist Hans Krebs (1900-1981), who received the 1953 Nobel Prize in Physiology or Medicine for his discovery of this chemical pathway.

The oxidative pathway involves three stages (figure 1.5). Stage 1 is the generation of a key two-carbon molecule, acetyl-CoA. Note that acetyl-CoA can be

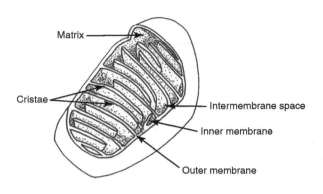

Figure 1.4 Structure of a mitochondrion.

Reprinted, by permission, from B.R. MacIntosh, P.F. Gardiner and A.J. McComas, 2006, *Skeletal muscle: Form and function*, 2nd ed. (Champaign, IL: Human Kinetics), 19.

formed from the breakdown of either carbohydrate, fat, or protein. Stage 2 is the oxidation of acetyl-CoA in the Krebs cycle. In this process, acetyl-CoA combines with oxaloacetate to form citrate. What follows is a series of reactions to regenerate oxaloacetate and two molecules of CO_2, and the pathway begins again. The primary function of the Krebs cycle is to remove hydrogens and associated energy from the various intermediates involved in the cycle using **nicotinamide adenine dinucleotide** (NAD) and **flavin adenine dinucleotide** (FAD) as hydrogen carriers. As a result, NADH and FADH are formed. Hydrogen removal is important in that hydrogen atoms, by virtue of the electrons they possess, contain the potential energy stored in the food molecules. Both NADH and FADH then proceed through a series of oxidative reactions collectively called the electronic transport chain, which is stage 3 of the oxidative pathway. In this process, energy stored in these molecules is used to combine ADP and P_i to form ATP. Oxygen does not participate in the reactions of the Krebs cycle but is the final acceptor of hydrogen at the end of electron transport chain to produce water. Because ATP is formed via the use of oxygen within mitochondria, this energy-yielding process is also termed **oxidative phosphorylation.** Using glucose as an example, this system can be summarized as follows:

$$C_6H_{12}O_6 + O_2 \rightarrow 32 \text{ ATP} + 6CO_2 + 6H_2O$$

The mechanism whereby the oxidation of NADH and FADH is coupled to the phosphorylation of ADP can be further explained by the chemiosmotic hypothesis advanced by Peter Mitchell in 1961. Mitchell

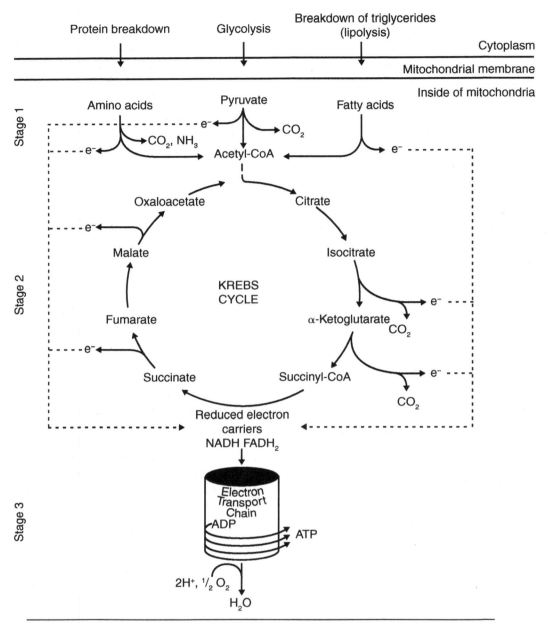

Figure 1.5 The three stages of the oxidative pathway of adenosine triphosphate (ATP) production.
MATHEWS, CHRISTOPHER K.; VAN HOLDE, KENSAL E.; AHERN, KEVIN G.; BIOCHEMISTRY, 3rd Edition, © 2000, Pgs. 485 & 747. Reprinted by permission of Pearson Education, Inc., Upper Saddle River, NJ.

proposed that electron transport and ATP synthesis are coupled by a proton gradient across the inner mitochondrial membrane (Stryer 1988). In this model, the energy that is released as electrons are transferred along the respiratory chain leads to the pumping of H+ protons from the matrix to the other side of the inner mitochondrial membrane. As a result, the concentration of H+ within the intermembrane space is higher than that in the matrix. This then generates an electrical potential that serves as a source of energy to be captured (figure 1.6). Mitchell proposed that it is

this proton-motive force that drives the synthesis of ATP. Mitchell's hypothesis that oxidation and phosphorylation are coupled by a proton gradient has been validated by a wealth of evidence. In 1978, he was awarded the Nobel Prize in Chemistry because of his extraordinary contribution to our understanding of the fundamental mechanisms of bioenergetics.

According to the chemiosmotic theory, cellular energy production takes place across the inner mitochondrial membrane. In this process, ADP is phosphorylated to ATP using energy associated with

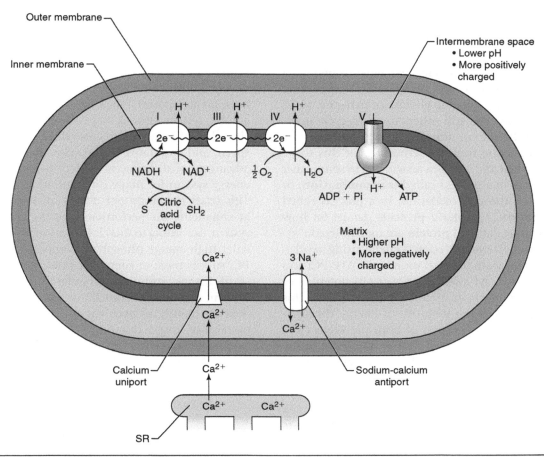

Figure 1.6 Electron transfer through the respiratory chain leads to the pumping of protons from the matrix to the intermembrane space of mitochondria. The H⁺ gradient serves as potential energy used to drive adenosine triphosphate (ATP) synthesis.

Reprinted, by permission, from M.E. Houston, 2006, *Biochemistry primer for exercise science*, 3rd ed. (Champaign, IL: Human Kinetics), 64.

a gradient of protons that is generated during electron transport. If protons leak back and abolish the gradient, heat is produced instead of useful energy. This disruption of the connection between food breakdown and energy production is known as "uncoupling." It was long thought that energy metabolism was fully coupled to production of ATP, which could then be stored or used in support of various cellular functions. However, with the discovery of **uncoupling proteins** (UCP), it is now known that this is untrue. The proton gradient can be diminished by the action of UCP. In fact, in living cells a significant proportion of mitochondrial respiration is normally not coupled to the phosphorylation of ADP, and energy that fails to be coupled to ATP synthesis is dissipated as heat.

Unlike the glycolytic pathway, which applies only to carbohydrate, the aerobic pathway allows oxidation of not only carbohydrate, but also lipid and protein. Lipids that normally participate in energy metabolism are known as triglycerides. Via **lipolysis,** triglycerides are broken down into fatty acids and **glycerol.** Fatty

acids can then undergo a series of reactions to form acetyl-CoA. Although glycerol can be converted in the liver into an intermediate of glycolysis, which later becomes pyruvate and then acetyl-CoA, this does not occur to a great extent in skeletal muscle. Therefore, glycerol is not an important muscle fuel source during exercise (Gollnick 1985; Holloszy and Coyle 1984). Fat oxidation is a rather slow process due to the complexity of its metabolism. Nevertheless, fat oxidation has the ability to yield a large amount of energy. For example, oxidation of the fatty acid palmitate, which contains 18 carbons, can liberate 129 ATP, nearly four times the amount of ATP produced from oxidation of a glucose molecule.

Protein is not considered a major energy source, since it contributes less than 15% of the energy produced during exercise (Dolny and Lemon 1988; Gollnick 1985; Lemon and Mullin 1980). However, it can be crucial in maintaining energy continuity and glucose homeostasis under special circumstances such as starvation or prolonged strenuous exercise in which bodily carbohydrate decreases significantly. Protein can enter bioenergetic pathways in many different places, but first it needs to be cleaved into amino acids. What happens next depends on which amino acids are involved. For example, some amino acids can be converted to glucose or pyruvate, some to acetyl-CoA, and still others to Krebs cycle intermediates. Before an amino acid can be used, the nitrogen residue must be removed. For this to be accomplished, the nitrogen is switched to some other compound, in a process called **transamination,** or is removed through oxidation, in a process called **deamination.** Chapter 2 presents details on how carbohydrate, fat, and protein are metabolized.

The oxidative system confers energy using oxygen, which differs from the operation of the ATP-PCr and anaerobic glycolytic systems. Due to its potential of extracting energy from all three macronutrients, this system produces the majority of energy throughout the day. The operation of both the Krebs cycle and the electron transport chain takes place in the mitochondria. Thus the ability to generate energy aerobically depends in part on the size and content of the mitochondria. Other factors such as myoglobin content and capillary density can also modulate the effectiveness of this system. The oxidative system is used primarily in sports emphasizing endurance, such as distance running ranging from 5 km (3.1 miles) to the marathon and beyond. Table 1.4 provides comparisons among the three energy systems with respect to their maximal power and capacities.

ENERGY TRANSFORMATION IN SPORT AND PHYSICAL ACTIVITY

The ATP-PCr, glycolytic, and aerobic pathways are the three energy systems functioning in each individual.

There are two inherent limits to energetic processes: the maximal rate (power) and the amount of ATP that can be produced (capacity) (Sahlin et al. 1998). Power and capacity vary drastically among the three systems; both the ATP-PCr and glycolytic systems have greater power but a lower capacity than the aerobic system. Brooks and colleagues (2005) have classified athletic activities into three groups: power, speed, and endurance. Such a classification has the advantage of allowing us to identify a predominant energy system for many different athletic activities. This leads to the proper design of training aimed at enhancing the performance of the given energy system. According to this classification, the intramuscular high-energy phosphate compounds ATP and PCr supply most of energy for power events such as short-distance sprinting and weightlifting. For rapid, forceful exercises that last about 1 min or so, muscle depends mainly on glycolytic energy sources. Intense exercise of longer duration (i.e., >2 min), such as middle-distance running and swimming, imposes a greater demand for aerobic energy transference. Table 1.5 illustrates energy sources of muscular work for various types of athletic activities.

> **KEY POINT**
>
> The oxidative system involves the breakdown of fuels with the use of oxygen. Compared with the ATP-PCr and glycolytic systems, the oxidative system is the slowest in generating ATP. However, it is the most capable of extracting energy stored in energy-containing nutrients.

One should note that the activities listed in table 1.5 are primarily track and swimming events in which exercise lasts continuously for a given time period. The fact that these individual events differ

Table 1.4   Comparisons of the Three Energy Systems

System	Maximal rate of energy production (kcal/min)	Maximal capacity of energy production (total kcal)
Immediate energy system (ATP-PCr)	36	11.1
Nonoxidative energy sources (anaerobic glycolysis)	26	15.0
Oxidative system (aerobic Krebs cycle)	10	2000*

*This value was estimated from muscle glycogen only.

Adapted from G.A. Brooks, T.D. Fahey and K. Baldwin, 2005, *Exercise physiology: Human bioenergetics and its applications,* 4th ed. (New York, NY: McGraw-Hill), 36, by permission of The McGraw-Hill Companies.

Table 1.5 Energy Sources of Muscular Work for Various Sporting Events

	Power	Speed	Endurance
Event	Shot put Discus Weightlifting High jump 40 yd dash Vertical jump 100 m run	200-800 m run 100-200 m swim	1500 m run 10K 400-800 m swim Cross country Road cycling Marathon
Duration of event	0-10 s	10 s-2 min	>2 min
Source of energy	APT PCr	ATP PCr Muscle glycogen	Muscle glycogen Liver glycogen Lipids
Energy system	ATP-PCr	Glycolysis	Aerobic pathway
Rate of process	Very rapid	Rapid	Slower
Oxygen involved	No	No	Yes

Adapted from G.A. Brooks, T.D. Fahey and K. Baldwin, 2005, *Exercise physiology: Human bioenergetics and its applications*, 4th ed. (New York, NY: McGraw-Hill), 32, by permission of The McGraw-Hill Companies.

only in duration has enabled us to estimate energy expenditure and fuel utilization using laboratory instrumentation. It is difficult to draw a general conclusion on energy metabolism in team sports such as soccer, field hockey, and lacrosse. The reason is that the energy and fuel requirements for these stop-and-go sports can vary depending on the field position and the duration of each burst of exercise. Using soccer as an example, it is likely that those who play midfield positions run longer and therefore derive proportionally more of their total energy from aerobic sources. Conversely, those who play forward positions often sprint and thus obtain a majority of their total energy from the ATP-PCr system.

The three energy systems can also be classified according to whether their operation requires a supply of oxygen. In this context, both the ATP-PCr and glycolytic systems are regarded as anaerobic in that they function outside of the mitochondria and the transfer of energy does not require oxygen. On the other hand, the oxidative system utilizes oxygen as the electron acceptor so that energy transfer can proceed. Sporting events and physical activities are often categorized as anaerobic or aerobic. This classification has made it easier to convey information to the public about the tolerability of various activities. Generally, activities that depend primarily on the aerobic system are less intense but longer lasting—for example walking, jogging, cycling, and swimming. Conversely, activities that

require anaerobic sources of energy are generally intense, fast moving, and more explosive—for example sprinting and jumping. These can also include resistance exercise in which muscle tension increases significantly once the muscle contracts.

How the three energy systems respond during exercise of changing intensity is a complex issue. This is so because as exercise intensity increases, a transition from one energy system to another takes place. It is important to keep in mind that for most activities, providing the energy needed is not a matter of simply turning on one energetic pathway. Rather, several energy systems operate in a sequential fashion, but with considerable overlap. Such a mixed use of energy systems may be manifested particularly during (1) the rest-to-exercise transition and (2) incremental exercise in which intensity rises progressively. In the transition from rest to light or moderate exercise, oxygen consumption increases progressively to reach a steady state within 1 to 4 min. The fact that oxygen consumption does not increase instantly to the desirable level suggests that energy systems other than the oxidative pathway contribute to the overall production of ATP at the beginning of exercise. There is evidence to suggest that from the onset of exercise, the ATP-PCr system is the first bioenergetic pathway activated, followed by glycolysis, and finally aerobic energy production. However, after a steady state is reached, the body's ATP requirement can be met primarily via aerobic metabolism.

It has long been thought that during incremental exercise in which intensity increases progressively, there are three phases that involve a transition of energy sources from aerobic to anaerobic (Skinner and McLellan 1980). First, with increasing levels of low-intensity exercise, a great amount of oxygen is being extracted by tissues while more carbon dioxide is being produced. In this phase, energy is primarily derived from the operation of the aerobic system because the intensity is quite mild. As intensity increases and reaches a point between 40% and 60% maximal oxygen uptake $\dot{V}O_2max$, there is an initial rise in blood lactate to approximately twice the resting values ($\sim$2 mmol $\cdot$ L^{-1}). This is indicative of increased involvement of the glycolytic system. During this second phase, blood lactate is successfully neutralized via respiratory compensation so that exercise can continue. With further increases in intensity to about 65% and 90% $\dot{V}O_2max$, exercise is mainly powered by energy derived from glycolysis. During this last phase, the blood lactate level begins at about 4 mmol $\cdot$ L^{-1} and increases much more rapidly because hyperventilation becomes incapable of compensating adequately. The transitions from the first to the second phase and from the second to the third are regarded as the aerobic threshold and the anaerobic threshold, respectively.

SUMMARY

Energy is defined as the ability to perform work. It is neither created nor destroyed, but instead is transformed from one state to another without being used up. The two major interchangeable forms of energy related to human movement are kinetic and potential energy.

In a living cell, ATP is the most important carrier of the energy necessary to perform many complex functions. Due to limited storage, ATP needs to be replenished instantly using the energy stored in foods in order to power most activities. Carbohydrate, fat, and protein are the three energy-containing nutrients people consume daily. Compared to fat, carbohydrate stored as glycogen is rather limited. However, it serves as a more immediate source of energy. Protein contains energy, but contributes little to energy metabolism.

Biosynthesis of ATP is accomplished via the operation of various energy-yielding systems. These systems consist of complex enzymatically controlled reactions that take place in or outside of mitochondria, with or without the use of oxygen. Three energy systems are responsible for the production of ATP: (1) the ATP-PCr system, (2) the glycolytic system, and (3) the oxidative system. Of these three, the oxidative system, including the Krebs cycle and the electron transport chain, is a "common" pathway shared by carbohydrate, fat, and protein. The three energy systems differ considerably in terms of their rate and their capacity to produce ATP, and their contribution varies depending on the intensity and duration of an activity. However, such differences among the three energy systems give the body the ability to derive energy under various circumstances, whether the person is generating explosive power, engaging in a long-distance event, or simply performing a household activity.

KEY TERMS

acetyl-CoA
adenosine diphosphate
adenosine monophosphate
adenosine triphosphatase
adenosine triphosphate
adenylate kinase
adipocytes
amino acids
biosynthesis
calorie
catabolism
creatine kinase
deamination
disaccharides
energy

fatty acids
flavin adenine dinucleotide
glucose
glycerol
glycogen
glycogenolysis
glycolysis
hepatic-portal vein
hydrolysis
joules
ketones
kilocalorie
kilojoules
kinetic energy
law of the conservation of energy

lipolysis
mechanical energy
mitochondria
monosaccharides
nicotinamide adenine
    dinucleotide
oxidative phosphorylation
phosphocreatine
phosphorylation
potential energy
transamination
triglycerides
uncoupling proteins
urea

REVIEW QUESTIONS

1. What is the law of the conservation of energy? Also define kinetic and potential energy and provide an example that illustrates the transformation between these two forms of energy.

2. What is the total energy stored in food containing 15 g of carbohydrate, 9 g of fat, and 4 g of protein?

3. Compare the three energy systems in terms of complexity, cellular location, end products, oxygen requirement, rate and capacity of ATP production, and sporting events supported by the system.

4. Define the terms *glycogenolysis, lipolysis, deamination*, and *transamination*.

5. State the chemiosmotic hypothesis.

CHAPTER
2

METABOLISM OF MACRONUTRIENTS DURING EXERCISE

This chapter illustrates how energy-containing nutrients and energy-yielding systems function and interact in support of physical activity and sport performance. It characterizes the patterns of energy expenditure and fuel utilization during various sporting events and physical activities that are aerobic, anaerobic, or both. Also included is a discussion of how these patterns may be influenced by extraneous factors such as mode, intensity, duration, training status, diet, and environment.

CARBOHYDRATE

Two main macronutrients provide energy for replenishing adenosine triphosphate (ATP) during exercise: (1) muscle and liver glycogen and (2) triglycerides within adipose tissue and exercising muscle. To a much lesser degree, protein or amino acids within skeletal muscle can donate carbon skeletons, thereby furnishing energy. During prolonged exercise, carbohydrates such as muscle glycogen and blood glucose derived from liver glycogenolysis are the primary energy substrates. Glycogen is a large polymer, comprised of many glucose residues and readily degradable upon the action of enzymes (figure 2.1). Glycogen undergoes a process of glycogenolysis that yields free glucose molecules (figure 2.2). This glucose can then enter the glycolytic pathway in which energy is transformed. The importance of glucose availability

during exercise is demonstrated by the observation that fatigue is often associated with muscle glycogen depletion or hypoglycemia or both (Coggan and Coyle 1987; Coyle et al. 1986; Sahlin et al. 1990). With respect to energy provision, carbohydrates are superior to fat in that (1) they can be used for energy with or without oxygen, (2) they provide energy more rapidly, (3) they must be present in order for fat to be used, (4) they are the sole source of energy for the central nervous system, and (5) they can generate 6% more energy per unit of oxygen consumed.

Carbohydrate Utilization at the Onset of Exercise

As mentioned in chapter 1, phosphocreatine (PCr) is the primary energy substrate available for replenishing ATP during very intense muscular exercise of short duration (i.e., ≤10 s). This idea was initially supported both by theoretical calculations of the energy required for production of muscle force and by the rapid decline in PCr observed during very intense exercise. Consequently, it was assumed that the provision of energy via a particular metabolic pathway occurs within a sequence—that is, that during intense exercise, PCr stores are almost depleted in the initial 10 s and further contractile activity is then sustained by the metabolism of muscle glycogen. However, more recent data from several laboratories do not

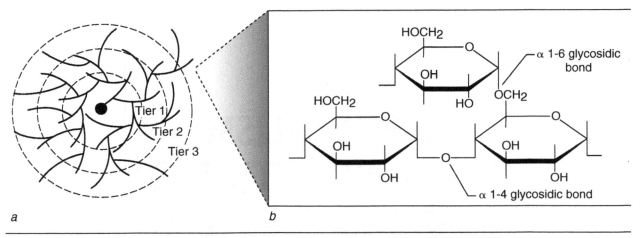

Figure 2.1 Structure of a glycogen particle.

Reprinted, by permission, from J. Shearer and T.E. Graham, 2002, "New perspectives on the storage and organization of muscle glycogen," *Can J Appl Physiol.* 27(2): 179-203.

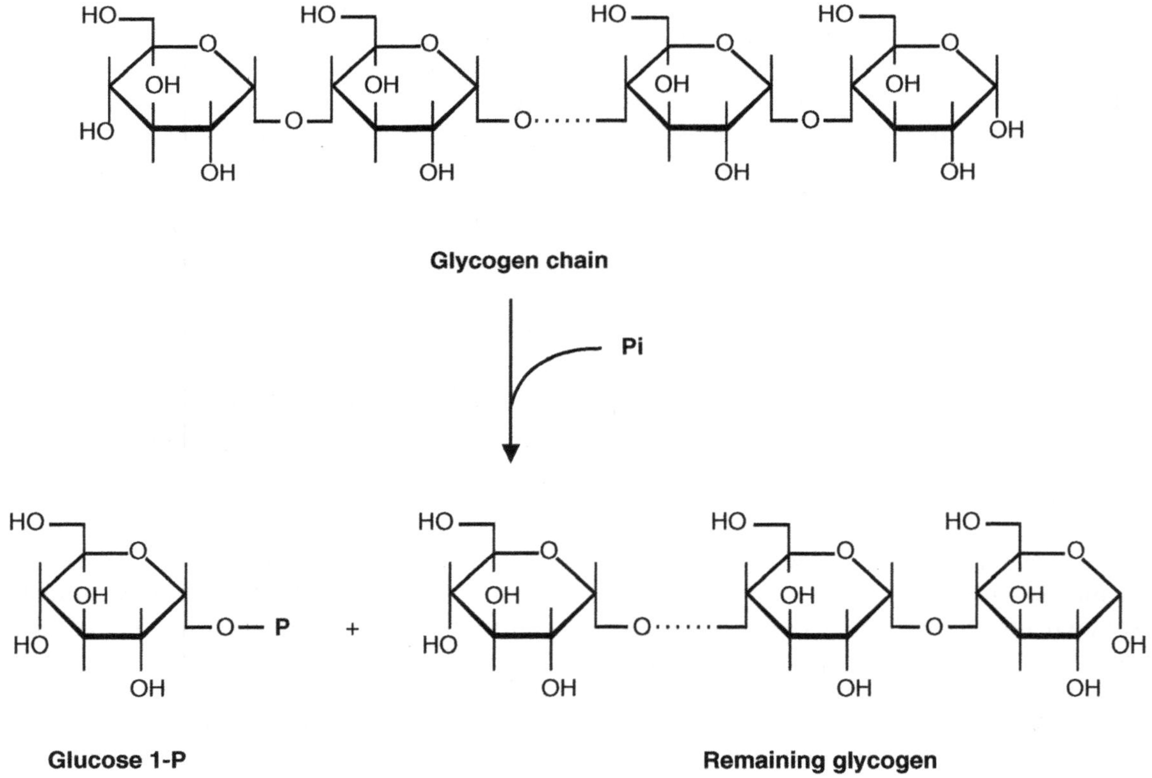

Glycogen chain

Pi

Glucose 1-P + **Remaining glycogen**

Figure 2.2 Degradation of glycogen.

Adapted, by permission, from P.C. Champe and R.A. Harvey, 1987, *Lippincott's illustrated reviews: Biochemistry* (Philadelphia, PA: Lippincott, Williams & Wilkins), 74.

support this hypothesis. Boobis and colleagues (1982) found that with all-out bicycle ergometer exercise aimed at accomplishing as much work as possible in 6 s, a 35% decrease in PCr occurred along with a 15% reduction in glycogen. When such exercise was performed for 30 s, a 65% decrement in PCr was found concomitant with a 25% reduction in glycogen. Similar results have been reported for short-duration maximal treadmill running and cycling (Cheetham et al. 1986; McCartney et al. 1986). Collectively, these studies indicate that PCr breakdown and glycogenolysis occur concomitantly from onset of the exercise.

> ▶ **K E Y P O I N T** ◀
>
> Carbohydrate in the form of glycogen is the most readily available energy source. Compared to fat, carbohydrate provides energy more quickly, can be used whether or not oxygen is present, and serves as the sole source of energy for the central nervous system. Also, it must be available in order for the body to use fat. Thus carbohydrate is the main source of fuel for most sporting events.

Influence of Exercise Intensity and Duration

During very intense exercise when oxygen consumption fails to meet energy demands, stored muscle and liver glycogen becomes the primary energy source, as energy transfer from carbohydrate can occur without oxygen. Since the introduction of the needle biopsy technique in the early 1960s, considerable effort has been devoted to the study of glycogen utilization during exercise as well as the reestablishment of glycogen stores after exercise. A landmark study by Gollnick and colleagues (1974) revealed that breakdown of muscle glycogen is most rapid during the early stages of exercise, with its rate of utilization exponentially related to exercise intensity. The authors also found that slow-twitch muscle fibers were the first to lose glycogen. This was followed by increased glycogen utilization in fast-twitch muscle fibers. As exercise continues, the rate of muscle glycogen utilization declines, and this is accompanied by an increased contribution of blood-borne glucose degraded from liver glycogen as a metabolic fuel. It is estimated that a 2 h vigorous workout just about depletes glycogen in the liver as well as in exercised muscle.

During moderate-intensity exercise, utilization of PCr as an energy source is relatively low even at the onset. It has been suggested that glycogen stored in active muscle supplies almost all of the energy in the transition from rest to exercise (McArdle et al. 2005). As soon as a steady state is attained, energy is provided through a mixed use of carbohydrate and lipids. Typically, liver and muscle glycogen supply between 40% and 50% of the energy requirement, whereas the remainder is furnished via oxidation of lipids. It is important to note that the energy mixture may vary depending on the intensity of exercise, although it can also be influenced by the training status of the individual and dietary intake of carbohydrate. As exercise continues, muscle glycogen stores diminish progressively. Consequently, as blood glucose from liver glycogen becomes the major supplier of carbohydrate energy, the contribution of fat relative to the total energy provision increases further. With glycogen depletion, the maximal steady-state exercise intensity that can be sustained decreases accordingly. The reduced power output allows exercise to be supported by oxidation of fat as a major fuel. As bodily glycogen decreases significantly, blood glucose also falls because the liver's glucose output fails to match the rate of glucose uptake by exercising muscles. **Hypoglycemia** is said to occur when the blood glucose concentration is lower than 2.5 mmol · L^{-1} or 45 mg · dL^{-1} (Powers and Howley 2001). This condition can occur during strenuous exercise that lasts for a duration close to or more than 2 h. Figure 2.3 illustrates the percentage of energy derived from the four major sources of fuel during moderate-intensity exercise (i.e., 65-75% $\dot{V}O_2$max) in trained individuals.

Liver Sources of Carbohydrate

During exercise when utilization of carbohydrate accelerates, an increased release of glucose from the liver is functionally important to maintain blood glucose homeostasis and to possibly attenuate muscle glycogen depletion. Debate continues as to whether the increased availability of blood glucose helps in sparing the use of muscle glycogen. However, it appears that once hypoglycemia is induced, muscle glycogen utilization is likely to accelerate. The increased contribution of liver glycogen is a universal phenomenon that was revealed more than 40 years ago. Using isotope dilution technique, Reichard and colleagues (1961) demonstrated that **hepatic glucose output** during muscular work reached three to six times that of resting values. This early

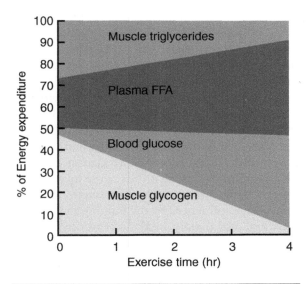

Figure 2.3 Percentage of energy derived from the four major sources of fuel during moderate-intensity exercise at 65% to 75% $\dot{V}O_2$max in trained individuals.

Reprinted from S.K. Powers and E.T. Howley, 2004, *Exercise physiology: Theory and applications to fitness and performance,* 5th ed. (New York, NY: McGraw-Hill), 64, by permission of The McGraw-Hill Companies.

finding was also confirmed by Rowell and colleagues (1965), who measured **splanchnic** glucose output in humans during exercise using arterial and hepatic venous catheterization. Over the next two to three decades, repeated evidence suggested that hepatic glucose output can increase by two- to threefold during moderate exercise and up to 10-fold during vigorous exercise.

The intensity and duration of an exercise are the factors that determine the source and quantity of glucose released by the liver. During moderate exercise (<60% $\dot{V}O_2$max), the blood glucose level remains relatively constant despite an increase in glucose utilization by exercising muscle (Kjaer et al. 1991). At this intensity, a drop in blood glucose will not occur unless exercise is prolonged for several hours (Ahlborg et al. 1974). Given that the level of blood glucose reflects a balance between hepatic glucose output and muscular glucose utilization, this finding indicates that at moderate intensity, an exercise-induced rise in hepatic glucose output is able to match the increased glucose utilization. In contrast, with more intense exercise (>60% $\dot{V}O_2$max), blood glucose has been found to increase especially during the early phase of the exercise, and the increase is more pronounced at higher exercise intensities (Kjaer et al. 1991; Hargreaves and Proietto 1994). This mismatch can be ascribed to the fact that hepatic glucose output exceeds peripheral glucose

uptake. It has been suggested that the production of glucose from the liver is not totally regulated by a feedback mechanism, which is fundamentally important in maintaining homeostasis. In this case of overproduction of hepatic glucose, the finding has been attributed to increased **efferent** signals of the central nervous system that regulate hepatic glucose metabolism (Kjaer et al. 1987).

An increase in liver glucose output can be brought about by an enhancement of glycogenolysis and **gluconeogenesis.** While glycogenolysis is a relatively simple process that involves glycogen breakdown into glucose (figure 2.2), gluconeogenesis entails a rather complex pathway using nonglucose molecules such as amino acids or **lactate** for the production of glucose in the liver. This "new" glucose can then be released into the blood and transported back to skeletal muscles to be used as an energy source. This internal production of glucose may be viewed as a secondary resort for the body with respect to obtaining glucose. Figures 2.4 and 2.5 illustrate the two separate processes of gluconeogenesis in which glucose is generated from lactate and alanine, respectively. In general, glycogenolysis appears to respond more quickly and to contribute more of the total hepatic glucose output during exercise. It has been demonstrated that during the first 30 min of exercise of either moderate or heavy intensity, most of the glucose released by the liver is derived from hepatic glycogenolysis. Gluconeogenesis, however, seems to be more responsive to the length of exercise and to play a more important role during the later phase of prolonged exercise. In a series of experiments using dogs, Wasserman and colleagues (1988) found that the relative contribution of gluconeogenesis to the total hepatic glucose output was only 15% during first 60 min of exercise. However, it reached 20% to 25% when exercise continued for another 90 min.

Exercise intensity can influence the type of gluconeogenic precursors used to produce glucose in the liver. Glycerol, lactate, and amino acid are the three major precursors that can be converted to glucose via gluconeogenesis. It is believed that when exercise is performed at an intensity level below **lactate threshold** (an intensity above which the production of lactate will increase sharply), glycerol is the primary molecule used for gluconeogenesis. As exercise intensity approaches and exceeds the lactate threshold, more lactate becomes available for producing glucose in the liver. Such different uses of gluconeogenic precursors at different exercise intensities make sense given that glycerol is a product of lipolysis and

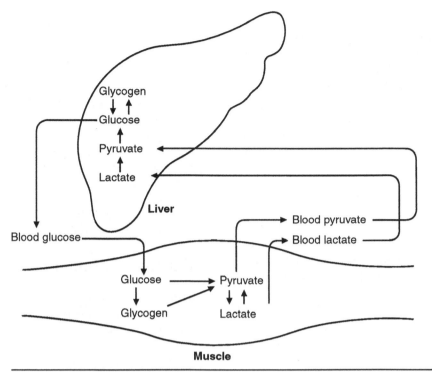

Figure 2.4 An example of gluconeogenesis. The lactate or Cori cycle involves the conversion in the liver of muscle-derived lactate to glucose, which can circulate back to muscle.

Adapted from G.A. Brooks, T.D. Fahey and K. Baldwin, 2005, *Exercise physiology: Human bioenergetics and its applications*, 4th ed. (New York, NY: McGraw-Hill), 87, by permission of The McGraw-Hill Companies.

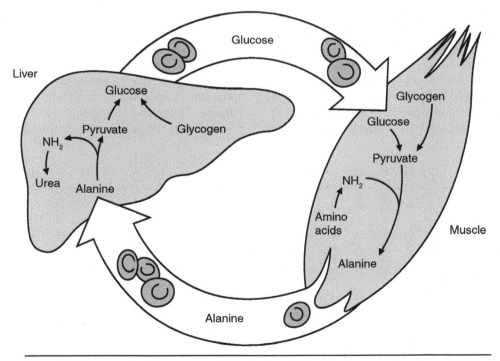

Figure 2.5 An example of gluconeogenesis. The alanine cycle involves the conversion of muscle-derived alanine to glucose, which can circulate back to muscle.

Adapted, by permission, from W.D. McArdle, F.I. Katch and V.L. Katch, 2001, *Exercise physiology*, 5th ed. (Philadelphia, PA: Lippincott, Williams & Wilkins), 39.

that fat utilization increases during exercise of low to moderate intensity. However, as intensity increases, more glycogen is degraded, thereby producing more lactic acid. An increased contribution of amino acids to gluconeogenesis would be seen particularly during vigorous exercise that lasts for a prolonged period of time. In this case, both muscle and liver glycogen stores decrease significantly, and more glucose must be produced in order to prevent the occurrence of hypoglycemia.

> ### ▶ K E Y   P O I N T ◀
>
> Muscle glycogen is an initial source of energy as strenuous exercise begins. As exercise continues, the contribution of degradation of liver glycogen increases as additional glucose is provided for use by muscle and to prevent hypoglycemia. The liver can also manufacture new glucose in an effort to maintain glucose homeostasis; this occurs under extreme conditions such as prolonged exercise or sustained starvation.

LIPID

Triglycerides are another major source of energy stored primarily in adipose tissue, although they are also found in muscle tissue. As shown in figure 2.6, a triglyceride molecule is composed of a glycerol and three fatty acids that can vary in the number of carbons they contain. Via lipolysis, a triglyceride is split to form a glycerol and three fatty acids (figure 2.7). These products can then enter metabolic pathways

for energy production. Despite the large quantity of lipid available as fuel, the processes of lipid utilization are slow to be activated and proceed at rates significantly lower than those of carbohydrate utilization. However, lipids are an important segment of energy substrates used during prolonged exercise or during extreme circumstances such as fasting or starvation when carbohydrate stores decline significantly. Even small increases in the ability to use lipid as fuel during exercise can help slow muscle glycogen and blood glucose utilization and delay the onset of fatigue. An increase in the ability to use lipid can be realized via an improved oxidative capacity of skeletal muscle following endurance training.

Energy Sources From Lipids

Three lipid sources supply energy: (1) fatty acids released from the breakdown of triglycerides, (2) circulating plasma triglycerides bound to lipoproteins, and (3) triglycerides within the active muscle itself. Unlike what happens with carbohydrate, which can yield energy without using oxygen, fat catabolism is purely an aerobic process that is best developed in the heart, liver, and slow-twitch muscle fibers. Most fat is stored in the form of triglycerides in fat cells or adipocytes, but some is stored in muscle cells as well. The major factor that determines the role of fat as an energy substrate during exercise is its availability to the muscle cell. In order for fat to be oxidized, triglycerides must first be cleaved to three molecules of free fatty acid (FFA) and one molecule of glycerol. This process, known as lipolysis, occurs through the activity of **lipase,** an enzyme found in the liver, adipose tissue, and muscle, as well as in blood vessels. Lipolysis is

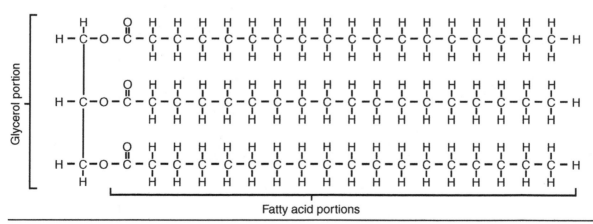

Figure 2.6   Structure of a triglyceride molecule.

Adapted from D. Shier, J. Butler and R. Lewis, 2004, *Hole's human anatomy and physiology*, 10th ed. (New York, NY: McGraw-Hill), 51, by permission of The McGraw-Hill Companies.

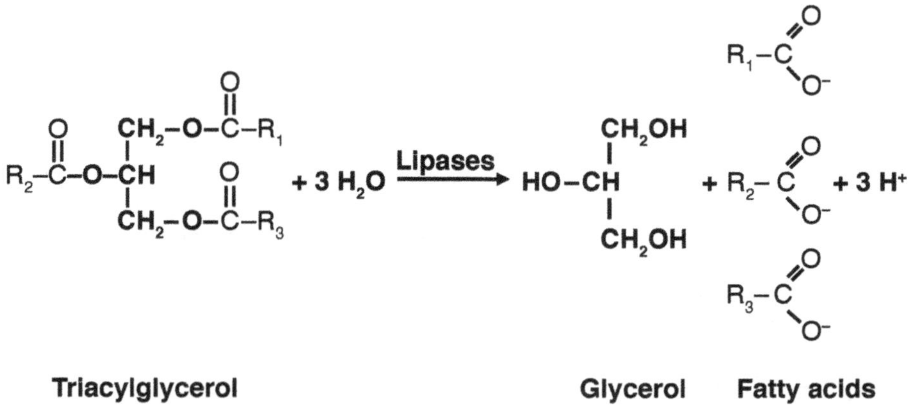

Figure 2.7 In lipolysis, a triglyceride molecule is cleaved into a glycerol and three fatty acids.
Reprinted from S.K. Powers and E.T. Howley, 2004, *Exercise physiology: Theory and application to fitness and performance*, 5th ed. (New York, NY: McGraw-Hill), 33, by permission of The McGraw-Hill Companies.

modulated by **catecholamines** (Arner et al. 1990). As such, this process is considered intensity dependent because the release of catecholamines increases as exercise intensity increases. It is important to note that lipolysis and oxidation are two separate processes of fat utilization. The latter process is facilitated during low- to moderate-intensity exercise in which production of lactic acid is low.

Preparatory Stages for Fat Utilization

Following lipolysis, two additional processes must also occur before FFA can undergo combustion: (1) transfer to the mitochondria and (2) β-**oxidation.** The oxidation of fatty acids occurs within the mitochondria. However, long-chain fatty acids are normally unable to cross the inner mitochondrial membrane due to their molecular size; thus a membrane transport system consisting of protein carriers is required. The carrier molecule for this system is **carnitine,** which is synthesized in humans from the amino acids lysine and methionine and is found in high concentrations in muscle. With the assistance of carnitine and an enzyme called **carnitine acyltransferase,** fatty acids can be brought from the cytoplasm into mitochondria. Upon entry into the mitochondria, they undergo another process called β-oxidation. β-oxidation is a sequence of reactions that reduce a long-chain fatty acid into multiple two-carbon units in the form of acetyl-CoA (figure 2.8). This process may be viewed as analogous to glycolysis, the first stage of the oxidative pathway for glucose in which a glucose molecule is converted into two molecules of acetyl-CoA. Once formed, acetyl-CoA becomes a fuel source for the Krebs cycle and leads to the production of ATP within the electron transport chain.

> **▷ K E Y   P O I N T ◁**
>
> Preparing carbohydrate and fat molecules for final entry into the metabolic pathway is an important step in energy metabolism. To be oxidized, both glucose and fatty acid need to be converted into acetyl-CoA. This occurs through glycolysis for glucose and β-oxidation for fatty acid. Glycolysis and β-oxidation are similar in that both pathways function to ultimately produce the molecule acetyl-CoA.

Influence of Exercise Intensity and Duration on Fat Utilization

Fat oxidation is influenced by exercise intensity and duration. Romijn and colleagues (1993) found that during exercise at 25% $\dot{V}O_2$max, 90% of the total energy was furnished via oxidation of plasma FFA and muscle triglycerides. The relative contribution of fat to total oxidative metabolism decreases as exercise intensity increases. However, this decrease in the relative contribution of fat is relatively minor compared with the increase in oxygen consumption. Therefore, despite a decrease in the relative contribution, the amount of fat being oxidized actually increases until the intensity reaches a value close to one's lactate threshold or ~60% $\dot{V}O_2$max. A number of recent studies have

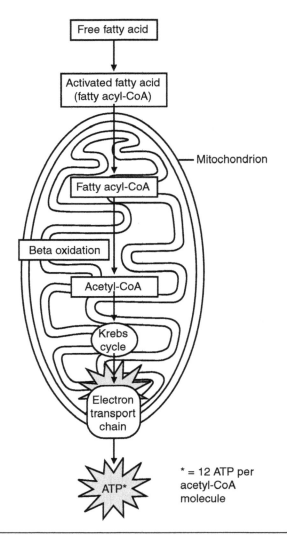

Figure 2.8 Illustration of β-oxidation.

Reprinted from S.K. Powers and E.T. Howley, 2004, *Exercise physiology: Theory and applications to fitness and performance*, 5th ed. (New York, NY: McGraw-Hill), 39, by permission of The McGraw-Hill Companies.

> **▶ KEY POINT ◀**
>
> The oxidation of fat depends on the level of exercise intensity. Contrary to what many believe, the maximal fat oxidation rate occurs at moderate rather than low intensity, because fat oxidation is also a function of absolute caloric expenditure. An intensity near one's lactate threshold, or around 60% to 65% $\dot{V}O_2$max, will elicit maximal fat oxidation.

As a steady-state exercise of light to moderate intensity continues, the contribution of fat to the total oxidative metabolism increases progressively (figure 2.3). Using a prolonged exercise protocol in which exercise lasted 4 h, earlier studies demonstrated a progressive decrease in the respiratory exchange ratio, signifying a steady increase in fat combustion (see chapter 4 for further details on the respiratory exchange ratio). It has been estimated that the relative contribution of fat can account for as much as 80% of the total energy expenditure during prolonged exercise. The progressive increase in fat utilization over time is due to a concomitant decrease in glycogen stores as a result of prolonged exercise. This reduction in carbohydrate energy substrates triggers a release of glucoregulatory hormones such as glucagon, cortisol, and growth hormone. These hormones function to stimulate the breakdown of lipid in response to reduced carbohydrate stores. Please refer to chapter 3 for more information about hormonal regulation of fuel utilization.

Interaction Between Carbohydrate and Fat Utilization

Utilization of carbohydrate and utilization of fat are not two separate processes. Instead, they are coordinated, and utilization of one substrate is affected by the availability of the other. This is attested to by the fact that fat utilization increases in accordance with a decrease in glycogen stores during prolonged exercise. It appears that carbohydrate availability during exercise modulates the level of lipolysis and fat oxidation. Studies have shown that ingesting high-glycemic carbohydrates prior to exercise significantly blunted the release of fatty acids from adipose tissue and thus oxidation of long-chain fatty acids by skeletal muscle (Coyle et al. 1997; De Glisezinski et al. 1998). As carbohydrate substrates decline, the suppressive effect of carbohydrate on fat utilization is withdrawn.

shown that the intensity at which the highest fat oxidation is observed ranges from 55% to 65% $\dot{V}O_2$max (Achten et al. 2002, Achten and Jeukendrup 2004). For example, by testing with multiple levels of intensity, Achten and colleagues (2002) found that exercise at 60% to 65% $\dot{V}O_2$max helped in eliciting the maximal rate of fat oxidation (figure 2.9). Their finding suggests that to obtain maximal fat oxidation, comparatively more intense exercise ought to be chosen. When exercise is performed at an intensity above lactate threshold, fat oxidation decreases significantly. This may result from increased carbohydrate utilization, or accumulation of lactic acid (which can serve to inhibit fat utilization), or both.

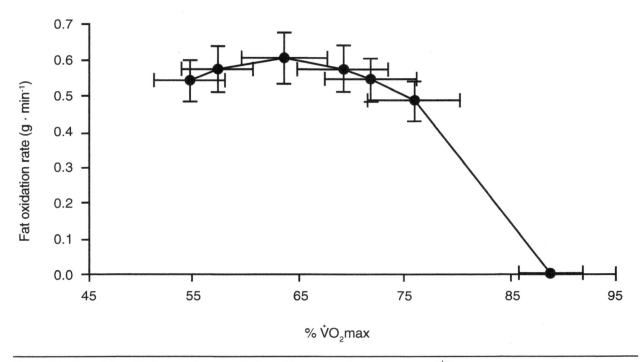

Figure 2.9 Rate of fat oxidation at various exercise intensities, expressed as a percentage of $\dot{V}O_2$max.

Reprinted, by permission, from J. Achten, M. Gleeson and A.E. Jeukendrup, 2002, "Determination of the exercise intensity that elicits maximal fat oxidation," *Medicine & Science in Sports & Exercise* 34(1): 92-97.

Similarly, it has long been known that elevated FFA levels are able to suppress glucose utilization. This can be explained by the glucose–fatty acid cycle theory advanced by British biochemist Philip Randle and colleagues (1964). According to this theory, increased availability of fatty acids slows down the utilization of glucose. Randle's theory has been used to explain how mitochondrial adaptations resulting from endurance training help promote lipid oxidation and thus spare glycogen utilization in skeletal muscle during exercise. Further details on this cycle and its function are presented in chapter 3.

PROTEIN AND AMINO ACIDS

Skeletal muscle constitutes approximately 40% of the body weight and is the second largest store of potential energy in the body, after fat. However, protein and amino acids are not commonly discussed as energy substrates, because amino acids contribute only a minor portion (i.e., 5-15%) of the total energy consumed during exercise. Unlike carbohydrate and fat, which can be stored as energy substrates, virtually no inert amino acids are used in this way. However, it must be recognized that during fasting and starvation, catabolism of proteins to amino acids and conver-

sion of amino acids into energy are very important processes in maintaining the levels of blood glucose essential for brain and kidney function. It has been reported that gluconeogenesis using amino acids increases every morning in response to a fall in glycogen stores (Brooks et al. 2005). In the last decade or so, researchers have realized that even a minor increase in protein consumption is important in conditions of high energy demands over a prolonged period of time. Growing evidence, especially from more recent research on the **branched-chain amino acids** leucine, valine, and isoleucine, suggests that protein serves as an energy fuel to a much greater extent than previously believed.

Protein Degradation and Synthesis

Metabolism of protein includes its degradation and synthesis. The magnitude of protein degradation is often smaller than that of carbohydrate and fat. However, it can increase significantly when exercise is performed at high intensity for a prolonged period of time. There are two classes of proteins in skeletal muscle: contractile and noncontractile. Contractile-related proteins are responsible for muscle contraction, whereas noncontractile-related proteins are those that are essential for other cellular functions.

In humans, contractile and noncontractile proteins compose 66% and 34% of the total muscle protein, respectively.

Remember that a protein molecule consists of chains of amino acids. Thus the amino acids tyrosine and phenylalanine have been used as indicators of noncontractile protein degradation (Graham et al. 1995). In an early study using an experimental protocol of 40 min of exercise performed at different intensities, Felig and Wahren (1971) demonstrated a greater release of tyrosine and phenylalanine, as well as alanine, during exercise compared with rest; this enhanced efflux of metabolites was greater at higher exercise intensities. Later, Babij and colleagues (1983) also observed a direct linear relationship between exercise intensity and oxidation of leucine, one of the three branched-chain amino acids. As to the metabolism of contractile proteins, measurement of 3-methyhistidine (3-MH) in the urine has been the approach most widely used to reflect contractile protein degradation, although this parameter can also be determined via the blood. After a thorough review of the literature, Dohm and colleagues (1987) concluded that the production of this catabolic index of contractile protein decreases during exercise. However, there are studies reporting an increase in the efflux of 3-MH during recovery. Taken together, these findings suggest that the integrity of contractile protein remains unaffected during exercise, when muscle contraction is demanded, but not during recovery. The mechanism responsible for the divergent response of 3-MH between exercise and recovery is unclear.

Assessment of protein degradation along with protein synthesis gives an indication of whether someone who performs exercise will need extra protein in order to prevent a loss in lean body mass. This type of assessment can be made via determination of **nitrogen balance.** As discussed in chapter 1, protein contains nitrogen, and the body cannot oxidize the nitrogen component. Consequently, nitrogen atoms combine with hydrogen to form urea to be excreted via the kidneys. Determining nitrogen balance involves assessing the relationship between the dietary intake of protein and the protein that is degraded and excreted. Nitrogen balance is said to occur when the amount of protein taken in equals the amount excreted. A positive nitrogen balance suggests that protein intake exceeds protein output and that the excessive protein may have been used to repair damaged tissue, synthesize new tissue, or both. On the other hand, a negative nitrogen balance indicates that protein loss is greater than intake and that the protein lost may have been used for energy or degraded due to exercise.

Protein synthesis decreases during exercise, and this finding has been universally demonstrated. The decreased protein synthesis, together with increased protein degradation, clearly suggests that those who are heavily trained will experience augmented protein loss and thus require a higher protein intake on a regular basis. Lemon and colleagues (1992) administered two levels of dietary protein in a group of novice bodybuilders who underwent a month of resistance training. They found that a majority of those on the lower protein intake (i.e., 0.99 g $\cdot$ kg^{-1} $\cdot$ day^{-1}) experienced a negative nitrogen balance, whereas all of those on the high protein intake (i.e., 2.62 g $\cdot$ kg^{-1} $\cdot$ day^{-1}) achieved a positive nitrogen balance. The authors further calculated that nitrogen balance occurs at 1.43 g $\cdot$ kg^{-1} $\cdot$ day^{-1}. The recommended daily allowance (RDA) for protein is 0.8 g $\cdot$ kg^{-1} $\cdot$ day^{-1} for a healthy adult. However, in light of the augmented protein catabolism associated with heavy exercise, it is suggested that people engaged in endurance training consume between 1.2 and 1.4 g $\cdot$ kg^{-1} $\cdot$ day^{-1} and that those who resistance train may benefit from consuming ~1.6 g $\cdot$ kg^{-1} $\cdot$ day^{-1} (Fielding and Parkington 2002).

Energy Metabolism of Amino Acids

The three principal sources of amino acids for energy metabolism are (1) dietary protein, (2) the free amino acid pool, and (3) **endogenous** tissue protein. Dietary protein is a relatively minor source of amino acids because it is not a common practice to consume a large protein meal prior to exercise. The free amino acid pool, which exists in muscle and blood, is also very small compared to the amounts of amino acids derived from degradation of tissue protein. It has been estimated that the intramuscular amino acid pool constitutes less than 1% of the total of metabolically active amino acids (Smith and Rennie 1990). Consequently, the most important source of amino acids comes from endogenous protein breakdown (Dohm 1986).

As mentioned in chapter 1, the catabolism of amino acids requires the removal of an amino group (the nitrogen-containing portion) by transamination or oxidative deamination. Transamination is a common route for the exchange of nitrogen in most tissues, including muscle, and involves the transfer of an amine from an amino acid to another molecule.

In a typical example, the amine is transferred from glutamate to pyruvate to produce alanine, which can then be utilized to produce glucose in the liver via the alanine cycle as shown in figure 2.5. The process of deamination occurs in the liver and is responsible for converting the nitrogen residue into the waste product urea, which can be excreted from the kidneys.

The remaining carbon skeleton can then be converted to various intermediates of the Krebs cycle, which is common to both carbohydrate and fat metabolism. As shown in figure 2.10, amino acids can give rise to pyruvate, acetyl-CoA, oxaloacetate, and so on, which can be either oxidized or converted to glucose. Another way in which amino acids contribute to energy metabolism is through conversion to glucose via gluconeogenesis; this glucose is then used to generate energy or prevent hypoglycemia. This process can be illustrated using the alanine cycle (figure 2.5). In this cycle, alanine is first produced from pyruvate via transamination in active skeletal muscle and then travels to the liver via the circulation. Upon entry into the liver, alanine becomes pyruvate through deamination. Gluconeogenesis then converts the remaining carbon skeleton of alanine into glucose, which then enters the blood for use by active muscle. This gluconeogenic process helps in maintaining blood glucose homeostasis during fasting and starvation. It also assists in prolonged exercise by providing additional energy fuel. It is estimated that the alanine-glucose cycle can generate up to 15% of the total energy requirement during prolonged exercise (Paul 1989).

> **▶ K E Y   P O I N T ◀**
>
> Protein does not normally participate in energy metabolism, and therefore there is no storage form of protein used for energy as there is for glycogen. However, when bodily carbohydrate decreases significantly, protein can be used as a fuel. Protein contributes to energy provision first through degradation into amino acids; the amino acids are then converted into glucose or various intermediates of the Krebs cycle in order to fulfill their energetic role.

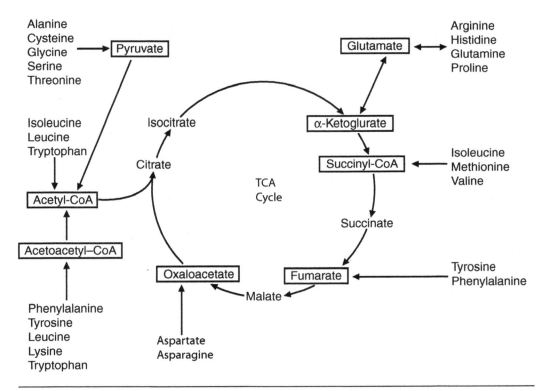

Figure 2.10 Major metabolic pathways for various amino acids following removal of the nitrogen group by transamination or deamination. The tricarboxylic acid, or TCA, cycle is also known as the Krebs cycle.

Reprinted, by permission, from A.L. Lehninger, 1970, *Biochemistry: The molecular basis of cell structure and function* (New York, NY: Worth Publishers), 436.

Issues Related to Branched-Chain Amino Acids

Leucine, isoleucine, and valine, the three branched-chain amino acids (BCAA), have drawn a great deal of attention for their role in bioenergetics. These are essential amino acids that cannot be synthesized in the body and therefore must be replenished via the diet. Branched-chain amino acids are unique in that they are catabolized mainly in the skeletal muscle. As with other amino acids, the first step in the metabolism of BCAA is removal of the amino group so that the remaining carbon skeleton can be further utilized. This transamination results in the production of glutamate, which can then donate its nitrogen-containing portion to pyruvate to form alanine.

Supplementation of BCAA has been claimed to enhance exercise performance in a variety of ways. It is said that BCAA can (1) serve as additional energy fuel, (2) prevent or attenuate the excessive loss of protein, and (3) help in improving the function of neurotransmitters, thereby reducing the feeling of fatigue. However, these claims should be interpreted with caution. The notion that BCAA can act as a fuel was initially reported by Goldberg and Chang (1978), who found that these amino acids can be oxidized by skeletal muscle. However, more recent studies by Wagenmakers and colleagues (1989, 1991) suggested that the activity of enzymes necessary for oxidizing BCAA was too low to allow a major contribution of BCAA to the total energy expenditure during exercise. In addition, in studies comparing the **ergogenic** effect of ingesting carbohydrate with that of ingesting carbohydrate and BCAA combined, no added effect of BCAA was observed. The claim that BCAA reduces protein degradation appears to make sense. However, this conclusion was drawn from early in vitro studies using a control environment outside the living system. Several in vivo studies in healthy individuals failed to demonstrate an improved nitrogen balance associated with the consumption of BCAA during or after exercise (Nair et al. 1992; Frexes-Steed et al. 1992; Louard et al. 1990).

> ### ▶ KEY POINT ◀
>
> Leucine, isoleucine, and valine are the three branched-chain amino acids. Some claim that they can serve as additional energy fuel; prevent or attenuate excessive loss of protein; and help in improving function of neurotransmitters, thereby reducing the feeling of fatigue. In light of these claims, this group of amino acids has been added in many sport drinks.

Recently, Eric Newsholme, a biochemist at Oxford University, proposed the central fatigue hypothesis, postulating that high levels of serum free tryptophan (fTRP) in conjunction with low levels of BCAA, or a high fTPR:BCAA ratio, may be a major factor causing fatigue during prolonged endurance exercise. This contention was developed based on the fact that fTRP is used for the production of serotonin, which is believed to play a key role in the onset of fatigue. Branched-chain amino acids, on the other hand, can compete against fTRP for carrier-mediated entry into the central nervous system, thus mitigating the production of serotonin. The central fatigue hypothesis also predicts that ingestion of BCAA will raise the plasma BCAA concentration and thus reduce transport of fTRP into the brain. The subsequent reduced formation of serotonin may alleviate sensations of fatigue, thereby improving endurance performance. If the central fatigue hypothesis is true, then it would also follow that the consumption of tryptophan before exercise decreases the time to exhaustion, thereby impairing performance. Nevertheless, a study by Stensrud and colleagues (1992) demonstrated no differences in exhaustive running performance between those who were on tryptophan and those who were on placebo. It appears that further research is still needed in order to substantiate the claims associated with BCAA.

SUMMARY

Carbohydrate and fat are the two primary sources of energy used in events that require replenishment of ATP in order to meet the energy demand. While carbohydrate is stored as glycogen in muscles and the liver, fat is stored in adipose tissue and muscles. Glycogenolysis and lipolysis are the two processes by which stored carbohydrate and fat are degraded into simple forms of molecules such as glucose and fatty acids, respectively. Compared to fat, carbohydrate provides energy more quickly, can be used whether or not oxygen is present, and serves as the sole source of energy for the central nervous system. It must also be available in order for the body to use fat. On the other hand, fat, virtually an unlimited source of energy, could be of importance under conditions in which carbohydrate stores decrease significantly.

Utilization of carbohydrate and fat is influenced by exercise intensity and duration. It is generally true that the greater the intensity of exercise, the greater the utilization of glycogen. However, this rather simple relationship does not hold for fat. Although the percent contribution of fat decreases as intensity increases, the actual amount of fat being oxidized is maximized at moderate intensities (i.e., 60-65% $\dot{V}O_2$max). As exercise continues, a gradual decrease in carbohydrate utilization is accompanied by a corresponding increase in fat utilization.

Protein or amino acids are rarely used for energy in most circumstances. However, under conditions in which bodily carbohydrate stores decrease significantly, they can furnish energy by donating their carbon skeletons via the Krebs cycle or through conversion into glucose via gluconeogenesis.

KEY TERMS

β-oxidation

branched-chain amino acids

carnitine

carnitine acyltransferase

catecholamines

efferent

endogenous

ergogenic

gluconeogenesis

hepatic glucose output

hypoglycemia

lactate

lactate threshold

lipase

nitrogen balance

splanchnic

REVIEW QUESTIONS

1. Why is carbohydrate often referred to as the most preferable source of energy?

2. How is the use of energy fuels influenced by exercise intensity?

3. How does the duration of exercise affect the use of energy substrates?

4. Define *gluconeogenesis*. How is this process related to the Cori cycle and the glucose-alanine cycle?

5. What is β-oxidation? How does this process differ from lipolysis?

6. Describe how protein participates in energy metabolism. Why are branched-chain amino acids considered ergogenic and used widely in sport?

3

REGULATION OF ENERGY METABOLISM

The basics of the metabolic pathways, as well as patterns of energy substrate utilization during exercise, are covered in chapters 1 and 2. One must keep in mind that an increased rate of energy metabolism takes place only if energy demand increases. In this context, questions remain with respect to how the demand-driven energy metabolism comes about and how the body modulates the rate of energy metabolism. This chapter deals with issues related to the intrinsic regulation of energy metabolism using carbohydrate, fat, and protein, with particular emphasis on how the energy metabolism of each of the three macronutrients is influenced by exercise-induced alterations in energy substrate and subsequent hormonal secretion.

OVERVIEW OF A BIOLOGICAL CONTROL SYSTEM

Muscular exercise can be considered a dramatic test of the body's homeostatic control systems. This is so because exercise has the potential to disturb many homeostatic variables. For example, heavy exercise results in large increases in the muscle oxygen (O_2) requirement and in the production of large amounts of carbon dioxide (CO_2). These changes must be corrected through increases in breathing and blood flow to increase O_2 delivery to the exercising muscle and remove metabolically produced CO_2, which otherwise would increase the body's acidity. In addi-

tion, as heavy exercise begins, there is an immediate increase in the use of adenosine triphosphate (ATP). As a result, ATP storage decreases. The body's energy systems must respond rapidly to replenish ATP from substrates such as phosphocreatine (PCr) and carbohydrate so that a continuous energy supply and thus energy homeostasis can be maintained.

Homeostasis and Steady State

French physiologist Claude Bernard was the first to recognize—in 1857—the central importance of maintaining a stable internal environment. This concept was further elaborated and supported in 1932 by the American physiologist Walter Cannon, who emphasized that such stability could be achieved only through the operation of a carefully coordinated physiological process. The activities of cells, tissues, and organs must be regulated and integrated with one another in such a way that any change in the internal environment initiates a reaction designed to minimize the change. Cannon used the term **homeostasis** to denote the maintenance of a constant or unchanging internal environment. Of course, changes in the composition of the internal environment do occur; but the magnitudes of those changes are small and are kept within narrow limits via multiple coordinated homeostatic processes.

A similar term, **steady state,** is often used by exercise scientists to denote a steady physiological environment. Although the terms steady state and homeostasis are often used interchangeably and both conditions result from compensatory regulatory responses, homeostasis generally refers to a relatively constant environment during unstressful conditions such as rest; in contrast, a steady state does not necessarily mean that the internal environment is completely normal, but simply that it is unchanging (Vander et al. 2001). In other words, a steady state reflects only a stability of the internal environment that is achieved via a balance between the demands placed on the body and the body's responses to those demands. An example that helps in distinguishing between these two terms is oxygen consumption during exercise. As shown in figure 3.1, at the beginning of moderate-intensity exercise, oxygen uptake reaches a plateau level within a few minutes. This plateau of oxygen uptake is a steady-state metabolic rate specific to the exercise. However, this constant oxygen uptake occurs at a rate greater than that of the resting level of metabolism and thus does not reflect a true homeostatic condition.

The fact that the internal environment, including such factors as body temperature, blood pressure, plasma glucose, or acidity, is kept relatively constant in most circumstances suggests that the body operates many control systems that maintain homeostasis on a regular basis. Indeed, every one of the fundamental processes performed by any single cell must be carefully regulated. What determines how much glucose enters a cell? Once inside the cell, what determines

how much of this glucose is used for energy and how much is stored as glycogen? To answer these questions, it is important to understand not only the metabolic processes but also the mechanisms that control them.

Control System and Its Operation

The body has hundreds of different control systems that regulate physiological variables. A control system within the organism can be defined as a series of interconnected components that maintain physiological and chemical parameters of the body at near-constant value. The general components of the system are (1) receptors, (2) the afferent pathway, (3) an integrating center, (4) the efferent pathway, and (5) effectors. Figure 3.2 presents a schematic for such a control system. A receptor is capable of detecting an unwanted change or disturbance in the environment and sends a message to the integrating center, which assesses the strength of the stimulus and the amount of the response needed to correct the disturbance. The pathway traveled by the signal between the receptor and the integrating center is known as the afferent pathway. The integrating center then sends an appropriate output message to an effector, which is responsible for correcting the disturbance and causes the stimulus to be removed. The pathway along which this output message travels is known as the efferent pathway.

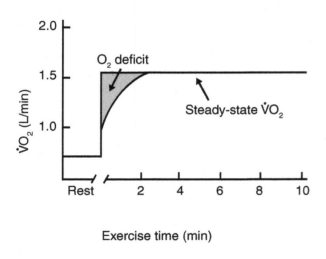

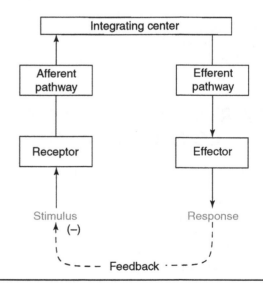

Figure 3.1 The time course of oxygen uptake ($\dot{V}O_2$max) in the transition from rest to submaximal exercise.

Adapted from S.K. Powers and E.T. Howley, 2004, *Exercise physiology: Theory and applications to fitness and performance*, 5th ed. (New York, NY: McGraw-Hill), 51, by permission of The McGraw-Hill Companies.

Figure 3.2 Schematic illustration of a biological control system.

Adapted from A.J. Vander, J. Sherman and D.S. Luciano, 1985, *Human physiology: the mechanisms of body function*, 4th ed. (New York, NY: McGraw-Hill), 156, by permission of The Mc-Graw-Hill Companies.

Most control systems of the body operate via **negative feedback.** Negative feedback is defined as the working process whereby a change in the variable being regulated brings about responses that tend to push the variable in the direction opposite to the change. An example of negative feedback can be seen in the respiratory control of CO_2 concentration in the **extracellular** fluid. An increase in extracellular CO_2 above the normal level triggers a chemical receptor, which sends information to the respiratory control center in the brainstem to increase breathing. The effectors in this example are respiratory muscles, and an increase in their contraction will reduce extracellular CO_2 concentration back to normal, thereby reestablishing homeostasis. There is, however, another type of feedback, known as positive feedback, whereby an initial disturbance in a system sets off a series of events that further increases the disturbance. Positive feedback does not favor the maintenance of the internal environment.

KEY POINT

The body's internal environment remains relatively constant despite changes in the external environment. The internal environment is maintained by control systems. A control system operates via negative feedback, the process whereby a change in a regulated variable brings about responses that tend to push the variable in the opposite direction. A control system detects, processes, and makes appropriate adjustments to correct the change.

Traditionally, the concept of a control system was restricted to situations in which the first four components were all parts of the nervous system. However, the term is no longer so narrowly focused, reflecting the recognition that the principles are essentially the same when blood-borne messengers such as hormones, rather than nerve fibers, serve as the afferent or (much more commonly) the efferent pathway, with an endocrine gland serving as the integrating center. In the case of thermoregulation, for example, when body temperature drops, the integrating centers in the brain not only send signals by way of nerve fibers to muscles to trigger contraction, but also cause the release of hormones that travel via the blood to many target cells, producing an increase in thermogenesis. Although hormones play an integral role in maintaining homeostasis, a control system that involves hormones may lack a receptor and an afferent pathway. For example, the release of **parathyroid hormone** is triggered by a fall in plasma calcium concentration. This hormone then functions to increase the release of calcium from bone into the blood. Likewise, the release of **insulin** is caused by a rise in plasma glucose concentration. This hormone then functions to increase cellular glucose uptake from blood. In both examples, the objective of the control system is to maintain a normal plasma concentration of calcium or glucose. However, neither control process involves a receptor or an afferent pathway. This is the case because glandular cells themselves are sensitive to changes in the chemical concentration of the blood that supplies them (Vander et al. 2001).

NEURAL AND HORMONAL CONTROL SYSTEMS

In light of the previous discussion, it is clear that both the nervous and endocrine systems are involved in the control and regulation of various functions in order to maintain homeostasis. Both are structured to be able to sense information, organize an appropriate response, and deliver the message to the proper organ or tissue. The two systems often work together to maintain homeostasis. However, they differ in that in order to deliver the output message the endocrine system relies on hormonal release, whereas the nervous system uses **neurotransmitters.**

With respect to nervous control, the autonomic nervous system is the efferent branch of the nervous system and is most directly related to the regulation of the internal environment (Brooks et al. 2005). The autonomic nerves innervate glands, blood vessels, cardiac muscle, and smooth muscle found in the respiratory and gastrointestinal systems. As such, the system operates below the conscious level. The autonomic nervous system can be further separated into the **sympathetic** and **parasympathetic** divisions. The parasympathetic division controls resting functions and has effects such as slowing the heart rate and stimulating digestion. It is composed of neurons that release **acetylcholine** (ACh). On the other hand, the sympathetic division controls fight-or-flight responses. Unlike the parasympathetic division, this division is composed of two types of neurons. The first type of neuron releases ACh, but the second type, which directly innervates the cell, releases **norepinephrine.** These neurotransmitters bind to receptors in the cell membranes of target tissues, altering the **membrane permeability** to certain ions. For example, in the heart, ACh promotes entry of Cl^{-1} to deter the occurrence of an action potential, whereas norepinephrine stimulates the entry of Na^+ and Ca^{++} to facilitate the production of an action potential. Consequently, ACh slows the heart rate, whereas norepinephrine speeds up the heart rate.

The endocrine glands release hormones directly into the blood, which carries the hormone to a tissue to exert an effect. The hormone exerts its effect by binding to a specific protein receptor. In doing so, the hormone can circulate to all tissues but affect only the tissues that have the specific receptor. As mentioned earlier, hormonal secretion from endocrine glands is regulated by feedback mechanisms. That is, a hormone is released in response to a change in the internal environment. However, the secretion of the hormone will diminish and eventually stop if a particular end result of the hormonal action is achieved. Of the many endocrine glands, the pancreas and adrenal glands are perhaps the most relevant to exercise metabolism.

> ### ▶ K E Y   P O I N T ◀
>
> Both the nervous and endocrine systems are involved in the control and regulation of various functions in order to maintain homeostasis. The two systems often work together as part of a control system. However, they differ in that in order to deliver the output message, the endocrine system relies on hormonal release, whereas the nervous system uses neurotransmitters.

Pancreatic Hormones

The pancreatic hormones are proteins secreted by the **islets of Langerhans,** clusters of endocrine cells in the pancreas. The name recognizes the work of German pathologist Paul Langerhans (1847-1888), who discovered these specialized cells in 1869. Islets of Langerhans contain several distinct types of cells, among which the α-cells and β-cells have been extensively investigated. The β-cells secrete insulin, which stimulates glucose and amino acid uptake by many cells; of these, muscle and adipose tissue cells are quantitatively the most important. Uptake is followed by increased synthesis of gly-

cogen and protein in muscle and of triglycerides in adipose tissue. High levels of circulating insulin also inhibit hepatic glucose output and thus promote glycogen as well as triglyceride synthesis in the liver. The α-cells of the pancreas secrete **glucagon.** While insulin promotes removal of glucose from the blood if the level is too high, glucagon functions to raise the blood glucose level if it is too low. Unlike insulin, glucagon exerts its effect primarily on the liver. It enhances both glycogenolysis and gluconeogenesis, the two processes that generate free glucose. An increase in gluconeogenesis is achieved via glucagon's role of stimulating hepatic amino acid uptake. Figure 3.3 illustrates a negative feedback control process in which insulin and glucagon function together to help maintain a relatively stable blood glucose concentration.

Adrenal Hormones

The **adrenal gland** has two sections: the adrenal **medulla** and the adrenal **cortex.** The adrenal medulla releases both **epinephrine** and norepinephrine, which

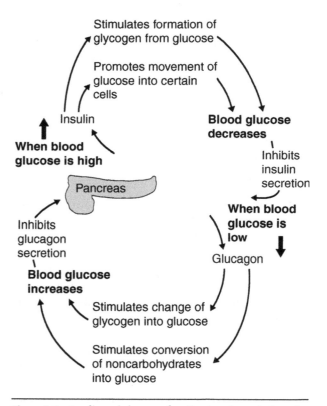

Figure 3.3 Insulin and glucagon function together to help maintain a relatively stable blood glucose concentration.

Adapted from D. Shier, J. Butler and R. Lewis, 1993, *Hole's human anatomy and physiology*, 6th ed. (New York, NY: McGraw-Hill), 489, by permission of The McGraw-Hill Companies.

are known as catecholamine. These two hormones are involved not only in activating energy metabolism in order to meet the demand of exercise, but also in maintaining blood glucose concentrations. They are also important in regulating cardiovascular and respiratory responses in an effort to facilitate energy homeostasis. The adrenal medulla is innervated by the sympathetic nervous system. As such, sympathetic activity stimulates the secretion of these hormones from the adrenal medulla. The adrenal cortex, the outer part of the adrenal gland, produces **cortisol, aldosterone,** and sex hormones; among these, only cortisol is directly related to energy metabolism. Cortisol contributes to the maintenance of plasma glucose by stimulating lipolysis in adipose tissue and gluconeogenesis in the liver. Unlike the release of catecholamines, which is controlled by sympathetic nerves, cortisol secretion is subject to the action of the stimulating hormones secreted by the hypothalamus and is regulated by a negative feedback mechanism (figure 3.4). Table 3.1 lists selected hormones and their catabolic roles in maintaining energy homeostasis.

Working Mechanisms

To produce their action, the catecholamines interact with two receptors, referred to as α and β receptors, that are located on the cell membrane surface.

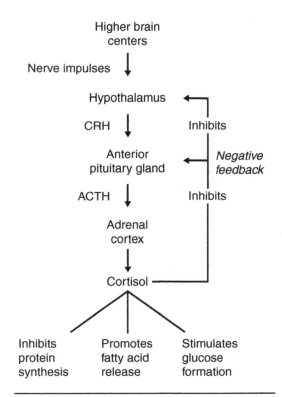

Figure 3.4 Regulation of cortisol secretion by a negative feedback mechanism.

Reprinted from D. Shier, J. Butler and R. Lewis, 1993, *Hole's human anatomy and physiology*, 6th ed. (New York, NY: McGraw-Hill), 486, by permission of The McGraw-Hill Companies.

Table 3.1 Selected Hormones and Their Catabolic Role in Maintaining Energy Homeostasis

Endocrine gland	Hormone	Catabolic action	Controlling mechanism	Stimuli
Anterior pituitary gland	Growth hormone	Mobilization of FFA Gluconeogenesis	Hypothalamic GH-releasing hormone	Exercise stress Low plasma glucose
Pancreas β-cells	Insulin	Uptake of glucose, amino acids, FFA into tissues	Plasma glucose concentration Autonomic nervous system	Elevated plasma glucose Decreased epinephrine and norepinephrine
Pancreas α-cells	Glucagon	Mobilization of FFA, glucose Gluconeogenesis	Plasma glucose concentration Autonomic nervous system	Low plasma glucose Elevated epinephrine and norepinephrine
Adrenal cortex	Cortisol	Mobilization of FFA Gluconeogenesis	Hypothalamic adrenal cortex stimulating hormone	Exercise stress Low plasma glucose
Adrenal medulla	Epinephrine Norepinephrine	Glycogenolysis Mobilization of FFA	Autonomic nervous system	Exercise stress Low plasma glucose

FFA = free fatty acids. GH = growth hormone.

Adapted from S.K. Powers and E.T. Howley, 2004, *Exercise physiology: Theory and applications to fitness and performance*, 5th ed. (New York, NY: McGraw-Hill), 88-89, by permission of The McGraw-Hill Companies.

Norepinephrine affects mainly the α receptors, whereas epinephrine affects both α and β receptors. The β receptors can be further subdivided into β1 and β2 receptors. The actions of these receptors are listed in table 3.2.

Both α and β receptors are sometimes called adrenergetic because they can be activated by epinephrine and norepinephrine. Once bound to either hormone, these receptors cause changes in cellular activity by increasing or decreasing cyclic adenosine monophosphate (AMP) or Ca^{++}, which are often referred to as **second messengers.** Second messengers can be viewed as intracellular molecules or ions that are regulated by extracellular signaling agents such as neurotransmitters and hormones (first messengers). The second messenger then activates another set of enzymes called protein kinases, which trigger various cellular events in response to the original stimulus. Unlike catecholamines, which are composed of **peptides,** cortisol is a lipid hormone that can diffuse easily through the cell membrane and become bound to a protein receptor in the cytoplasm of the cell. The hormone–receptor complex enters the nucleus and binds to a specific protein linked to deoxyribonucleic acid (DNA). This then leads to the synthesis of proteins necessary to alter the metabolism. This process does not involve the production of second messengers. It takes longer for the action of cortisol to be turned on, but its effect lasts longer than that of catecholamines.

> **KEY POINT**
>
> Hormones involved in energy metabolism first combine with protein receptors and then activate enzymes necessary to catalyze the intended chemical reactions. Peptide hormones such as catecholamines bind with receptors on the cell membrane, which triggers the production of a second messenger within the cell that helps complete the action of the hormone. In contrast, lipid-like hormones such as cortisol diffuse across the cell membrane and bind to a receptor within the cell before exerting their action.

REGULATION OF SUBSTRATE METABOLISM DURING EXERCISE

In response to an increase in exercise intensity, utilization of muscle glycogen increases. As a bout of exercise continues over a prolonged period, there is a gradual shift toward an increased use of liver glycogen—a change that is needed to maintain energy supply and blood glucose homeostasis. As carbohydrate energy stores diminish, both degradation and oxidation of triglycerides increase to ensure energy homeostasis. This pattern of substrate utilization was outlined in chapter 2. However, what has not yet been discussed

Table 3.2   Interaction of Epinephrine and Norepinephrine With Adrenergic Receptors

Receptor type	Intracellular mediator	Effect
α	Cyclic AMP and Ca^{++}	Vasoconstriction Intestinal relaxation
β1	Cyclic AMP	Increased heart rate Increased cardiac contraction Increased lipolysis Increased glycogenolysis
β2	Cyclic AMP	Vasodilation Bronchodilation

Adapted from J. Tepperman and H.M. Tepperman, 1987, *Metabolic and endocrine physiology*, 5th ed. (Chicago, IL: Year Book Medical Publishers).

is how these changes come about in terms of both an increase in utilization and a shift between different pools of energy substrates. Chapter 1 covered the various metabolic pathways through which carbohydrate and fat are degraded to yield the energy necessary to support the increased energy demand imposed by exercise. Note that the speed at which these pathways operate is regulated directly by a series of enzymes that catalyze reactions. Moreover, the extent to which these enzymes are activated is in turn controlled by the release of certain catabolic hormones. The following sections deal more specifically with the factors thought to control metabolic processes during exercise.

Regulation of Muscle Glycogen Degradation

Muscle glycogen is the primary carbohydrate fuel for most types of exercise; and the heavier the exercise intensity, the faster glycogen is degraded. Interestingly, an increase in plasma epinephrine, a hormone that facilitates glycogenolysis, has been found to correlate positively with exercise intensity (Kjaer 1989). It is generally thought that epinephrine plays the most important role in mediating glycogen degradation. This catabolic effect is believed to be initiated by second messengers, which activate protein kinases needed for glycogenolysis; and plasma epinephrine is responsible the formation of cyclic AMP when bound to β-adrenergic receptors (Hargreaves 2006).

Muscle glycogenolysis is also regulated by the activity of **glycogen phosphorylase,** the key enzyme for this process. Glycogen phosphorylase catalyzes the first step of glycogen breakdown and is responsible for supplying individual glucose molecules to the glycolytic pathway for producing ATP. In the resting state, this enzyme exists primarily in the inactive *b* form, whose activity can be stimulated by an increase in adenosine diphosphate (ADP) and AMP or inhibited by an increase in ATP. In response to muscle contraction or stimulation by epinephrine, the phosphorylase *b* inactive form is converted to the phosphorylase *a* active form. However, under the influence of insulin, activated phosphorylase can become deactivated (Johnson 1992). This then reduces the availability of glucose.

Substrate availability also affects the rate of glycogen degradation. Early studies showed that increases in preexercise muscle glycogen result in enhanced muscle glycogen utilization during exercise (Richter and Galbo 1986). It was proposed that glycogen can bind to glycogen phosphorylase and, in doing so,

increase the enzyme's activity (Johnson 1992). There is also evidence that alterations in the availability of blood-borne substrates such as glucose may influence muscle glycogenolysis. Coyle and colleagues (1991) found that an increase in blood glucose as a result of intravenous infusion of glucose resulted in a decrease in muscle glycogen utilization. However, when the blood glucose level was brought down, no alteration in glycogen utilization was observed.

> **▶ K E Y   P O I N T ◀**
>
> A key enzyme that regulates muscle glycogen degradation is glycogen phosphorylase. This enzyme is influenced by the level of ATP, epinephrine concentration, and glycogen stores. Increases in ADP and AMP levels, epinephrine release, and muscle glycogen concentration have been found to stimulate glycogen phosphorylase and thereby glycogenolysis.

Regulation of Glycolysis and the Krebs Cycle

Glycolysis and the aerobic pathway involving the Krebs cycle are the two possible pathways in which glucose is further metabolized with or without oxygen. Each process consists of a series of sequential chemical reactions, and each reaction is catalyzed by a specific enzyme. The rate of glycolysis is controlled by the activity of **phosphofructokinase** (PFK). Phosphofructokinase catalyzes the third step of glycolysis. When exercise begins, increases in ADP and P_i levels activate PFK, thereby accelerating glycolysis. Phosphofructokinase is also activated by an increase in cellular levels of hydrogen ions and ammonia. The Krebs cycle, like glycolysis, is subject to enzymatic regulation. Among the numerous enzymes involved, **isocitrate dehydrogenase** (IDH) is thought to be the rate-limiting enzyme in the aerobic pathway (Stanley and Connett 1991). This enzyme catalyzes a reaction during the early phase of the Krebs cycle. Like PFK, it is stimulated by ADP and P_i and inhibited by ATP. Isocitrate dehydrogenase is also sensitive to changes in cellular levels of calcium. McCormack and Denton (1994) found that an increase in Ca^{++} levels in mitochondria stimulated IDH. This finding is congruent with the concept that an increase in Ca^{++} in muscle is essential to initiate muscle contraction requiring energy. It is intriguing to note that these two rate-limiting enzymes are found early in the metabolic pathway. This is an important phenomenon because unwanted products of a pathway could accumulate if these enzymes were located near its end.

> ▶ **K E Y   P O I N T** ◀
>
> Phosphofructokinase (PFK) and isocitrate dehydrogenase (IDH) are the rate-limiting enzymes that control glycolysis and the Krebs cycle, respectively. The rate-limiting enzymes are found early in the metabolic pathway, which is important for preventing an accumulation of unwanted metabolic by-products.

Effect of Fatty Acid Availability on Carbohydrate Utilization

In light of the preceding discussion, one can conclude that carbohydrate utilization is regulated by the energy needs of the exercising muscle. Carbohydrate utilization can also be influenced by the availability of energy substrates, and it has been found that an increase in fatty acids reduces carbohydrate utiliza-

tion. By infusing triacylglycerol (Intralipid), Costill and colleagues (1977) found that muscle glycogen breakdown was reduced with elevated plasma fatty acids following 60 min of moderate-intensity exercise. A similar finding was also obtained in a study in which subjects were fed fat in conjunction with heparin, which helps to facilitate lipolysis (Vukovich et al. 1993).

As briefly mentioned in chapter 2, this interaction between carbohydrate and fat may be explained by the classical glucose–fatty acid cycle delineated by British biochemist Philip Randle over 40 years ago. As shown in figure 3.5, with an increase in plasma fatty acid concentration, there is an increase in fatty acid entry into the cell and a subsequent increase in β-oxidation whereby fatty acids are broken down to acetyl-CoA. An increased concentration of acetyl-CoA then inhibits the **pyruvate dehydrogenase complex**, which breaks down pyruvate to acetyl-CoA. In addition, the increased production of acetyl-CoA raises the concentration of citrate, an intermediate of the Krebs cycle. This increased citrate level then inhibits PFK, a rate-limiting enzyme of glycolysis, as well as **hexokinase**, which regulates cellular glucose uptake. Together, these reduced enzymatic activities decrease

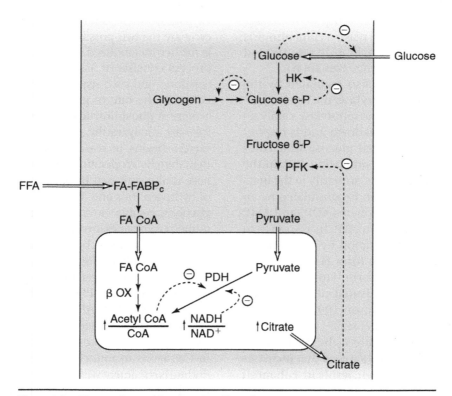

Figure 3.5   Glucose–fatty acid cycle or Randle cycle.

Reprinted, by permission, from M.E. Houston, 2006, *Biochemistry primer for exercise science*, 3rd ed. (Champaign, IL: Human Kinetics), 174.

carbohydrate utilization. An increase in fat utilization also helps decrease cellular levels of AMP and P_i and therefore mitigates carbohydrate breakdown (Dyck et al. 1993, 1996). As discussed earlier, increases in cellular AMP and P_i reflect a low energy state and as a result stimulate the activity of PFK in an attempt to maintain energy homeostasis.

> **KEY POINT**
>
> An increase in fatty acid availability can reduce dependence on carbohydrate energy sources. This is significant in prolonged exercise in that an increased use of fat helps spare muscle and liver glycogen. The inhibitory effect of fat on carbohydrate utilization can be explained by the Randle cycle theory, which suggests that an increase in acetyl-CoA derived from fat degradation will suppress a number of enzymes necessary for converting glucose into energy.

Regulation of Hepatic Glucose Output

Glucose released from the liver plays an important role in maintaining blood glucose during exercise. As utilization of blood glucose increases in exercising muscle, glucose output from the liver increases concurrently so that symptoms associated with hypoglycemia are prevented. In principle, the regulation of hepatic glucose output during exercise is believed to be accomplished through a negative feedback mechanism. Via glucose infusion, Jenkins and colleagues (1985, 1986) found that an increase in blood glucose concentration inhibited hepatic glucose production during moderate-intensity exercise. As a change in blood glucose is linked to the production of insulin and glucagons, it appears that this feedback mechanism may be hormone related. Indeed, both insulin and glucagon can influence the activity of the liver as the target organ. By manipulating insulin and glucagon levels with the infusion technique, researchers have demonstrated favorable responses of hepatic glucose output (Wasserman et al. 1984, 1989). In other words, when plasma insulin was made to decrease or when plasma glucagon was made to increase, an increase in hepatic glucose production resulted. The greatest effect on hepatic

glucose uptake was observed with a simultaneous decrease in glucagons and an increase in insulin (Marker et al. 1991).

Hepatic glucose output can outpace the rate at which blood glucose is taken up by the tissue, thereby causing **hyperglycemia.** Various studies have shown an increase in plasma glucose during intense exercise (Kjaer et al. 1986, 1987, 1991). The authors ascribed the increase in plasma glucose to an increase in hepatic glucose output that exceeded the rise in glucose utilization by exercising tissues. As this mismatch between hepatic glucose production and peripheral glucose utilization is more pronounced with increasing exercise intensity, Kjaer (1995) suggested that the regulation of glucose production is in part mediated by activity in motor centers in the central nervous system in a feedforward fashion; that is, the greater the exercise intensity, the greater the central command and thus the greater hepatic glucose output. Glucose production from the liver may also be subject to control by adrenergetic activities. In a study using leg exercise as well as combined arm and leg exercise, Kjaer and colleagues (1991) observed a positive correlation between plasma catecholamines and hepatic glucose output. In addition, in an animal study in which the adrenal medulla was removed, Richter and colleagues (1981) and Sonne and colleagues (1985) found reduced liver glycogenolysis and hepatic glucose output. These findings suggest that epinephrine and norepinephrine also play a role in the exercise-induced increase in glucose output from the liver.

Exercise-induced hyperglycemia is a transitory phenomenon. Upon completion of exercise, blood glucose is quickly restored to normal levels, although this may not be the case for people with diabetes who have diminished insulin secretion.

> **KEY POINT**
>
> The primary function of glucose output from the liver is to maintain normal blood glucose concentration, although some glucose can also be taken in by muscle tissue for use as energy. Hepatic glucose output is well regulated by insulin and glucagon so that under normal circumstances there is not a mismatch between glucose output and utilization. However, during high-intensity exercise, hepatic glucose output can outpace utilization and thereby cause hyperglycemia.

Regulation of Lipolysis

To be metabolized, storage fat must first undergo lipolysis, whereby triglycerides are degraded to fatty acids and glycerol. The resulting fatty acids are then converted into acetyl-CoA and enter the Krebs cycle for further metabolism. Adipose tissue lipolysis is controlled by hormone-sensitive lipase, which breaks the bonding of triglycerides to form fatty acids and glycerol. This enzyme is regulated by hormones—insulin, glucagon, catecholamines, and cortisol, as well as **growth hormone**, which is released from the **pituitary gland.** Of these hormones, insulin is the only one that inhibits lipolysis; each of the others functions as a stimulator. Among the stimulatory hormones, catecholamines are the most potent stimulator of lipolysis during exercise. Catecholamines are bound to the β-adrenergic receptors, causing production of the second messenger cyclic AMP. This latter product then triggers a series of chemical events whereby hormone-sensitive lipase becomes activated (Fain and Sainz-Garcia 1983). On the other hand, insulin can reverse the effects of lipolytic hormones. The mechanism underlying the antilipolytic action of insulin remains uncertain. It is thought that insulin suppresses lipolysis either by decreasing cyclic AMP concentration or by inhibiting the enzymes needed to activate hormone-sensitive lipase (Gabbay and Lardy 1984).

Just as fatty acid availability affects carbohydrate utilization, plasma glucose concentration can influence adipose tissue lipolysis. Wolfe and colleagues (1987) found that in healthy subjects, glucose infusion suppressed the rate of appearance of glycerol, a common measure of lipolysis. From that study, however, it is difficult to discern whether the decreased lipolysis was caused by an increase in plasma glucose or by a resultant increase in plasma insulin. In a later study in which insulin concentration was kept unchanged, Carlson and colleagues (1991) showed that the appearance of glycerol decreased by 32% due to hyperglycemia, with plasma glucose raised to ~10 mM. This finding suggests that plasma glucose regulates fat mobilization independently of changes in hormones.

Regulation of Triglyceride Utilization

Regulation of fat utilization requires the transport of fatty acids into the mitochondria where fat can be oxidized. In this context, the mass of mitochondria and oxygen delivery are very important factors in determining rates of fat utilization. Tissues such as the heart and liver are highly adapted for fat utilization, whereas brain and red blood cells rely almost exclusively on glycolysis for energy. In skeletal muscle, the ability to use fat as an energy source varies according to muscle fiber type. Fast-twitch muscle fibers are limited in their fat utilization because of their low volume of mitochondria, as well as a less than optimal blood supply. In contrast, slow-twitch muscle fibers are highly capable of oxidizing fat because they are rich in mitochondria and capillaries.

The process by which fatty acids are transported across the mitochondrial membrane is mainly controlled by the activity of carnitine palmitoyl transferase I (CPT-I). As discussed in chapter 2, CPT-I is part of a transport system needed in order for long-chain fatty acids to enter the mitochondrion, where oxidation takes place. In this context, it makes sense that CPT-I plays an important role in the control of fat oxidation (Winder 2001). Carnitine palmitoyl transferase I is regulated by **malonyl-CoA,** an intermediate in fatty acid synthesis. An increase in malonyl-CoA content inhibits the activity of CPT-I (McGarry and Brown 1997, Rudderman et al. 1999), thereby reducing fat utilization. During high-intensity exercise, the high rate of glycogenolysis increases the amount of acetyl-CoA in the muscle cell, and some of this acetyl-CoA is converted to malonyl-CoA. The increased malonyl-CoA then suppresses CPT-I and thus reduces the transport of fatty acids into the mitochondria. Conversely, as carbohydrate energy sources are depleted, the inhibitory effect of malonyl-CoA on CPT-I is attenuated due to reduced glycogenolysis. Consequently, the activity of CPT-I is augmented and fat utilization is enhanced. Apart from malonyl-CoA, Starritt and colleagues (2000) suggested that a reduction in pH (increased acidity) associated with high-intensity exercise would also serve as an inhibitor of the activity of CPT-I.

In most situations, carbohydrate and fat together confer most, if not all, of the energy to be expended. However, compared with regulation of carbohydrate utilization, control of fat utilization is not as tight. For example, as mentioned earlier, the rate of carbohydrate utilization is closely related to changes in markers such as ADP and AMP that reflect the energy needs of the exercising muscle. This is not the case with fat utilization. The rate of fat oxidation appears to be influenced mainly by the oxidative capacity of the tissue, the availability of fatty acids, and the rate of carbohydrate utilization (Jeukendrup and Gleeson 2004).

> **▶ K E Y P O I N T ◀**
>
> Fat utilization is not as tightly controlled as carbohydrate utilization. While carbohydrate utilization is closely regulated by changes in AMP and ADP that reflect the energy needs of the cell, the rate of fat oxidation appears to be influenced mainly by the oxidative capacity of the tissue, the availability of fatty acids, and the rate of carbohydrate utilization.

Regulation of Protein Synthesis and Degradation

Protein is synthesized in order to provide architectural support, enzymes to catalyze metabolic reactions, and signaling intermediates within and between cells. However, protein can also be metabolized as a vital fuel when the body faces the need for additional energy, as during fasting or prolonged exercise. Protein synthesis and degradation are two opposite process that take place concurrently. Amino acids are constantly incorporated into proteins, and at the same time proteins are constantly broken down to make free amino acids available. Consequently, a balance between protein synthesis and degradation becomes a very important measure that reflects whether the body is retaining an adequate amount of lean tissue.

Protein Degradation

Protein synthesis and degradation are mainly regulated by hormonal secretion, which can be influenced by many circumstances such as exercise, stress, and dietary feeding. Glucagon, cortisol, and catecholamines are mainly associated with protein degradation, whereas insulin, growth hormones, and testosterone are primarily linked to protein synthesis. Thus far, cortisol has been considered the most potent stimulator of protein catabolism or degradation (Graham and MacLean 1992; Rooyackers and Nair 1997). Cortisol degrades tissue protein to yield amino acids for glucose synthesis in the liver via gluconeogenesis. As this action helps to generate new glucose units, its role during exercise has been extensively investigated. Cortisol secretion does not increase much during the early phases of exercise even when the exercise is strenuous (i.e., >75% $\dot{V}O_2$max). However, as highly intense exercise persists, the blood cortisol level begins to increase; this has been reported to occur 30 min or longer into exercise. The delayed rise in cortisol appears to occur just in time, because muscle glycogen is likely to decrease significantly during strenuous exercise that lasts for more than 30 min. In conditions in which the muscle glycogen level is low, an increase in cortisol secretion helps to maintain a continuous energy supply throughout the entire period of exercise.

Protein Synthesis

Insulin, growth hormone, and **testosterone** appear to be the most influential hormones mediating the anabolic process. Insulin is known for its role in regulating plasma glucose homeostasis in response to hyperglycemia. However, in addition to its effect on stimulating blood glucose uptake by peripheral tissues, insulin also promotes the entry of circulating amino acids into certain cells, such as skeletal muscle fibers, where protein synthesis takes place. Thus insulin is also regarded as an anabolic hormone in terms of protein synthesis. Insulin release can be blunted during exercise when intensity and duration exceed certain thresholds. This is an appropriate response, in that a decrease in insulin favors the mobilization of glucose from the liver and of fatty acids from adipose tissue, both of which are necessary to maintain the plasma glucose concentration. If exercise were associated with an increase in insulin, blood glucose would be taken up into tissues at a faster rate, leading to immediate hypoglycemia.

The effect of growth hormone on protein synthesis occurs via an increase in membrane transport of amino acids into cells, synthesis of ribonucleic acid (RNA) and ribosomes, and activity of ribosomes, as well as all other events essential to protein synthesis. Recent literature has also revealed a linkage between growth hormone administration and diminished amino acid oxidation (Rooyackers and Nair 1997). Growth hormone can act indirectly via enhanced hepatic release of **somatomedins,** which are carried by the blood to target tissues where they induce growth-promoting effects, particularly in cartilage and bone (Vander et al. 2001). Because the somatomedins are structurally and functionally similar to insulin, they are referred to as **insulin-like growth factors.** It is particularly intriguing that blood levels of growth hormone increase during vigorous exercise and remain elevated for some time period after exercise. Elevated growth hormone during exercise has been found to function similarly to cortisol as a lipolytic hormone in the maintenance of blood glucose homeostasis (Powers and Howley 2001).

Growth hormone stimulates fat breakdown and indirectly suppresses carbohydrate utilization. A low plasma glucose concentration can serve to stimulate the release of growth hormone by the anterior pituitary gland.

Testosterone and testosterone analogs such as anabolic steroids are well known for their anabolic effect on protein metabolism. However, the mechanism whereby testosterone regulates protein metabolism remains elusive. After administering testosterone over a 12-week period, Griggs and colleagues (1989) reported a 27% increase in protein synthesis. It remains doubtful whether this increased protein synthesis was brought about by an increase in amino acid uptake. It has been argued that testosterone increases protein synthesis mainly by enhancing the utilization of amino acids from the intracellular pool (Wildman and Miller 2004). Thus far, little is known about the effects of estrogen and progesterone on protein metabolism, although it is evident that menopause in women is associated with accelerated muscle loss (Forbes 1987).

 K E Y   P O I N T

Protein synthesis and degradation are mainly regulated by hormones. Glucagon, cortisol, and catecholamines are associated primarily with protein degradation. Insulin, growth hormones, and testosterone are the most influential hormones mediating the anabolic process, although growth hormone can function catabolically during exercise. Both cortisol and growth hormone stimulate lipolysis and gluconeogenesis during prolonged exercise when carbohydrate stores decrease significantly, helping to prevent hypoglycemia and muscle glycogen depletion.

SUMMARY

The term *homeostasis* is defined as the maintenance of a constant internal environment. It differs from the term *steady state* in that the latter refers to a constant internal environment achieved under stressful conditions such as exercise. The maintenance of a constant internal environment is achieved by many biological control systems that operate mainly in a negative feedback manner and are capable of detecting, processing, and making appropriate adjustments to correct changes. The nervous and endocrine systems often work together as part of a control system. They are structured to be able to sense information, organize an appropriate response, and deliver messages via neurotransmitters or hormones to the proper organ or tissue.

Muscular exercise can be a dramatic challenge to the homeostasis of energy. However, with the operation of control systems, in most cases individuals are able to respond to the challenge by augmenting energy provision. For example, exercise increases the use of ATP initially. By-products such as ADP, AMP, and P_i can serve as stimuli to cause the release of catabolic hormones, such as catecholamines, that are necessary to activate metabolic pathways so that ATP can be replenished almost instantaneously. Likewise, as muscle glycogen stores decrease, the production of glucagon, cortisol, and growth hormones increases correspondingly. These hormones can then function to stimulate lipolysis, as well as liver glycogenolysis and gluconeogenesis, so that additional energy substrates can be available for ATP synthesis.

KEY TERMS

acetylcholine
adrenal gland
aldosterone
cortex
cortisol
epinephrine
extracellular
glucagon
glycogen phosphorylase
growth hormone
hexokinase
homeostasis

hyperglycemia
insulin
insulin-like growth factors
islets of Langerhans
isocitrate dehydrogenase
malonyl-CoA
medulla
membrane permeability
negative feedback
neurotransmitters
norepinephrine
parasympathetic

parathyroid hormone
peptides
phosphofructokinase
pituitary gland
pyruvate dehydrogenase complex
second messengers
somatomedins
steady state
sympathetic
testosterone

REVIEW QUESTIONS

1. Define the term *homeostasis*. How does it differ from the term *steady state*?

2. What are the components of a biological control system? Give an example of the operation of a control system.

3. Briefly describe the theory of the glucose–fatty acid or Randle cycle.

4. Epinephrine, insulin, glucagon, cortisol, and growth hormone are the major hormones involved in energy metabolism. Which endocrine gland is each hormone secreted from? What specific role does each hormone play with regard to energy metabolism during exercise?

5. Blood glucose concentration is generally maintained throughout exercise. How does this come about? When can hypoglycemia and hyperglycemia occur?

II

APPLICATION OF BIOENERGETICS IN PHYSICAL ACTIVITY

As a result of cellular respiration, mechanical work is accomplished. This process of energy transformation is made possible by the consumption of oxygen, which plays a key role in ensuring efficient energy transfer. In this context, measurement of the amount of oxygen consumed enables us to indirectly track energy expenditure. Such a quantitative approach is imperative because it allows us not only to characterize metabolic profiles for various physical activities, but also to design exercise programs that can bring about desirable metabolic benefits.

The chapters in part II describe practical ways of using information pertaining to bioenergetics. These three chapters are arranged so as to give readers ideas about how they may apply what they learn. Chapter 4 demonstrates developments in the technology used to assess energy metabolism. Chapter 5 provides empirical data related to energy metabolism measured during physical activities and sports. Chapter 6 presents ideas, principles, and methodologies that can be used to formulate exercise programs aimed at maximizing energy expenditure.

4

MEASUREMENT OF ENERGY METABOLISM

This chapter reviews various techniques that are currently available for assessing energy expenditure and fuel utilization. It covers measurement techniques used in both laboratory and field settings. For each technique, the rationale and principle relating to the development of the technique are explained. This discussion is followed by details about the technique and an explanation of how the technique should be carried out. For field-based methodology, further information is provided concerning the validity and reliability of each technique presented.

LABORATORY APPROACHES

Interest in monitoring energy expenditure and respiratory gas exchange during physical activity can be traced back to almost 100 years ago when Haldane and Douglas, in preparation for the 1911 Anglo-American expedition to Pike's Peak in Colorado, developed the "Douglas bag," a method that measures oxygen consumption and carbon dioxide production. Since then a wide range of electronic and computer-assisted metabolic systems have been developed that allow precise quantification of energy expenditure during various physical activities. In today's world, with physical inactivity often playing a contributing role in the development of many chronic diseases, such technological advancement has enabled us not only to quantify the energy cost of an activity but also to assess the association between physical activity and health as well as the effectiveness of interventions aimed at increasing physical activity in order to treat or prevent diseases. In general, in the laboratory setting, two techniques are employed to measure energy expenditure and substrate utilization: direct **calorimetry** and indirect calorimetry.

Direct Calorimetry

Energy metabolism can be defined as the rate of heat production. This definition recognizes the fact that when the body uses energy to do work, heat is liberated. The production of heat occurs through cellular respiration and mechanical work. Direct calorimetry involves measurement of the heat produced during metabolism. The technique works similarly to the bomb calorimeter discussed in chapter 1, but it uses a chamber large enough to allow an individual to remain in the space and work for an extended period of time. Figure 4.1 depicts a room calorimeter designed on the basis of early experiments of Atwater and Benedict (1903). One can measure energy expenditure during muscular exercise by installing an exercise device such as a treadmill or bicycle ergometer within the chamber.

In this insulated calorimeter, a thin copper sheet lines the interior wall to which heat exchangers are attached. A known amount of water circulates through the heat exchange regularly, absorbing the heat radiated from the subject in the chamber; this reflects the person's metabolic rate. Insulation protects the entire chamber so that any change in water temperature relates directly to the individual's energy metabolism. The air is recirculated, and carbon dioxide and water are filtered out

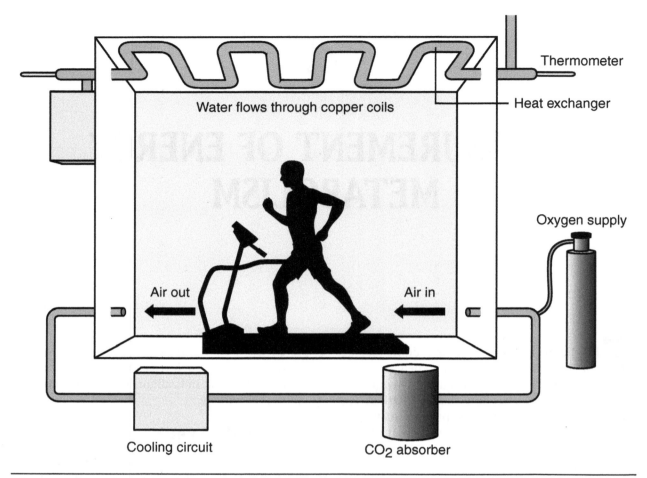

Figure 4.1 Direct calorimetry chamber.

Reprinted, by permission, from A. Jeukendrup and M. Gleeson, 2004, *Sports nutrition: An introduction to energy production and performance* (Champaign, IL: Human Kinetics), 66.

of the air before it reenters the chamber together with added oxygen. Direct measurement of heat production in humans has proven to be very precise based on the well-defined concept that 1 calorie is equivalent to the amount of heat needed to raise the temperature of 1 gram of water by 1 °C. However, application is limited because considerable time, expense, and engineering expertise are required for operating and maintaining the equipment (Ravussin and Rising 1992). In addition, though these factors are much less of a limitation, the results can be affected in that not all the heat produced is liberated to the environment so that it can be captured, and extra heat is produced due to the operation of the exercise equipment itself rather than resulting from metabolism.

Indirect Calorimetry

In indirect calorimetry, measurement of whole-body respiratory gas exchange is used to estimate the amount of energy produced though the oxidative process. The rationale behind indirect calorimetry is that virtually all bioenergetic processes are oxygen dependent (Ravussin and Rising 1992). Indirect calorimetry differs from direct calorimetry in that the indirect method determines how much oxygen is required for biological combustion to be completed, whereas the direct method measures the heat that is produced as a result of metabolism. The principle of indirect calorimetry can be explained by the following relationship:

$$\text{Substrate} + O_2 \rightarrow \text{heat} + CO_2 + H_2O$$

In light of the direct relationship between oxygen consumption and the amount of heat produced, it is logical that measuring the amount of oxygen consumed can replace measuring the heat produced as a result of biological oxidation.

To convert the amount of oxygen consumed into heat equivalents, one must know the type of energy substrate that is being metabolized, that is, carbohydrate or fat, since protein rarely participates in energy metabolism. The energy liberated when fat is the only substrate being oxidized is 4.7 kcal or 19.7 kJ per liter of oxygen used. However, the energy released when carbohydrate is the only fuel being oxidized is 5.05 kcal or 21.1 kJ per liter of oxygen used. Another measure often implemented, although it is less accurate, estimates the energy expenditure of exercise on the basis of 5 kcal or 21 kJ per liter of oxygen used. Therefore, a person exercising at an oxygen consumption of 2.0 L · min^{-1} would expend approximately 10 kcal or 42 kJ of energy per minute. Despite the simplicity of this method, this assumed value of energy equivalency should be used with caution because it implies that over 95% of energy comes from the oxidation of carbohydrate. In reality, this may not always be the case, as the energy contribution from carbohydrate should be much lower when exercise is performed within a low-intensity range. Consequently, the use of an energy equivalent of 4.825 kcal per liter of oxygen has been suggested for circumstances such as resting or steady-state exercise at mild intensities.

> **KEY POINT**
>
> Two laboratory procedures are commonly used to assess human energy expenditure: direct calorimetry and indirect calorimetry. Direct calorimetry measures the heat produced as a result of metabolism, whereas indirect calorimetry measures the amount of oxygen being used during energy transformation.

Closed-Circuit Spirometry

Oxygen uptake ($\dot{V}O_2$) and carbon dioxide production ($\dot{V}CO_2$) can be measured with either closed-circuit or open-circuit **spirometry.** With the closed-circuit method, the subject breathes through a mouthpiece into a spirometer that is filled with 100% oxygen. During each **inspiration,** some of the oxygen in the spirometer is consumed. Gas exhaled during **expiration** passes through a filter where carbon dioxide is removed. The residual oxygen in the spirometer is available for the next inspiration. As oxygen is consumed, the volume of oxygen in the spirometer decreases, and this change in volume is measured. Measuring $\dot{V}O_2$ using closed-circuit spirometry can be a potential problem during exercise because this method requires subjects to be attached to the equipment while exercising. The technique can also produce **airway resistance** at high flow rates. In addition, during strenuous exercise in which the pro-

duction of carbon dioxide is high, some of the expired carbon dioxide can be trapped in the system and be subsequently inhaled. This can reduce the amount of oxygen being taken in or the amount of carbon dioxide being eliminated, or both, and therefore impair exercise performance.

Open-Circuit Spirometry

The technique most commonly used to measure oxygen consumption during exercise is open-circuit spirometry. This method allows subjects to inhale ambient air with a constant composition of 20.93% oxygen, 0.03% carbon dioxide, and 79.04% nitrogen. $\dot{V}O_2$ is determined as the difference between the volume of oxygen inspired and the volume of oxygen expired:

$$\text{volume of } O_2 \text{ consumed} = \text{volume of } O_2 \text{ inspired} - \text{volume of } O_2 \text{ expired}$$

Laboratory equipment used to measure oxygen consumption is illustrated in figure 4.2. The volume of air inspired and expired is measured with a gas meter that is attached to a subject by means of a flexible hose and a face-fitting mask. The expired gas from the

Figure 4.2 An open-circuit indirect calorimetry system.
© Human Kinetics

subject is analyzed for fractions of oxygen and carbon dioxide by electronic gas analyzers. Results are then sent to a computer via an analog–digital converter, and the computer is programmed to perform the necessary calculations of $\dot{V}O_2$ and other metabolic parameters.

Respiratory Chamber

The technology of indirect calorimetry has become increasingly sophisticated in recent years. A respiratory chamber based on the principle of open-circuit indirect calorimetry appears similar to the chamber used for direct calorimetry but does not have the coils for measuring heat exchange. This room-sized chamber has the basic furniture and utilities needed for carrying out various daily functions so that measurements can be made in a real-life situation. The chamber is equipped with instrumentation that can measure oxygen uptake and carbon dioxide, as well as **pulmonary ventilation;** these measurements can be performed continuously for an extended period of time. With use of this large chamber, all components of daily energy expenditure can be assessed. These include **basal metabolic rate** (BMR), sleeping metabolism, energy cost of arousal (BMR minus sleeping metabolism), **thermic effect of food,** and energy cost of physical activities.

Although indirect calorimetry does not involve the direct measurement of heat production, this technique is still considered quite accurate in reflecting energy metabolism and substrate oxidation. In fact, the technique has been extensively used as a criterion measure in validation studies aimed at developing a new field-based method for quantifying energy expenditure. Compared to direct calorimetry, the indirect approach is relatively simpler to operate and less expensive to maintain and staff. With the recent emergence of portable versions, this technique can also be used under many free-living conditions such as during the performance of common household and garden tasks and leisure physical activities.

Measurement of Substrate Oxidation

In addition to quantifying energy expenditure, indirect calorimetry provides a means of estimating the composition of fuels oxidized. To derive this estimation, one determines the ratio of $\dot{V}CO_2$ produced to $\dot{V}O_2$ consumed, which is referred to as the **respiratory quotient** (RQ). Because of structural differences in the composition of carbohydrate, lipid, and protein, complete oxidation of the vari-

ous nutrients requires different amounts of oxygen and produces different amounts of carbon dioxide. For example, the oxidation of 1 g of glucose requires 0.746 L O_2 and produces 0.743 L CO_2 (Jeukendrup and Gleeson 2004), and as a result, RQ is close to 1. On the other hand, the oxidation of 1 g of free fatty acid (palmitic acid) requires 2.009 L O_2 and produces 1.414 L CO_2 (Jeukendrup and Gleeson 2004), and as a result, RQ is close to 0.7. Differences in RQ caused by carbohydrate and fat oxidation are also illustrated in the following oxidative chemical reactions:

Glucose $C_6H_{12}O_6 + 6O_2 \rightarrow 6\ CO_2 + 6\ H_2O$
$$RQ = 6CO_2 \div 6O_2 = 1$$

Palmitic acid $C_{16}H_{32}O_2 + 23O_2 \rightarrow 16\ CO_2 + 16\ H_2O$
$$RQ = 16CO_2 \div 23O_2 = 0.7$$

The RQ serves as a convenient measure to provide quantitative information on the relative contributions of energy nutrients to the total energy provision at rest and during steady-state exercise. If RQ is found to be equal to 1, then all energy is derived from the oxidation of carbohydrate energy substrates. If RQ is found to be equal to 0.7, then all energy is derived from oxidation of fat energy substrates. It is, however, very unlikely in most circumstances that fat or carbohydrate will be the only fuel used. In fact, during rest and submaximal exercise, RQ is often found to be somewhere between 0.7 and 1.0. Table 4.1 lists a range of RQ values and corresponding percentages of energy derived from fat or carbohydrate oxidation, which were first published by American nutrition scientist Graham Lusk (1924). This table also illustrates the **caloric equivalent of oxygen** corresponding to each RQ value. Caloric equivalent of oxygen is defined as the number of calories produced for each liter of oxygen used, and as shown in table 4.1, this parameter is subject to changes in the composition of carbohydrate and fat oxidation. To determine energy expenditure accurately, we need to know not only the level of oxygen uptake, but also the composition of energy fuels being utilized.

The RQ represents gas exchange across the blood–cell barrier within an organ or tissue bed. Thus, direct measurement of this parameter can be difficult because it requires an invasive medical procedure. However, this methodological limitation is overcome by determining $\dot{V}O_2$ and $\dot{V}CO_2$ at the lungs using indirect calorimetry. To take into account the fact that $\dot{V}O_2$ and $\dot{V}CO_2$ are measured at the lungs, the

Table 4.1 Thermal Equivalents of Oxygen for the Nonprotein Respiratory Quotient (RQ) and Percentages of Calories From Carbohydrate and Fat

Nonprotein RQ	Kcal per liter of O_2 consumed	% Carbohydrate	% Fat
0.70	4.686	0.0	100.0
0.71	4.690	1.1	98.9
0.72	4.702	4.8	95.2
0.73	4.714	8.4	91.6
0.74	4.727	12.0	88.0
0.75	4.739	15.6	84.4
0.76	4.750	19.2	80.8
0.77	4.764	22.8	77.2
0.78	4.776	26.3	73.7
0.79	4.788	29.9	70.1
0.80	4.801	33.4	66.6
0.81	4.813	36.9	63.1
0.82	4.825	40.3	59.7
0.83	4.838	43.8	56.2
0.84	4.850	47.2	52.8
0.85	4.862	50.7	49.3
0.86	4.875	54.1	45.9
0.87	4.887	57.5	42.5
0.88	4.889	60.8	39.2
0.89	4.911	64.2	35.8
0.90	4.924	67.5	32.5
0.91	4.936	70.8	29.2
0.92	4.948	74.1	25.9
0.93	4.961	77.4	22.6
0.94	4.973	80.7	19.3
0.95	4.985	84.0	16.0
0.96	4.998	87.2	12.8
0.97	5.010	90.4	9.6
0.98	5.022	93.6	6.4
0.99	5.035	96.8	3.2
1.00	5.047	100.0	0.0

Adapted, by permission, from W.D. McArdle, F.I. Katch and V.L. Katch, 2001, *Exercise physiology*, 5th ed. (Philadelphia, PA: Lippincott, Williams & Wilkins), 182.

respiratory exchange ratio (RER) is used instead of RQ. The RQ and RER depict the same ratio, but the RER can be interpreted as the ratio of $\dot{V}CO_2$ to $\dot{V}O_2$ corresponding to metabolism of the overall body rather than a specific tissue bed and is typically determined using open-circuit spirometry.

▶ KEY POINT ◀

The respiratory quotient (RQ), which indicates gas exchange across the blood–cell barrier within tissue beds, is a key parameter used in determining the composition of substrate utilization. Oxidation of carbohydrate versus fat requires different amounts of oxygen and produces different amounts of carbon dioxide. Under most circumstances, this parameter can be estimated via measurement of gas exchange occurring at the lungs, and the ratio $\dot{V}CO_2/\dot{V}O_2$ obtained is referred to as the respiratory exchange ratio (RER).

One needs to address certain issues when measuring RER but using an RQ table to quantify substrate utilization. First, the RQ table was developed based on the assumption that the amount of protein oxidized is negligible or can be corrected for the oxidation of protein computed from nitrogen excretion in urine and sweat (Ferrannini 1988; Frayn 1983). Secondly, application of the RQ assumes that oxygen and carbon dioxide exchange measured at the lungs reflects the actual gas exchange from macronutrient metabolism within tissues. In this regard, exercise intensity could be relevant. It has been found that this assumption works well during exercise of light to moderate intensity. During heavy exercise, however, $\dot{V}CO_2$ measured at the lungs represents not only that produced during energy metabolism but also that derived from the buffering of metabolic acid, which increases at a greater rate during-high intensity exercise. Consequently, use of the RER would no longer accurately reflect the pattern of substrate utilization. Finally, use of the RER may not be adequate for those with pulmonary disorders because the pattern of gas exchange at the lungs can be altered due to obstructive ventilation.

FIELD-BASED TECHNIQUES

The traditional chamber and calorimetry technology can be inadequate for reasons such as cost, instru-

mentation, and time necessary for running tests. It is very difficult to use these sophisticated approaches to capture the complexity of activities that people engage in as they go about their daily lives. Consequently, there have been many attempts to develop relatively simple and more convenient methods to allow the determination of energy demands associated with free-living activities. Of these methods, doubly labeled water technique, motion sensors, heart rate monitoring, and physical activity questionnaires or logs are perhaps the most common methods for which validity and reliability have been extensively investigated. These field-based approaches enable us to track our physical activity participation in many free-living conditions. Accurate measurement of energy expenditure due to physical activity is critical for several reasons: (1) to quantify participation in physical activity, (2) to monitor compliance with physical activity guidelines, (3) to enable understanding of the dose–response relationship between physical activity and health, and (4) to assess the effectiveness of intervention programs designed to improve physical activity levels. Although field-based techniques provide a convenient approach to the measurement of energy expenditure, they generally do not measure or distinguish metabolism of energy substrates such as carbohydrate and fat.

Doubly Labeled Water Technique

The **doubly labeled water** method is a relatively new approach for estimating energy expenditure. The use of this technique for assessing energy expenditure in humans was first reported by Schoeller and van Santen (1982). The technique requires the subject to consume a quantity of water containing a known concentration of the stable **isotopes** of hydrogen (^2H or deuterium) and oxygen (^{18}O or oxygen-18). The term *isotope* refers to one of two or more species of the same chemical element that have different atomic weights (Shier et al. 1999). Isotopes have nuclei with the same number of protons but varying numbers of neutrons. Stable isotopes are those whose nuclei will not emit radiation and thus are not radioactive. Both ^2H and ^{18}O are used as tracers, as they are slightly heavier than regular H and O and can be measured within various body compartments. For example, through oxidative metabolism, labeled hydrogen is lost as ^2H_2O in sweat, urine, and water vapor during respiration, while labeled oxygen leaves as $H_2^{18}O$ in water and as $C^{18}O_2$ in expired air. A **mass spectrometer** is used to determine the difference in excretion rates between the two tracers, and the difference then represents the rate of carbon dioxide production. Oxygen uptake is further estimated from $\dot{V}CO_2$ as well as RQ, which is often assumed to be 0.85 (Black et al. 1986).

The primary advantage of the doubly labeled water technique for measuring total energy expenditure is that it does not interfere with everyday life and thus can be used in a variety of free-living settings. The fact that the technique is not constrained by time allows acquisition of the typical daily energy expenditure. To date, the technique has been used in circumstances such as bed rest and during prolonged activities like climbing Mt. Everest, cycling the Tour de France, rowing, and endurance running and swimming (Hill and Davis 2002; Stroud et al. 1997; Mudambo et al. 1997). The potential drawbacks include the high cost of the ^{18}O and both the expense and specialized expertise required for analysis of the isotope concentrations in body fluids by a mass spectrometer. As measurements are often taken for a long period of time, no information is obtained about brief periods of peak energy expenditure.

Motion Sensors

Motion sensors are mechanical and electronic devices that capture motion or **acceleration** of a limb or the trunk, depending on where the device is attached to the body. There are several different types of motion sensors, ranging in cost and complexity from the pedometer to the triaxial accelerometer. The pedometer is a relatively simple device primarily used to measure walking distance. It can be clipped to a belt or worn on the wrist or ankle. Pedometers do not operate on the vertical pendulum principle. Rather, they count motions by responding to vertical acceleration. The early version of this instrument was merely mechanical in that it had a lever arm attached to a gear, which rotated each time the lever arm clicked. The horizontal spring-suspended lever arm moved vertically up and down as a result of each step. More sophisticated pedometers are now commercially available. These rely on the use of a **micro-electromechanical system** (MEMS) composed of inertia sensors and computer software to detect steps. These battery-operated pedometers have digital readouts that can display not only total steps and distance, but also values in calories. Some can be adjusted for stride length so that walking distance can be more precisely calculated. Some newly developed electronic pedometers, for example, the Yamax DW500 pedometer, have proven useful in tracking physical activity such as overground walking, which accounts

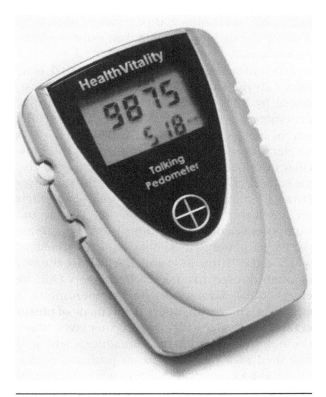

Figure 4.3 An example of a digital pedometer.
A. Wilson/Custom Medical Stock Photography

for a substantial fraction of the energy expended in physical activity (Bassett et al. 1996).

Pedometers are typically small and low in cost, and they can be used in epidemiological studies in large-scale populations (figure 4.3 and table 4.2). However, the pedometer has a number of limitations when used as a research tool. It is unable to distinguish vertical accelerations above a certain threshold, and thus cannot discriminate walking from running or different levels of exercise intensity (Bassett 2000). In terms of converting steps into energy expenditure, the device works on the assumption that a person expends a constant amount of energy per step. In Yamax pedometers, for example, this constant is assumed to be 0.55 cal $\cdot$ kg^{-1} $\cdot$ step^{-1} regardless how fast the person is moving (Hatano 1993). Also, to date, most instruments have not been equipped with the ability to store data over a specific time period. Thus they generally cannot furnish information on the duration, frequency, and intensity of physical activity. It is also important to note that for activities that do not involve locomotion, such as cycling or upper body exercise, the unit may need to be attached to the body part that is moving. Saris and Binkhorst (1977) and Washburn and colleagues (1980) found that ankle-mounted pedometers were more accurate during bicycle riding than waist-mounted pedometers.

Accelerometers are more sophisticated electronic devices that measure acceleration produced by body movement (figure 4.4). Unlike pedometers, accelerometers are able to detect the rate of movement or the intensity of exercise, as acceleration is directly proportional to the muscular force being exerted. Accelerometers can also measure acceleration in one (uniaxial) or three (triaxial) planes. Although a variety of models are now commercially available, the Caltrac and Computer Science Applications (CSA) models are the two commonly used uniaxial accelerometers, whereas the Tritrac R3D and Tracmor are the two commonly used triaxial accelerometers (Ainslie et al. 2003). Structurally, the accelerometer is equipped with a **transducer** that is made of **piezoceramic material** with a brass center layer. When the body accelerates, the transducer, which is mounted in a cantilever beam position, bends, producing an electrical charge that is proportional to the force being exerted by the subject. This creates an acceleration–deceleration wave; the area under this wave is summed and converted to digital signals referred to as "counts." Results can be displayed on a screen as an accumulated total

or downloaded as raw data to be further analyzed. Most current models also have the ability to display the level of accumulated energy expenditure for an extended period of time. This is done by means of a microprocessor that utilizes an activity–energy conversion factor as well as prediction equations for BMR based on age, body size, and gender as independent variables (Washburn et al. 1989).

A notable advantage of accelerometers is that they can detect the rate of movement and thus the intensity of exercise (table 4.2). Together with the use of an internal clock, this intensity-discriminating feature helps to characterize the intensity and duration of the physical activity being performed. Thus a dose–response relationship between physical activity and health and fitness outcome can be assessed. Other advantages are that accelerometers are small, can be worn without interfering with normal movement, and record data for extended periods of time. The instrument also seems to be reliable. By having a subject wear two Caltrac devices, one on each side of the body, Sallis and colleagues (1990) observed that the interinstrument reliability reached 0.96.

A number of studies have shown significant correlations between energy expenditure estimated by accelerometers and by other methods that have proven accurate, such as indirect calorimetry and the doubly labeled water technique (Haymes and Byrnes 1993; Heyman et al. 1991; Wong et al. 1981). However, an equal number of studies showed that the accelerometer underestimated energy expenditure (Bray et al. 1994; Klesges et al. 1985; Miller et al. 1994; Montoye et al. 1983). The high validity of this instrument for assessing physical activity appears to have been demonstrated primarily in studies that employed level walking and running (Hendelman et al. 2000). Questions remain as to whether accelerometers are able to accurately assess energy expenditure during leisure activities such as household and occupational activities, weight-bearing and static exercises such as cycling and load carriage, and walking or running on soft or graded terrain (Hendelman et al. 2000; Sherman et al. 1998; Bouten et al. 1994).

Given that a triaxial accelerometer combines three independent sensors to detect acceleration in three-dimensional space, it seems logical that it would be more accurate in capturing physical activities than a single-plane accelerometer. However, validation results supporting this contention are mixed. Welk and Corbin (1995) reported that

Figure 4.4 An example of a uniaxial accelerometer.

Photo courtesy of Caltrac

Table 4.2   Advantages and Disadvantages of Various Field Methods for Assessing Physical Activity and Energy Expenditure (EE)

Method	Advantages	Disadvantages
Pedometers	Small in size Low cost Suitable for epidemiological studies	Unable to detect acceleration Unable to quantify intensity, duration, and frequency Unable to detect certain movements such as in weightlifting, cycling, and upper body exercise
Accelerometers	Small in size Detect the rate of movement Able to provide information on intensity, duration, and frequency	Questionable for converting motion data into EE Unable to detect certain movements such as in weightlifting, cycling, and upper body exercise Unable to discriminate walking or running performed on soft or graded terrain
HR monitors	Correlate closely with $\dot{V}O_2$ Measure all movements including those that can't be detected by motion sensors Able to provide information on intensity, duration, and frequency	Weak relation with $\dot{V}O_2$ in low-intensity domain Require individual calibration curves for accurate estimates of EE HR subject to change in stress, body posture, dehydration, environmental temperature
Motion sensors + HR monitors	Overcome major weaknesses associated with motion sensors and HR monitors in addition to retaining their advantages	Data analysis time-consuming Problematic if multiple sensors are used Validation of use of algorithm needed to estimate EE using large and heterogeneous samples

the Caltrac and Tritrac accelerometers were similar in reflecting aspects of lifestyle activities. On the other hand, Bouten and colleagues (1996) and Eston and colleagues (1998) found that a three-dimensional monitor was better in predicting oxygen consumption during physical activities than a uniaxial monitor. In light of the advantages and concerns associated with accelerometers, there is still a need to continue improving not only the hardware in order to better track motion, but also the software so that conversion of activity counts into energy values can become more accurate.

Heart Rate Monitoring

Due to the difficulties encountered in measuring $\dot{V}O_2$ and thus energy expenditure in the field, there has been steady interest in developing less direct methods of recording physiological responses associated with $\dot{V}O_2$. Among these physiological parameters are heart rate (HR), pulmonary ventilation,

and body temperature. However, monitoring HR appears to be the most popular technique. There is a fairly close and linear relationship between HR and $\dot{V}O_2$ or energy expenditure during dynamic exercise involving large muscle groups; that is, the greater the HR, the greater the $\dot{V}O_2$. This is especially the case when HR ranges from 110 to 150 beats $\cdot$ min^{-1}; here the relationship between the two parameters has been found to be linear. Thus it is reasonable to use HR as a physiological marker of $\dot{V}O_2$ to assess physical activity and its associated energy expenditure. This technique has also been validated against **electrocardiography** (ECG) monitoring in both the laboratory and the field (Karvonen et al. 1984; Leger and Thivierge 1988; Treiber et al. 1989). Measurement of HR is relatively low cost, noninvasive, and easy. With today's technology, HR can be monitored and recorded with the use of a chest strap transmitter and a small receiver watch. A typical HR monitor transmits the **R-R waves** of ECG into a receiver where ECG signals can be digitized and displayed. Many advanced models also have an internal clock that allows sampling over different time intervals and can store data over a period of days or weeks, thereby providing information on various components of physical activity including intensity, duration, and frequency.

Earlier studies suggested using average pulse rate as a predictor of daily energy expenditure (Payne et al. 1971). However, this approach was criticized in that it could not account for the fact that the $HR-\dot{V}O_2$ relationship can be altered by age, gender, fitness level, psychological stress, and substances and medications known to affect HR. To address this concern, Goldsmith and Hale (1971), and later Washburn and Montoye (1986), proposed the use of net HR (exercise HR – resting HR) to estimate energy expenditure, as they found a much smaller error in estimating energy expenditure using this approach. This method helps in correcting for variations in resting HR and thus the state of training or level of fitness. However, even with this modification, it still assumes that the slopes of the $HR-\dot{V}O_2$ curve are the same for various individuals, which is often not the case.

More recently, Spurr and colleagues (1988) developed another method that involves the determination of FlexHR. With this method, an individual's resting metabolic rate (RMR) and $HR-\dot{V}O_2$ curve across exercises of varying intensities are first determined. The FlexHR is then calculated as the mean of the highest HR achieved at rest and the lowest HR achieved during exercise. If a given HR observed during field activity is below the FlexHR, then energy expenditure is determined based on RMR. If a given HR is above FlexHR, then energy expenditure is determined based on the individual's $HR-\dot{V}O_2$ curve. Even though this method takes into account individual variability in the slope of the $HR-\dot{V}O_2$ relationship, it is time-consuming to establish individualized calibration curves. Although the use of a generalized curve in order to avoid the excess time has been suggested, Rowlands and colleagues (1997) reported that this strategy would increase the error in estimation of energy expenditure, especially in children, whose HR can vary widely. It should also be noted that in the Spurr and colleagues study, the $HR-\dot{V}O_2$ curve was developed based on stationary cycle exercises performed in the laboratory. It remains questionable how accurate the curve is in reflecting the energy cost of activities in a free-living condition.

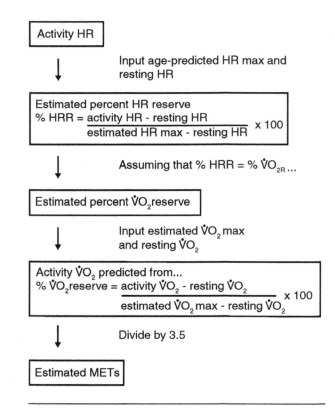

Figure 4.5 Computation process demonstrating the use of activity HR, age-predicted %HR reserve, and estimated %$\dot{V}O_2$ reserve to calculate energy expenditure. HR = heart rate. HRR = heart rate reserve.

Reprinted, by permission, from S.J. Strath et al., 2000, "Evaluation of heart rate as a method for assessing moderate intensity physical activity," *Medicine and Science in Sports and Exercise* 32(Suppl.): S465-S470.

Strath and colleagues (2000) validated the use of HR monitors with a different approach. First, the study employed free-living activities such as mowing, trimming, and gardening. Secondly, HR was converted into energy expenditure based on findings reported by Swain and Leutholtz (1997) and Swain and colleagues (1998) showing that %HR reserve is equivalent to %$\dot{V}O_2$ reserve. Figure 4.5 demonstrates a step-by-step process for obtaining a given level of activity $\dot{V}O_2$. In the study by Strath and coworkers, while HRmax was calculated as 220 – age, $\dot{V}O_2$max was obtained using the nonexercise prediction equation developed by Jackson and colleagues (1990). In addition, the resting $\dot{V}O_2$ was assumed to be equal to $3.5 \text{ ml} \cdot \text{kg}^{-1} \cdot \text{min}^{-1}$. With this new approach, Strath and colleagues found that the estimated energy expenditure correlated strongly with energy expenditure measured by portable indirect calorimetry. It appears that HR could be a strong predictor of energy expenditures in free-living activity if the $HR-\dot{V}O_2$ converting process is adjusted for individual differences not only in the resting HR, but also in maximal aerobic capacity or $\dot{V}O_2$ max.

With the HR monitoring technique, the key issue involves the precision of converting HR into energy expenditure. The reason is that HR in relation to energy cost can be affected by many factors other than physical activity per se. Factors such as age, fitness, and resting metabolism may be accounted for with the use of individualized HR–$\dot{V}O_2$ curves or with the use of measures of relative intensity, for example, %HR reserve, that adjust for age and fitness. However, there are still some doubts deserving of attention (table 4.2). For example, a question remains as to whether HR is a valid indicator of energy expenditure during low-intensity activities because of its weak relationship with $\dot{V}O_2$ within the low-intensity domain (Freedson and Miller 2000). This is pertinent in that the intensity at which many daily activities are performed ranges from low to moderate. Heart rate is also susceptible to emotional stress, which would result in a disproportional rise in HR for a constant $\dot{V}O_2$. In addition, HR can vary due to changes in stroke volume, and this latter parameter is influenced by body posture, exercise modes, and heat stress and dehydration.

▶ KEY POINT ◀

Accelerometry and HR monitoring are the two techniques most commonly chosen to assess energy expenditure, but each has limitations. The errors associated with accelerometry appear to be mechanical in that these devices generally cannot detect increases in energy expenditure due to movement up inclines, increases in resistance, or static exercise. The limitations of HR monitoring are primarily biological. For example, the HR–$\dot{V}O_2$ relation can be affected by age, gender, fitness, stroke volume, psychological stress, and environmental temperature.

Heart Rate and Motion Monitoring Combined

Accelerometry and HR monitoring are the field-based methods most often chosen for assessing physical activity and energy expenditure. However, there are limitations associated with each method when used alone. It appears that the limitations of HR monitoring are primarily due to biological variance. For example, as mentioned earlier, the HR–$\dot{V}O_2$ relation has been found to be affected by age, gender, fitness, stroke volume, and psychological stress. Responses of HR can also be influenced by ambient temperature, hydration status, and the quantity of muscle mass involved in the activity (Haskell et al. 1993; Brage et al. 2003). On the other hand, the limitations of accelerometry are mainly biomechanical, as the technique is generally unable to adequately detect increases in energy expenditure due to (1) movement up inclines, (2) an increase in resistance to movement, or (3) static exercise. In addition, a single sensor cannot identify movement that involves various parts of the body. As errors associated with the two methods are not inherently related, the combination of HR and accelerometry should in theory yield a more precise estimate of physical activity and energy expenditure than either method used alone.

Over the past decade or so, researchers have studied the validity of the simultaneous heart rate–motion sensor technique for measuring energy expenditure during exercise in the laboratory and the field setting (Haskell et al. 1993; Luke et al. 1997; Rennie et al. 2000; Strath et al. 2001a, 2001b; Brage et al. 2003). In general, these studies have shown that measuring HR and movement concurrently is a better approach for estimating oxygen consumption and energy expenditure than using either technique alone. In these studies, HR and movement counts were recorded at the same time. $\dot{V}O_2$ was estimated through use of data on HR, motion, and both HR and motion. Each estimated $\dot{V}O_2$ was then compared against a criterion $\dot{V}O_2$ measured with a standard technique, that is, indirect calorimetry. As information on both HR and motion was available, these studies were able to use the motion data to exclude HR that were increased due to nonexercise reasons or to use the HR data to capture an increase in energy metabolism that was not detected by motion sensors. Some studies recorded HR and motion using two or more devices or sensors attached to different body parts, an approach that may be problematic in a field setting over an extended period of time. A single unit that detects HR and motion with only one sensor has now been developed for use in tracking physical activity. This miniature device also uses a sophisticated **algorithm** in determining energy expenditure and can record data for as long as 11 days.

Apparently this combined approach has many advantages. It overcomes the major weaknesses associated with HR monitoring or motion sensing alone (table 4.2). For example, HR monitoring is not subject to error in movement sensing such as in detecting the level of activity during resistance exercise, swimming, and cycling. Likewise, movement sensing complements HR monitoring as it allows differentiation between increased HR caused by physical activity and that caused by nonexercise-related influences. Nevertheless, this method still requires the establishment of individualized calibration curves for both HR and movement counts, which can be time-consuming. Perhaps because of such technical difficulties associated with multiple sensing, most validation studies have used only a relatively homogenous, small sample size. It is hoped that future efforts will be directed toward examining whether this combined approach can be made precise enough to preclude the need for individual calibration.

▶ K E Y   P O I N T ◀

Because errors associated with the accelerometer and HR monitoring are not inherently related, a simultaneous heart rate–motion sensor technique has been developed. This combined approach has proven better for estimating energy expenditure than the use of either technique alone.

Multisensor Monitoring System

Recently, a new device for assessing energy expenditure, called the SenseWare armband (SWA), came on the market. The device is worn on the right upper arm over the triceps muscle. It has an ergonomic design and can be easily slipped on and off, and it does not interfere with day-to-day activities or sleeping. The device has multiple sensors that can measure various physiological and movement parameters simultaneously, including body surface temperature, skin vasodilation, and rate of heat dissipation, as well as a two-axis accelerometer. Data on these parameters, together with demographic information including gender, age, height, and weight, are used to estimate energy expenditure with a generalized algorithm. The principal difference between this system and the devices discussed previously is the inclusion of a heat flux sensor.

This allows the system to detect a change in heat produced as a result of metabolism.

The SWA is still in the earlier stages of development, and data on its validity and reliability are very limited. It appears that the system is accurate in tracking resting energy expenditure (Fruin and Walberg Rankin 2004). However, its ability to detect energy expenditure during exercise remains questionable. In recent studies comparing the SWA with indirect calorimetry, Fruin and Walberg Rankin (2004) found that the SWA overestimated the energy expenditure of flat walking and underestimated inclined walking, although it was found to be reasonably accurate for exercise performed on a cycle ergometer. Jakicic and colleagues (2004) also observed differences between the SWA and indirect calorimetry during treadmill walking, cycling, stair stepping, and arm cranking. Interestingly, in this latter study, no difference was observed when the authors adopted a series of mode-specific algorithms. While this finding is appealing, the use of multiple algorithms would require the development of a mechanism that allows the device to switch between algorithms so that minimal burden is placed on the user.

SUBJECTIVE MEASURES

A range of subjective methods are available for the assessment of energy expenditure in humans. These methods may be classified into (1) questionnaires and (2) activity logs or diaries. In general, questionnaires can be viewed as primarily recall based, and subjects are required to provide information on the pattern of daily activities that have already occurred. Questionnaires have been used to assess physical activities for the previous 24 h, the previous week, and the previous year. Some questionnaires are structured to specifically focus on occupational or leisure-time activities. Others are quite general, eliciting information on activities that have occurred both on and off the job. Questionnaires can be further divided into those that are self-reported and those that are interviewer administered.

Activity logs or diaries involve recording of subjects' activities periodically, at a frequency ranging from every minute (Riumallo et al. 1989) to every several hours (LaPorte 1979). This method often includes a standardized form, used to facilitate both compliance and quality. Some forms require subjects to record activities minute by minute, whereas others require them to record changes in activity. In these forms, names of activities are often abbreviated to make recording easier. In some activity logs, activities to be recorded are specific enough to include not only the type but also the intensity and duration. This method demands a greater level of attention from subjects in maintaining their diary. To alleviate this burden, some researchers have used a wristwatch that alerts the subject to record activities at specified times (Riumallo et al. 1989).

Both questionnaires and activity logs and diaries have been frequently used in large population surveys and epidemiological studies, mainly because these techniques are relatively inexpensive as compared to HR or motion monitoring. With these techniques, specific activities can be identified together with information on intensity, duration, and frequency. In addition, data can be collected by many subjects simultaneously. However, there are some limitations to these methods. Quantifying various aspects of physical activity such as type, intensity, duration, and frequency, which are important in estimating energy expenditure, is difficult. For example, with questionnaires, subjects may not necessarily recall all the activities they have performed, or they may overestimate or underestimate intensity and duration for some activities. These shortcomings could be particularly relevant when a questionnaire is administered to children who have lower cognitive functioning (Janz et al. 1995). There is also evidence that physical activity was underestimated in women when household chores were not included in the surveys (Ainsworth and Leon 1991; Shaw 1985). Additionally, with the activity log or diary technique, recording activities frequently for a long period of time can be difficult. It has been suggested that the longer the period of data collection, the less accurate the results may be (Montoye et al. 1996). This technique can also influence the pattern of activity of subjects who record ongoing activities: They may purposely modify behaviors due to the survey. Readers interested in searching for a particular survey instrument may refer to the work of Montoye and colleagues (1996) and several reviews that provide many survey samples and analysis of a variety of questionnaires and activity logs and diaries (Shephard 2003; Ainslie et al. 2003; Sirard and Pate 2001; Tudor-Locke and Myers 2001; Sallis and Saelens 2000; Pereira et al. 1997).

Determining exercise intensity appears to be the most challenging task. Questionnaires or activity logs often express the intensity of physical activity semantically. However, perceptions of intensity or exertion depend largely on the subject's prior experience and exercise tolerance. There have been attempts to anchor semantic descriptions of exercise intensity using ratings of perceived exertion or the onset of sweating and breathlessness. However, these approaches seem to help only in distinguishing light from heavy efforts (Shephard 2003). Many questionnaires have not included items related to lower levels of physical activity (i.e., less intense than brisk walking), which could be sufficient to produce positive health outcomes in extremely sedentary or older populations (Hakim et al. 1999; Rakowski and Mor 1992).

> ### ▶ K E Y   P O I N T ◀
>
> Subjective measures of physical activity and energy expenditure mainly involve questionnaires and activity logs or diaries. Questionnaires are often answered in a recall fashion, and subjects provide information on activities that have already occurred. Activity logs or diaries are completed by people periodically during the day. These subjective methods provide the least expensive approach to tracking physical activity and energy expenditure, and this advantage can be important in studies using a large-scale populations.

Ainsworth and colleagues (1993, 2000) have developed the Compendium of Physical Activity, which provides a comprehensive list of physical activities and their associated level of exercise intensity. This standardized classification system should be used in the construction of a questionnaire or activity log so that results are comparable across studies. This system also furnishes information on intensity for each activity listed, so energy expenditure can be calculated if necessary. In this compendium, the intensity associated with each activity is expressed as a **metabolic equivalent** (MET), which is the ratio of work metabolic rate to a standard resting metabolic rate equal to a resting $\dot{V}O_2$ of 3.5 ml $\cdot$ kg^{-1} $\cdot$ min^{-1}. The MET is an absolute measure of intensity of effort. To obtain the relative intensity imposed by a given activity, the subject's fitness or **maximal aerobic capacity** (or $\dot{V}O_2$max) needs to be known. Then a given $\dot{V}O_2$ can be expressed as a percent of $\dot{V}O_2$max. Expressing intensity relatively is important because for individuals with varying fitness, a given activity can produce different physiological responses. For practical purposes, measurement of $\dot{V}O_2$max may be accomplished via a nonexercise approach that utilizes information on the subject's age, body mass, body composition, and physical activity participation (Jackson et al. 1990).

SUMMARY

The ability to assess energy expenditure has been a topic of scientific interest for more than a century. With physical inactivity emerging as an important risk factor for many chronic diseases, the topic becomes even more important. The potential negative impact of a sedentary lifestyle on risk of cardiovascular and other diseases has prompted governments and scientific organizations to develop physical activity guidelines. In this context, being able to accurately determine energy expenditures of daily living, including physical activity, is critical for several reasons: (1) to quantify participation in physical activity, (2) to monitor compliance with physical activity guidelines, (3) to understand the dose–response relationship between physical activity and health, and (4) to assess the effectiveness of intervention programs designed to improve physical activity levels.

Two laboratory procedures are commonly used to assess human energy expenditure: direct calorimetry and indirect calorimetry. Direct calorimetry measures the heat produced as a result of metabolism, whereas indirect calorimetry measures the amount of oxygen being used during energy transformation. The initial intent of developing these laboratory techniques may have been to satisfy scientific curiosity. However, because of their accuracy, these methods using sophisticated instrumentation have been frequently employed as criterion measures in studies aimed at developing less technical and more user-friendly alternatives applicable outside the laboratory setting.

Among field-based techniques, motion sensing and HR monitoring appear to be the two most commonly chosen to assess energy expenditure in daily living. These systems are able to provide energy expenditure estimates with reasonable accuracy. In light of the limitations associated with either technique, recent efforts have been devoted to either combining the two monitoring techniques or adding other parameters such as heat flux to the existing configuration. In light of the limitations associated with virtually all field-based techniques, the decision about what technique to use often involves an issue of trade-off. For example, subjective measures of physical activity using questionnaires or activity logs can be affected by human error, but these methods are most suitable for epidemiological studies using large sample sizes.

KEY TERMS

acceleration

airway resistance

algorithm

basal metabolic rate

caloric equivalent of oxygen

calorimetry

doubly labeled water

electrocardiography

expiration

inspiration

isotopes

mass spectrometer

maximal aerobic capacity

metabolic equivalent

micro-electromechanical system

piezoceramic material

pulmonary ventilation

respiratory exchange ratio

respiratory quotient

R-R waves

spirometry

thermic effect of food

transducer

REVIEW QUESTIONS

1. How is the term calorie defined? What is the caloric equivalent of oxygen?

2. What is the difference between direct and indirect calorimetry? What is the difference between closed-circuit and open-circuit spirometry?

3. What is the respiratory quotient? How is this parameter used to determine the pattern of substrate utilization?

4. If someone walks on a treadmill and is consuming oxygen at 1.5 L/min and expiring carbon dioxide at 1.2 L/min, what is the rate of energy expenditure of this individual? What are the percentages of carbohydrate and fat being used? How much of the energy expended comes from oxidizing carbohydrate and how much comes from fat oxidation?

5. Describe how pedometers, accelerometers, and heart rate monitors work in tracking energy expenditure of daily living.

6. What are the advantages and disadvantages associated with subjective measures of physical activity using questionnaires or activity logs?

5

ENERGY COST OF PHYSICAL ACTIVITIES AND SPORTS

This chapter provides an in-depth analysis of energy expenditure during movements such as walking, running, cycling, and resistance exercise. It also covers the effect of extrinsic factors such as body mass, body size, weight distribution, exercise surface, and stride pattern on energy expenditure during exercise. In addition, the chapter reviews methodology involving the use of mathematical equations for estimating energy expenditure for common activities such as walking, running, cycling, and bench stepping.

PRINCIPLES OF ENERGY UTILIZATION DURING EXERCISE

Physical activity is a powerful metabolic stressor (Coyle 2000). It stimulates chemical processes in which the potential energy stored in energy substrates is converted to the type of energy that cells can utilize, namely adenosine triphosphate (ATP). This biologically usable energy can then serve to power various cellular activities necessary to maintain homeostasis. Adenosine triphosphate can be produced without the use of oxygen. This occurs via a process in which energy from substrates is harnessed from the degradation of glucose or glycogen in the cytoplasm of cells. Adenosine triphosphate can also be produced with the use of oxygen. This takes place via a process in which energy from substrates is sequentially captured

from a series of chemical reactions in the mitochondria of cells. As mentioned in chapter 4, in an aerobic event, the energy utilization associated with a particular exercise can be quantified through determination of the amount of oxygen used, with the oxygen later converted into calories. Determination of energy cost for an anaerobic event is much more complex and may be achieved via measurement of the production of muscular force and power, utilization of energy substrates (i.e., ATP, phosphocreatine, glucose), or energy utilization following the completion of exercise. Knowing the energy cost of exercise can be important if a comparable nutritional requirement needs to be provided or if the efficiency of the body during performance of exercise is to be calculated.

Gross and Net Energy Expenditure

The principle of using oxygen uptake to estimate energy expenditure was discussed in previous chapters. The following example pertaining to a physical activity uses this estimation process. A 40-year-old female, jogging for 30 min at a steady pace that requires 1.5 L of oxygen uptake every minute, consumes a total of 45 L of oxygen. To convert oxygen uptake to energy values, we need to know the caloric equivalent of oxygen, which is influenced by the pattern of substrate utilization as discussed in chapter 4. This parameter can be estimated to be 5 kcal per liter of oxygen. Thus, the jogger expends about 225 kcal (45 L $\times$ 5 kcal $\cdot$ L^{-1}) during her workout. It is particularly noteworthy that this value is not the energy cost induced by jogging per se. Rather, it is the measured total or **gross energy expenditure** that also includes the energy the woman would expend at rest over the same time period. To obtain the **net energy expenditure,** we must subtract a resting value from the measured total. For example, if over a 30 min period the woman consumes 7 L of oxygen or 35 kcal (7 L × 5 kcal · L^{-1}) at rest, her net energy expenditure will be 190 kcal (225 – 35 kcal). Such a net value allows more precise assessment of the energetic cost of an exercise and its impact on weight loss (Swain 2000). In fact, the seventh edition of the American College of Sports Medicine (ACSM) guidelines recommends that net caloric expenditure be used in all body weight management recommendations (ACSM 2006).

Slow Component of Oxygen Uptake

Metabolic reactions always proceed at the rates required to maintain ATP homeostasis. Therefore, an increase in exercise intensity will dictate increases in substrate utilization and oxygen consumption. As shown in figure 5.1a, as power output increases, steady-state oxygen uptake increases correspondingly (Åstrand and Rodahl 1996). When the steady-state $\dot{V}O_2$ values are plotted against their respective power outputs, there is a linear increase in $\dot{V}O_2$ with increases in power output. The steady-state $\dot{V}O_2$ is typically regarded as the energy cost for that specific power output. In recent studies, it has been recognized that the $\dot{V}O_2$ response at higher work rates does not follow the steady-state pattern as shown in figure 5.1a, but rather follows the pattern illustrated in figure 5.1b. It appears that at power outputs above lactate threshold, a minimal intensity above which blood lactate increases significantly, $\dot{V}O_2$ will increase to a level greater than its steady-state value. This additional increase in energy cost, which has been referred to as the **slow component of oxygen uptake,** occurs during high-intensity steady-state exercise (Gaesser and Poole 1996). The term "slow" is in recognition that an increase in the oxygen component is usually small and gradual. The slow component of $\dot{V}O_2$ has been attributed to greater than expected increases in pulmonary ventilation, body temperature, and recruitment of fast-twitch muscle fibers (Poole and Richardson 1997).

> ## ▶ KEY POINT ◀
>
> Once exercise is performed at an intensity near or above lactate threshold, a gradual increase in oxygen uptake occurs despite the maintenance of the workload. This excess in oxygen uptake above what is typically required for performing the workload is referred to as the oxygen slow component and should be considered when one is determining the total energy cost of an activity.

Exercise Efficiency

The energy cost of performing an activity can increase significantly if an individual is less skillful or inexperienced. In this case the individual can be characterized as inefficient or less economical. **Efficiency** of human movement has to do with how much of the energy liberated is utilized to accomplish mechanical work such as moving the body over a certain distance. It is determined as a ratio, or the actual work accomplished over the amount of energy required to perform the work. Exercise efficiency is typically quantified by computing mechanical efficiency, which reflects the fraction of the total energy expended that contributes to the external work, with the remainder being lost as heat. The greater the proportion of energy that contributes to the external work, the more efficient the person is. The mathematical expression of this efficiency parameter is as follows:

mechanical efficiency = external work accomplished ÷ energy expenditure × 100%.

Using the cycle ergometer as an example, one can derive the external work accomplished by computing the total power output, which is equal to brake resistance in kilograms (kg) times pedaling speed in meters per minute. If someone pedals against a resistance of 2 kg and at a speed of 300 m/min, the total power output will be 600 kgm/min (2 kg × 300 m/min) or ~6 kJ/min (1 kJ ≈ 100 kgm). The energy expenditure is obtained from measurement of oxygen consumption. Assuming that the oxygen uptake for this person during cycling is 1.5 L/min or 31.5 kJ/min (1 L of O_2 ≈ 21 kJ), the mechanical efficiency will be 19% (6 kJ/min ÷ 31.5 kJ/min × 100%).

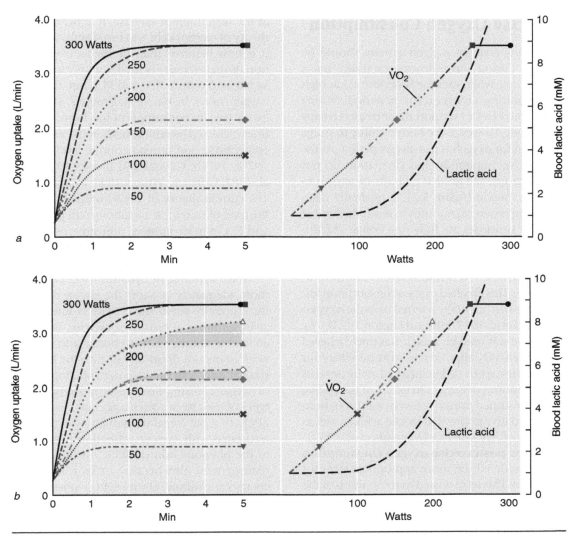

Figure 5.1   Increase in oxygen uptake with increasing exercise intensity *(a)* as originally proposed by Åstrand and Rodahl (1986) and *(b)* as redrawn by Gaesser and Poole (1996).

Reprinted, by permission, from G.A. Gaesser and D.C. Poole, 1996, "The slow component of oxygen uptake kinetics in humans," *Exercise and Sport Sciences Reviews* 24: 36.

Exercise efficiency is often used to assess athletic performance for a prolonged event in which preserving energy is imperative. For example, a cross country runner who uses less oxygen than another for the same running pace will certainly have an advantage during competition. As with all machines, the efficiency of the human body for mechanical work falls considerably below 100%. This is mainly due to the energy required to overcome internal and external friction. On average, efficiency ranges between 20% and 25% for walking, running, and stationary cycling (Cavanagh and Kram 1985). Moreover, exercise performed with upper body musculature has been shown to be less efficient than that performed with the lower body (Sawka 1986; Kang et al. 1997).

> **KEY POINT**
>
> Efficiency of human movement concerns how much of the energy liberated is utilized to accomplish mechanical work. As with all machines, the efficiency of the human body for mechanical work is well below 100%, mainly due to the energy required to overcome internal and external friction. On average, efficiency in humans ranges between 20% and 25%. Being more efficient is an advantage for athletes who perform long-distance events.

Postexercise Oxygen Consumption

The total energy cost for a given activity should be assessed both during and following exercise. This is especially the case when exercise is performed at high intensities, requiring a longer recovery period. During exercise, $\dot{V}O_2$ increases to support the increased energy need of the body. However, the body's ability to gauge such a demand for oxygen is not always perfect. At the onset of exercise, respiration and circulation do not immediately supply the needed quantity of oxygen to the exercising muscle (figure 5.2). It normally takes the oxygen supply several minutes to reach the level at which aerobic processes are fully functional. The difference in oxygen requirement and oxygen supply is known as the **oxygen deficit.** After exercise, $\dot{V}O_2$ does not return to resting levels immediately, but does so rather gradually. This elevated oxygen consumption following exercise was originally referred to as the **oxygen debt** (Cerretelli et al. 1977; Hill and Lupton 1923). The term *oxygen debt* was used because it was once believed that the elevated $\dot{V}O_2$ after exercise was necessary for repayment of the oxygen deficit incurred at the onset of exercise. The amount of oxygen in excess of the resting level that is consumed during recovery was thought to be used for oxidative removal of lactate acids as well as replenishment of ATP and PCr (Margaria et al. 1933).

Today **excess postexercise oxygen consumption** (EPOC) is considered the more appropriate term for the oxygen debt. This term is used in recognition of the fact that the excess oxygen consumption can actually continue much longer than once believed and that its magnitude often exceeds the oxygen deficit incurred

at the beginning of exercise. It appears that the early theory of oxygen debt was too simplistic. For example, it is now believed that the excess oxygen consumption during recovery is used not only to remove blood lactate and restore ATP and PCr, but also to replenish oxygen stores in blood and muscle. In addition, the excess oxygen consumption has been found to result from increased breathing, body temperature, and blood epinephrine and norepinephrine levels (Brooks et al. 1971a, 1971b; Gaesser and Brooks 1980; Gladden et al. 1982; Gore and Witters 1990; Hagberg et al. 1978). The current theory of EPOC reflects two factors: (1) the level of anaerobic metabolism in previous exercise and (2) exercise-induced adjustments in respiratory, circulatory, hormonal, and thermal function that still exert their influences during recovery.

Understanding the dynamics of EPOC enables us to more adequately quantify the energy cost of an activity, especially when the activity is intense and brief, as in sprinting or resistance exercise. Because of their anaerobic nature, such strenuous and short-duration work bouts can drastically disturb the body's homeostasis throughout exercise, necessitating a greater level of oxygen consumption during recovery. As such, these types of exercise are often associated with a fairly large EPOC that can constitute a majority of the total energy associated with exercise. Attention has been brought to the phenomenon of EPOC in the area of weight management also because of its role in facilitating energy expenditure. Overweight or obesity is the result of a positive energy balance, and EPOC can contribute to the opposite when exercise is undertaken regularly. Nevertheless, it has been suggested that in order for

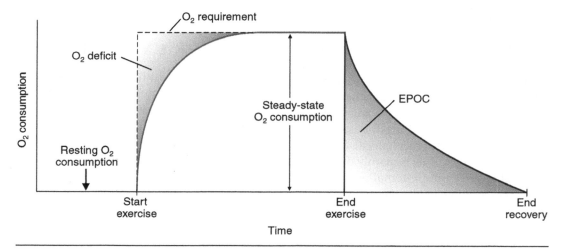

Figure 5.2   Response of oxygen uptake during steady exercise and recovery.

EPOC to be effective, one would have to exercise at an intensity exceeding 70% $\dot{V}O_2$max for more than an hour three times a week (Borsheim and Bahr 2003). This volume of exercise is impractical for many overweight or obese individuals.

> ### ▶ K E Y P O I N T ◀
>
> Upon completion of an exercise, $\dot{V}O_2$ does not return to resting levels immediately, but does so rather gradually. This elevated oxygen consumption following exercise is referred to as excess postexercise oxygen consumption (EPOC). The EPOC is the energy necessary to restore homeostasis disrupted during the preceding exercise, and its quantity is proportional to the level of exercise intensity. Computation of the total energy expenditure of a single exercise session should take EPOC into account.

The EPOC may have greater implications for athletes who often perform two or more bouts of exercise on the same day or follow an interval training program that uses preestablished spacing of exercise and rest intervals. In this case, the objective is to prompt a speedy recovery and thus produce a smaller EPOC. Bahr and colleagues (1991) found that an elevated $\dot{V}O_2$ from previous exercise can reduce mechanical efficiency during the next workout. While performing no activity during the rest interval helps preserve energy, growing evidence suggests that performing low-intensity aerobic exercise during the recovery or rest interval following a strenuous non-steady-state workout accelerates lactate removal despite an associated increase in $\dot{V}O_2$ (Ahmaidi et al. 1996; Choi et al. 1994).

ENERGY COST DURING VARIOUS EXERCISES

As the body shifts from rest to exercise, energy transformation increases. This increase in energy is reflected by measurement of oxygen consumption, as oxygen is what makes energy transformation possible. Even with energy transformation that proceeds without oxygen initially, oxygen is eventually necessary for the restoration of homeostasis. The preceding discussion introduced the physiological principles associated with energy expenditure during exercise; this section further and more specifically elucidates energy cost and its determinants during various types of common exercise. From a practical standpoint, we must know the energy costs of specific activities and how they can be influenced by extraneous factors in order to be able to choose the appropriate exercise format, which is an integral part of exercise prescription.

Walking and Running

Walking and running are the two principal forms of human locomotion. They share the same basic feature: Each step consists of one phase of stance and one phase of swing. However, the two activities differ in terms of the timing of these phases. For example, the stance phase for each foot is longer in walking and shorter in running, whereas the opposite is true for the swing phase. In addition, during each walking cycle, at least one foot is always on the ground, whereas there is a period during each running cycle when both feet are off the ground. From an energetic standpoint, walking is an energy-cheap activity; its energy cost is generally no more than three times the resting metabolic rate. On the other hand, running can be very demanding metabolically as it engages more muscles that contract forcefully. On average, the energy expenditure during running can be as high as 10 times the resting metabolic rate (table 5.1).

For most people, walking is their major daily physical activity. Literature on energy cost during walking can be traced back to more than a half-century ago. In walking, there is a curvilinear relationship between energy expenditure and speeds ranging from 1.5 to 9.5 km · h^{-1} (0.9 to 5.9 miles · h^{-1}) (Fellingham 1978). As shown in figure 5.3, the relationship between oxygen consumption and speed appears to be linear between speeds of 3.0 and 5.0 km · h^{-1} (1.9 to 3.1 miles · h^{-1}). However, at faster speeds, a steeper curve in the relationship suggests a decrease in walking economy. In fact, 5.0 km · h^{-1} or 3 miles · h^{-1} has been viewed as an optimal walking speed that is often naturally chosen (Saibene and Minetti 2003). According to figure 5.3, this speed elicits an energy expenditure of 3 to 4 kcal · min^{-1}, although this value can change depending on body mass. In an early study of race walkers, Menier and Pugh (1968) reported that the average walking speed during competition walking was about 13.0 km · h^{-1} or 8 miles · h^{-1}. At high speeds, walking can be more costly than running. In the same study, the authors compared the energy costs of walking and running. As shown in figure 5.4, at speeds faster than 8 km · h^{-1} or 5 miles h^{-1}, a greater oxygen consumption was elicited during walking than running.

Table 5.1 Energy Expenditure During Various Physical Activities

Activity	Energy expenditure (kcal · min⁻¹)	
	Men	Women
Sitting	1.7	1.3
Standing	1.8	1.4
Sleeping	1.2	0.9
Walking (3.5 mi · h⁻¹, 5.6 km · h⁻¹)	5.0	3.9
Running (7.5 mi · h⁻¹, 12.0 km · h⁻¹)	14.0	11.0
(10.0 mi · h⁻¹, 16.0 km · h⁻¹)	18.2	14.3
Cycling (7.0 mi · h⁻¹, 11.2 km · h⁻¹)	5.0	3.9
(10.0 mi · h⁻¹, 16.0 km · h⁻¹)	7.5	5.9
Weightlifting	8.2	6.4
Swimming (3.0 mi · h⁻¹, 4.8 km · h⁻¹)	20.0	15.7
Basketball	8.6	6.8
Handball	11.0	8.6
Tennis	7.1	5.5
Wrestling	13.1	10.3

mi = mile.

Adapted, by permission, from J.H. Wilmore and D.L. Costill, 2004, *Physiology of sport and exercise*, 3rd ed. (Champaign, IL: Human Kinetics), 148.

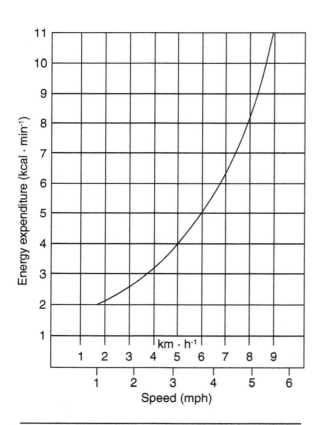

Figure 5.3 Energy expenditure during walking on a level surface at various speeds.

Adapted, by permission, from W.D. McArdle, F.I. Katch and V.L. Katch, 2001, *Exercise physiology*, 5th ed. (Philadelphia, PA: Lippincott, Williams & Wilkins), 204.

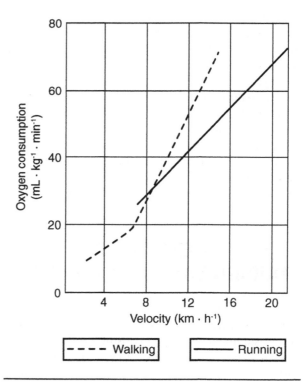

Figure 5.4 Comparisons of oxygen uptake between horizontal walking and running in competition walkers.

Reprinted, by permission, from W.D. McArdle, F.I. Katch and V.L. Katch, 2001, *Exercise physiology: Energy, nutrition and human performance*, 5th ed. (Philadelphia, PA: Lippincott, Williams & Wilkins), 206; adapted, by permission, from D.T. Menier and L.G. Pugh, 1968, "The relation of oxygen intake and velocity of walking and running, in competition runners," *The Journal of Physiology* 197(3): 717-721.

> ▶ **K E Y   P O I N T** ◀
>
> Walking and running are the two principal forms of human locomotion. From an energetic point of view, walking is less demanding metabolically than running. However, walking can impose a relatively high energy cost at speeds faster than a natural walking pace. For example, at a speed of 8 km · h^{-1} or 5 miles · h^{-1}, it is more efficient to run than to walk.

Effects of Body Mass

Energy cost during walking or running can also be affected by factors such as body mass, body mass distribution, load carriage, gradient, and terrain. As walking and running are activities in which the body mass is supported by the lower extremities, it follows that at any given speed, the greater the body mass of the person, the greater the energy cost incurred. It is important also to recognize that a major portion of the metabolic demand during walking and running is associated with acceleration and deceleration of the limbs with each stride. Thus, the mass of the limbs can be considered inertia that muscles must overcome, and the greater the inertia, the greater the muscular effort necessary to move the limbs. Furthermore, the inertia can increase if more limb mass is distributed away from the center of the rotational axis. Therefore, assuming that other factors are equal, those who have more mass concentrated in the limbs or have limb mass concentrated more distally require more energy to perform a walking or running task (Martin and Morgan 1992). This mechanical principle is relevant to the situation in which people walk with handheld weights. Miller and Stamford (1987) found that the energy expended during walking at 4 miles · h^{-1} (6.4 km · h^{-1}) with ankle and hand weights was comparable to that used during running at 5 miles · h^{-1} (8 km · h^{-1}). Among other examples of utilization of this mechanical rationale to maximize energy expenditure are exercises that use walking poles (to simulate arm action in cross-country skiing) or power belts (worn around the waist and equipped with resistance cords with handles for arm action), and upper body exercise such as swinging the arms during walking.

Effects of Terrain

Walking and running can take place on many different types of terrain, including pavement, grass, and sand and in snow. Because of the differences in hardness among these surfaces, there are differences in the energy required to move the body forward. The energy cost during walking has been reported to be similar on paved road and grass (McArdle et al. 2001). However, more energy is expended during walking or running on a soft surface such as sand or in snow. Lejeune and colleagues (1998) found that walking on sand required 2.5 times more energy expenditure than walking on a hard surface at the same speed. On the other hand, energy utilization during running on sand was only about 1.6 times more than during running on a hard surface at the same speed. Similarly, Smolander and colleagues (1989) observed that walking in snow triples the metabolic cost of walking on a treadmill. This added energy cost during locomotion on a soft terrain like sand or in snow has been ascribed to the additional external work that is done in response to the loss of vertical velocity (Lejeune et al. 1998). The loss of vertical velocity appears to be smaller as velocity increases, suggesting that people may choose jogging or running to minimize the energy discrepancy between hard and soft surfaces. Overall, it has been suggested that walking or running along a beach or in snow can provide an excellent opportunity to burn additional calories (Semih and Feluni 1998).

> ▶ **K E Y   P O I N T** ◀
>
> The energy cost during walking and running increases if body mass increases or if more of the mass is distributed away from the center of the body. This mechanical principle applies when people walk with handheld weights, which help in maximizing energy expenditure. Energy cost also increases if exercise is performed on a soft terrain such as sand or in snow; this gap diminishes as velocity increases. Walking or jogging on such surfaces can be an excellent opportunity to burn additional calories.

Running Economy

Research on the energetics of running has frequently dealt with **running economy** and its impact on endurance performance. Running economy is typically defined as the energy demand for a given velocity of submaximal running. It is determined through measurement of steady-state oxygen uptake during running at different speeds (Daniels 1985). A runner with greater oxygen uptake is said to be less able to use stored energy efficiently. Efficient utilization of available energy is an important approach for those running a long-distance event because this will help preserve energy substrates and reduce the production of performance-inhibiting substances such as lactic acid. Numerous studies have demonstrated a strong relationship between running economy and distance running performance (Costill 1967; Costill et al. 1973; Conley and Krahenbuhl 1980; Di Prampero et al. 1993).

Measurement of running economy has been typically made in the laboratory as the subject runs on a motorized treadmill. In this indoor environment, air and wind resistance are absent. Several studies have yielded estimations that 2% to 8% of the total energy cost during outdoor running is expended to overcome air resistance (Pugh 1970; Davies 1980); this is a rather small quantity. However, the effect of air resistance becomes more pronounced as running speed increases. Running economy can also be influenced by biomechanical factors such as body physique and running mechanics, as well as physiological factors like ventilation, body temperature, and substrate utilization (Anderson 1996). Running economy decreases during prolonged running (Collins et al. 2000; Kyrolainen et al. 2000). This reduction is thought to be mainly due to altered physiological responses such as increases in ventilation, body temperature, and fat utilization, all of which are associated with increased oxygen consumption (Morgan et al. 1990; Nichol et al. 1991).

Cycling

Cycling is the major means of transportation in many countries and is also used for recreation and competitive sport. During the last decade or so, off-road cycling has enjoyed an exponential growth in popularity. In the fitness and rehabilitation industry, cycling is often performed on a stationary ergometer, which allows regulation of exercise intensity. Stationary cycling differs from outdoor cycling in that it provides a broad range of exercise intensities that can be adjusted via the manipulation of either pedal speed or brake resistance. Cycling uses less muscle mass than running and as a result requires less energy (table 5.1). However, because of its non-weight-bearing character, cycling is more tolerable for and commonly chosen by people who are extremely sedentary or obese or those who have difficulties with **gait.**

The stationary cycle ergometer can be either mechanically or electronically braked, and the brake resistance is measured in kilograms. Brake resistance may be viewed as somewhat similar to the gear ratio used by cyclists during road cycling. To maintain a given velocity, a lower gear ratio needs to be accompanied by a high pedal cadence or vice versa. When mechanically braked ergometers are used, the work rate or power output depends on the maintenance of a particular pedaling frequency. On the other hand, electronically braked ergometers are designed such that at any given selected work rate, the brake resistance will automatically increase or decrease depending on the pedaling rate selected. In other words, work rate can be kept constant independent of the change in pedal frequency. These devices provide a stability in work rate that is less dependent on subject motivation as compared to the mechanically-braked cycle ergometer. For a cycle ergometer, the work rate is measured in either watts (W) or kilogram meters per minute (kgm · min^{-1}), with 1 W equal to 6 kgm · min^{-1} (or, more precisely, 6.12 kgm · min^{-1}). As mentioned earlier, work rate or power output can be calculated as the product of brake resistance (kg) and pedaling speed (m · min^{-1}). To obtain a pedaling speed, a pedaling frequency in revolutions per minute is multiplied by linear distance traveled per revolution, which is 6 m with use of a Monark cycle ergometer (ACSM 2006).

> ### ▶ K E Y P O I N T ◀
>
> Most cycle ergometers on the market are electronically braked, meaning that the brake resistance automatically increases or decreases depending on the pedaling rate chosen. This so-called **servomechanism** allows an individual to vary pedal rate but still maintain a desirable workload, so adherence to a target intensity depends less on the individual's motivation.

There is a linear relationship between energy expenditure and work rate. In fact, one can predict energy expenditure from work rate using a regression equation, as discussed later in this chapter. As mentioned earlier, work rate during cycling is a function of brake resistance and pedal frequency. Thus a change in brake resistance or pedal frequency can affect energy expenditure. Many researchers have examined the effect of pedal frequency per se on energy expenditure. This is an interesting question, in that pedal frequency can influence neuromuscular fatigue in working muscles (Takaishi et al. 1994, 1996). Pedal frequency can also affect the recruitment pattern of muscle fiber types (Ahlquist et al. 1992). To examine the impact of cycling cadence adequately, the work rate needs to be kept constant. With a constant work rate, however, a change in pedal rate is always accompanied by a change in brake resistance in the opposite direction. We recently demonstrated greater energy expenditure during cycling at 80 as compared to 40 rev · min^{-1} when exercise was performed at a low intensity (i.e., 50 W or 300 kgm · min^{-1}) (Kang et al. 2004). This finding is consistent with results of Londeree and colleagues (1997), who showed a positive relationship between oxygen consumption and pedal rate. But riding at a given work rate with a higher cadence may not necessarily be more costly during exercise at high intensity. In the same study, Londeree and colleagues found that the difference in energy expenditure between cadences decreased as the work rate increased. Professional cyclists often choose to pedal at a rather fast rate, close to 100 rev · min^{-1}, during competition. It has been found that for athletes who are competing, riding at a high cadence tends to be more efficient metabolically (Palmer et al. 1999; Gueli and Shephard 1976).

During cycling outdoors, the energy cost is greater due to air resistance. This is especially the case when cycling is performed at high velocities. It has been reported that aerodynamic resistance can be over 90% of the total resistance during cycling at speeds >30 km · h^{-1} (>19 miles · h^{-1}) (Kyle 1991). Considerable effort has been made to improve the configuration of bicycles and their components to minimize extra use of energy. For example, some of the clip-on handlebars invented in the 1980s have allowed cyclists to assume a forward-crouched upper body position while competing. This change in upper body position has been found to result in as much as a 35% decrease in air drag due to a reduction in the frontal area of the cyclist (Capelli et al. 1993; Kyle 1989). Interestingly, the aerodynamic position itself has been associated with increased $\dot{V}O_2$ as well as heart rate and respiratory exchange ratio (Gnehm et al. 1997). However, the magnitude of this increase in energy cost is considered too small to cancel out the reduction in energy cost due to the improved aerodynamic position (Faria et al. 2005).

Resistance Exercise

Resistance exercise or weightlifting has been widely used in various sport training programs as well as in health- and fitness-related exercise interventions. Every activity, including activities of daily living, requires a certain percentage of an individual's maximum strength and endurance. Regular resistance exercise can serve as a potent stimulus to the musculoskeletal system that is necessary to bring about gains in muscle size and function. It also helps in enhancing bone mass and the strength of connective tissue. A training routine that combines both aerobic and resistance exercises has been highly recommended because the resulting improvement in cardiorespiratory and muscular function can allow individuals not only to reduce their risks for chronic diseases related to physical inactivity, but also to perform activities of daily living comfortably and safely.

Considerably less information is available concerning the energy cost of resistance exercise than for other forms of exercise. A possible reason is that this type of exercise is typically performed in an intermittent fashion so that the accumulated exercise time is relatively short. Although exertion can be quite strenuous at times during resistance exercise, these strenuous phases usually do not last for more than 1 min. The energy cost during resistance exercise is generally higher than the cost during walking and can be compared with that during cycling (table 5.1).

Energy cost during resistance exercise is typically measured using net $\dot{V}O_2$max, calculated as exercise $\dot{V}O_2$ + recovery $\dot{V}O_2$ − resting $\dot{V}O_2$ for the period of exercise and recovery. This formula recognizes the fact that oxygen deficit during resistance exercise can be so significant that inclusion of recovery $\dot{V}O_2$ provides a more accurate measure of the energy cost of the exercise. Table 5.2 provides results on net $\dot{V}O_2$ from studies on the energy cost of resistance exercises (Halton et al. 1999; Hunter et al. 1988, 1992; Olds and Abernethy 1993; Willoughby et al. 1991; Wilmore et al. 1978). It appears that the energy cost is greater during a **circuit weight training** routine comprising a larger volume of exercise than during a regular resistance training workout. Energy cost during resistance exercise has also been reported using gross $\dot{V}O_2$: The resting component is not subtracted from the total value (Burleson et al. 1998; Phillips and Ziuraitis 2003). With this approach, $\dot{V}O_2$ was found to be near 1 to 1.5 L · min^{-1}, values higher than those in table 5.2. When the total oxygen consumed is accumulated over an entire exercise session, it ranges from 30 to 45 L · session^{-1} or 150 to 225 kcal · session^{-1}. This level of oxygen uptake is considered mild given that the gross caloric expenditure can reach 400 to 500 kcal during a typical endurance exercise of moderate intensity that lasts for ~45 min.

Table 5.2   Comparisons of Studies on the Energy Cost of Resistance Exercise

Studies	Subjects	Volume/intensity	$\dot{V}O_2$ (L · min⁻¹)*
Willoughby et al. (1991)	10 men	Squat at 50% 1RM, 7 reps	0.18
		Squat at 70% 1RM, 6 reps	0.19
		Squat at 90% 1RM, 5 reps	0.24
Halton et al. (1999)	7 men	1 circuit of 8 exercises at 75% 20RM, 20 reps	0.29
Olds and Abernethy (1993)	7 men	2 circuits of 7 exercises at 60% 1-RM, 15 reps	0.78
		2 circuits of 7 exercises at 75% 1-RM, 12 reps	0.78
Hunter et al. (1988)	10 men 7 women	4 sets of bench press at 20% 1RM, 30 reps	0.14
		4 sets of bench press at 40% 1RM, 20 reps	0.23
		4 sets of bench press at 60% 1RM, 10 reps	0.30
		4 sets of bench press at 80% 1RM, 5 reps	0.51
Wilmore et al. (1978)	20 men 20 women	3 circuits of 10 exercises at 40% 1RM, 15-18 reps	1.00
Hunter et al. (1992)	14 men 8 women	4 sets of knee extension at 60% 1RM, 10 reps	0.46
		4 sets of knee extension at 80% 1RM, 5 reps	0.48
		4 sets of knee flexion at 60% 1RM, 10 reps	0.58
		4 sets of knee flexion at 80% 1RM, 5 reps	0.60

*Values are the net $\dot{V}O_2$ that is averaged over exercise and recovery periods. Net $\dot{V}O_2$ is computed as exercise $\dot{V}O_2$ + recovery $\dot{V}O_2$ – resting $\dot{V}O_2$ accumulated for the total period of exercise and recovery.

Although resistance exercise may not accumulate as much energy expenditure, it can disturb the body's homeostasis to a greater extent than aerobic exercise. The physiological strain that resistance exercise imposes can persist through a sustained period of recovery. As such, resistance exercise is often associated with a greater EPOC than aerobic exercise. For example, Burleson and colleagues (1998) found a greater EPOC following circuit weight training performed at 60% 1-repetition maximum (1RM) as compared to treadmill exercise that was matched for the same $\dot{V}O_2$ elicited during resistance exercise. The greater EPOC is attributable to the fact that a majority of the energy supporting the activity is derived from the use of anaerobic energy sources such as ATP and PCr, whose replenishment following exercise requires oxygen (Dudley 1988). The larger EPOC can also result from a greater change in heart rate, as well as increased concentrations of blood lactate and the release of selected hormones during exercise, which will elevate oxygen utilization once exercise is completed (Dudley 1988).

Many studies have addressed the effect of resistance exercise on EPOC (Binzen et al. 2001; Kang et al. 2005b; Thornton and Potteiger 2002; Melby et al. 1993; Melanson et al. 2002; Olds and Abernethy 1993; Schuenke et al. 2002). It is generally agreed that although the lifting mode and intensity (i.e., % 1RM) can be influential, it is the volume of the training, which is the total quantity of weights lifted, that is the most important factor determining the magnitude of EPOC. Using the respiratory exchange ratio, several studies also revealed a greater fat utilization in addition to the greater EPOC following resistance exercise (Binzen et al. 2001; Melby et al. 1993). Despite these favorable metabolic responses, the long-term potential of resistance training in managing body weight and body composition is still unclear.

METABOLIC CALCULATION

Researchers studying the energy cost of exercise have demonstrated that it is possible to estimate energy expenditure during selected physical activities without using sophisticated instrumentation. Walking, running, and cycling are forms of activity that have been studied in great detail, because the movement patterns in these activities are relatively simple and repetitive. A linear relation has been found between energy cost and intensity indices such as speed or cadence. In the case of stationary cycling, energy cost is affected by brake resistance in addition to cadence. With respect to walking and running, the energy requirements also depend on whether exercise is performed on a horizontal surface or against certain levels of grade. At any given speed, the greater the grade, the greater the involvement of muscle mass required to maintain balance and therefore the greater the energy expenditure.

Walking and Running

Walking and running differ in many respects. During walking, approximately 0.1 ml oxygen is needed for transporting each kilogram of body mass per meter of horizontal distance covered (Dill 1965). The oxygen demand of running is twice as great, or $0.2 \text{ ml} \cdot \text{kg}^{-1} \cdot \text{m}^{-1}$ (Margaria et al. 1963; Nagle et al. 1971). The oxygen demand in raising one's body mass against gravity is about 1.8 ml per kilogram body mass for each meter of vertical distance traveled during walking (Nagle et al. 1965, 1971). However, the oxygen cost of vertical ascent during running is half that of walking, or $0.9 \text{ ml} \cdot \text{kg}^{-1} \cdot \text{m}^{-1}$ (Margaria et al. 1963). In light of these differences, two separate equations have been developed to allow estimation of energy expenditure during walking and running (ACSM 2006). The equations are shown in figure 5.5.

Cycling

Estimation of energy expenditure for cycling is based on the use of a stationary cycle ergometer that allows the measurement of work rate or power output. As with walking and running, a linear relationship has been found between energy expenditure and work rate during stationary cycling. The total net energy cost of cycling entails the energy associated with unloaded cycling or cycling without brake resistance and the energy that is proportional to the external load being applied. At 50 to 60 $\text{rev} \cdot \text{min}^{-1}$, the oxygen cost of unloaded cycling is approximately $3.5 \text{ ml} \cdot \text{kg}^{-1} \cdot \text{min}^{-1}$ above that of rest (Lang et al. 1992; Latin and Berg 1994; Londeree et al. 1997). The oxygen cost of cycling in relation to the external load is about 1.8 ml per kilogram resistance for each meter of linear distance traveled (Nagle et al. 1971). Figure 5.5 shows the metabolic equation used to estimate the energy requirement of cycling.

Metabolic Equations for Other Activities

Equations are available for other popular activities such as arm cranking and bench stepping. Many sports, such as swimming, rowing, and kayaking, involve use of the arms. Thus arm cranking has been used in training or testing of athletes performing these events. In many rehabilitation settings, exercise with the arms is also frequently chosen as either a testing or a training mode for those with spinal cord injury or leg amputation. An ergometer similar to that for leg cycling is available for the arms so that exercise intensity can be regulated during arm cranking. In comparison with leg exercise, arm exercise at the same power output produces greater heart rate and blood pressure. The increment of $\dot{V}O_2$ in $\text{ml} \cdot \text{kg}^{-1} \cdot \text{min}^{-1}$ related to increased work rate during arm exercise was also found to be steeper than during leg cycling (Sawka 1986), which suggests that arm cranking can use up more oxygen at high intensities than leg cycling. This difference has been attributed to the increased metabolic demand for upper body stabilization during arm cranking at high intensities. See figure 5.5 for the formula for calculating net $\dot{V}O_2$ with arm exercise (ACSM 2006).

This formula reflects a steeper relationship between work rate and relative $\dot{V}O_2$ during arm cranking than

Walking

net $\dot{V}O_2$ (ml · kg^{-1} · min^{-1}) = horizontal $\dot{V}O_2$ + vertical $\dot{V}O_2$

horizontal $\dot{V}O_2$ = 0.1 ml · kg^{-1} · min^{-1}/ (m · min^{-1}) × speed (m · min^{-1})

vertical $\dot{V}O_2$ = 1.8 ml · kg^{-1} · min^{-1}/ (m · min^{-1}) × speed (m · min^{-1}) × %grade

Running

net $\dot{V}O_2$ (ml · kg^{-1} · min^{-1}) = horizontal $\dot{V}O_2$ + vertical $\dot{V}O_2$

horizontal $\dot{V}O_2$ = 0.2 ml · kg^{-1} · min^{-1}/ (m · min^{-1}) × speed (m · min^{-1})

vertical $\dot{V}O_2$ = 0.9 ml · kg^{-1} · min^{-1}/ (m · min^{-1}) × speed (m · min^{-1}) × %grade

Cycling

net $\dot{V}O_2$ (ml · kg^{-1} · min^{-1}) =

1.8 ml · kg^{-1} · m^{-1}× work rate (kg · m · min^{-1}) ÷ body mass (kg) + 3.5 ml · kg^{-1} · min^{-1}

Arm Exercise

net $\dot{V}O_2$ (ml · kg^{-1} · min^{-1}) =

3 ml · kg^{-1} · m^{-1} × work rate (kg · m · min^{-1}) ÷ body mass (kg)

Bench Stepping

net $\dot{V}O_2$ (ml · kg^{-1} · min^{-1}) =

0.2 ml · kg^{-1} · step^{-1} × step rate (step · min^{-1}) + 1.33 × 1.8 ml · kg^{-1} · m^{-1} × step rate (step · min^{-1}) × step height (m · step^{-1})

Figure 5.5 Metabolic equations for various activities.

during leg cycling. Unlike the formula for leg cycling, this formula does not include an unloaded component. It was thought that this component could be neglected because of the smaller muscle mass involved during arm exercise (Swain 2000).

Bench stepping was one of the earliest exercise modalities used to measure physical work capacity. To perform this type of exercise, still in use today, an individual simply steps up onto and down from a bench at a predetermined rate. It is particularly noteworthy that this activity contains a significant amount of **eccentric** work. Unlike a **concentric** contraction, in which muscle shortens, an eccentric contraction occurs when muscle lengthens because an outside force or gravity is greater than the force being exerted by the muscle. Eccentric contraction occurs primarily during the down step. A similar activity is stair climbing, which has emerged as a popular alternative to walking and jogging. With the recent advent of newly designed stair climbing machines, a wide range of exercise intensities can be performed effectively. See figure 5.5 for the equation used for computing energy cost during bench step exercise (ACSM 2006):

In this equation, 0.2 ml · kg^{-1} · step^{-1} is the oxygen cost of horizontal movement for each step

that occurs; 1.33 is a factor that accounts for the added energy cost due to eccentric work during each down step; and 1.8 ml · kg^{-1} · m^{-1} is the oxygen cost of vertical work per meter distance traveled, which is similar to the vertical factor in the walking formula shown earlier.

In general, metabolic equations are reasonably accurate in providing estimates of energy cost during exercise. They can allow fitness professionals to obtain exercise $\dot{V}O_2$ without having to utilize a metabolic system, which is often difficult and time-consuming. If the caloric cost of exercise is desired, it is necessary to convert the net $\dot{V}O_2$ to L · min^{-1} by multiplying the $\dot{V}O_2$ in ml · kg^{-1} · min^{-1} by body mass in kilograms and then dividing by 1000. $\dot{V}O_2$ in L · min^{-1} is then multiplied by 5 kcal · L^{-1} and by exercise duration to obtain the total calories expended during exercise. Energy expenditure can also be expressed in metabolic equivalents (METs). As discussed in chapter 4, a MET is the ratio of work metabolic rate to a standard resting metabolic rate that is equal to a resting $\dot{V}O_2$ of 3.5 ml · kg^{-1} · min^{-1}. In this case, the actual $\dot{V}O_2$ in ml · kg^{-1} · min^{-1} is simply divided by 3.5 ml · kg^{-1} · min^{-1} so that a MET value can be obtained. For example, for someone who consumes 35 ml · kg^{-1} · min^{-1} while running, the MET level is 10. Use of metabolic calculations is relatively simple and straightforward, especially for those who have experience with the formulas, though it can be a source of confusion for the novice user. Tables 5.3 and 5.4 provide precalculated energy expenditures in METs over a wide range of exercise intensities during walking, running, and cycling, which eliminate the need to use the metabolic equations.

Table 5.3   Estimated Energy Expenditure in METs During Level and Grade Walking and Running

Speed		Grade (%)						
mi · h^{-1}	m · min^{-1}	Level	2.5	5.0	7.5	10.0	12.5	15.0
Walking								
1.7	45.6	2.3	2.9	3.5	4.1	4.6	5.2	5.8
2.0	53.6	2.5	3.2	3.9	4.6	5.3	6.0	6.6
2.5	67.0	2.9	3.8	4.6	5.5	6.3	7.2	8.1
3.0	80.4	3.3	4.3	5.4	6.4	7.4	8.5	9.5
3.4	91.2	3.6	4.8	5.9	7.1	8.3	9.5	10.6
3.8	100.5	3.9	5.2	6.5	7.8	9.1	10.4	11.7
Jogging or running								
5.0	134.0	8.6	9.5	10.3	11.2	12.0	12.9	13.8
6.0	161.0	10.2	11.2	12.3	13.3	14.3	15.4	16.4
7.0	188.0	11.7	12.9	14.1	15.3	16.5	17.7	18.9
8.0	214.0	13.3	14.7	16.1	17.4	18.8		
9.0	241.0	14.8	16.3	17.9	19.4			
10.1	268.0	16.3	18.0	19.7				

mi = mile.

Adapted, by permission, from American College of Sports Medicine (ACSM), 2000, *ACSM guidelines for exercise testing and prescription*, 6th ed. (Philadelphia, PA: Lippincott, Williams & Wilkins), 307.

Table 5.4   Estimated Energy Expenditure in METs During Leg Cycle Ergometry

Power output		Body weight (kg)					
kgm · min^{-1}	Watts	50	60	70	80	90	100
300	50	5.1	4.6	4.2	3.9	3.7	3.5
450	75	6.6	5.9	5.3	4.9	4.6	4.3
600	100	8.2	7.1	6.4	5.9	5.4	5.1
750	125	9.7	8.4	7.5	6.8	6.3	5.9
900	150	11.3	9.7	8.6	7.8	7.1	6.6
1050	175	12.8	11.0	9.7	8.8	8.0	7.4
1200	200	14.3	12.3	10.8	9.7	8.9	8.2

Adapted, by permission, from American College of Sports Medicine (ACSM), 2000, *ACSM guidelines for exercise testing and prescription*, 6th ed. (Philadelphia, PA: Lippincott, Williams & Wilkins), 308.

SUMMARY

At the beginning of exercise, the body increases its oxygen uptake in order to match the demand imposed by the exercise. While this homeostatic state is important for the body to produce, the maintenance of oxygen uptake that would be expected to result is often not achieved when exercise is performed at intensities above lactate threshold. At these strenuous intensities, despite an unchanging workload, there is a gradual increase in oxygen uptake that can lead to fatigue. An increase in oxygen uptake without a concordant increase in workload decreases metabolic efficiency, and this is of interest to long-distance athletes, who typically seek to preserve energy while performing the activity. Any given exercise will also provoke a transitory elevation in oxygen uptake during the subsequent recovery; and the more strenuous the preceding activity, the greater and more sustained the postexercise oxygen consumption becomes.

Walking, running, and cycling are the exercise modes for which issues related to energy cost have been most extensively investigated. For walking and running, as weight-bearing activities, the energy cost has been found to be affected by extraneous factors such as body mass and its distribution, as well as road surface or terrain. Cycling, as a non-weight-bearing activity, is usually not as energetically costly as running. It is often performed on an ergometer, which allows one to set the workload using a combination of brake resistance and pedal speed. Resistance exercise, a popular activity among young adults, differs from these other activities in that it is performed in an intermittent fashion and entails multiple short bursts of intense muscle contraction. Despite the fact that the accumulated energy expenditure is often lower than for most aerobic exercises, calculation of energy cost must take into account the excess postexercise oxygen consumption incurred, especially if rest intervals are short.

In light of the complexity associated with the measurement of energy cost, formulas have been made available that allow us to obtain energy cost estimates without having to use sophisticated instrumentation. Understanding the energy cost of exercise is important both for designing an exercise program to maximize energy expenditure or efficiency and for determining appropriate nutritional requirements toward the maintenance of a desirable energy balance.

KEY TERMS

circuit weight training

concentric

eccentric

efficiency

excess postexercise oxygen consumption

gait

gross energy expenditure

net energy expenditure

oxygen debt

oxygen deficit

running economy

slow component of oxygen uptake

REVIEW QUESTIONS

1. Define the following terms related to oxygen uptake: *oxygen deficit, oxygen debt, excess postexercise oxygen consumption,* and *oxygen slow component.*

2. How does gross oxygen uptake differ from net oxygen uptake? How do you convert $\dot{V}O_2$ from $L \cdot min^{-1}$ to $ml \cdot kg^{-1} \cdot min^{-1}$? Why is there a need to do so?

3. What is the formula for computing the metabolic efficiency of human movement? Using cycling as an example, illustrate how metabolic efficiency is determined mathematically. Why is arm cranking generally less efficient than leg cycling?

4. Suppose that someone is prescribed exercise on a Monark cycle ergometer at a workload of 600 kgm/min. If the resistance is set at 2 kg, how fast should the person pedal to achieve this workload?

5. A client is walking on a treadmill at 3.4 mph (5.5 km/h) up a 5% grade. What is this person's oxygen consumption in $ml \cdot kg^{-1} \cdot min^{-1}$?

6. What is the relative oxygen consumption for running on a treadmill at 5.5 mph (9 km/h) and 12% grade? Provide the answer in $ml \cdot kg^{-1} \cdot min^{-1}$ and METs.

CHAPTER

6

EXERCISE STRATEGIES FOR ENHANCING ENERGY UTILIZATION

This chapter presents exercise guidelines that have been established to facilitate energy expenditure and reviews various prescriptive strategies that have proven effective in altering energy provision and expenditure. These strategies include continuous exercise, intermittent exercise, exercise with variable intensity, and resistance exercise. This chapter also discusses the efficacy of increasing physical activity alone as a weight management strategy.

PHYSICAL ACTIVITY AND ENERGY BALANCE

Chronic diseases and disabilities account for most of the health problems in highly developed countries today. Most of these health problems, such as coronary heart disease, hypertension, obesity, non-insulin-dependent diabetes, osteoporosis, and emotional disorders, have been ascribed to our habits of living, including low levels of physical activity. Obesity appears to play a central role in mediating many of these chronic conditions and has been strongly associated with a sedentary lifestyle. The prevalence of obesity has risen significantly in much of the world over the last several decades. For example, the prevalence of individuals with body mass index (BMI, weight ÷ height^2) ≥25 kg · m^{-2} in the United States increased from 44.3% to 52.6% from the late 1970s to the early 1990s, with the prevalence of those with a BMI ≥30 kg · m^{-2} increasing from 13.4% to 21.2%

during this same time period (Flegal et al. 1998). Obesity is the condition resulting from increased energy intake, reduced exercise expenditure, or a combination of the two.

Indeed, advances in technology have altered most of our occupations and modes of transportation so that we now need to expend much less energy in our daily activities. This reduced energy expenditure, coupled with unchanged or increased energy intake, can easily result in a positive energy balance that leads to weight gain. With the use of instruments that measure quantity of physical activity, numerous studies have demonstrated an inverse relationship between the level of physical activity and risk for developing overweight and obesity (Grundy et al. 1999; National Institutes of Health 1998; Williamson et al. 1993). In a comprehensive review of over 20 longitudinal studies, Ross and Janssen (2001) found that reduction in total adiposity was positively related to energy expenditure. The greater the energy expended in exercise and physical activity, the greater the fat loss. Physical activity has also been shown to play a role in preventing weight gain. Di Pietro (1999) found that men with increased cardiorespiratory fitness gained less body weight as compared to those who were less conditioned. This finding suggests that those who stay fit are less likely than others to become overweight. Some have conjectured that cardiovascular fitness is mainly genetically determined. However, this has been proven untrue, as Bouchard and colleagues (1992) found that heredity generally accounts for about 25% of variance in maximal aerobic capacity. This finding clearly supports the necessity and efficacy of pursuing an active lifestyle, which can be imperative in terms of treatment and prevention of obesity.

Physical activity is defined as any bodily movement produced by skeletal muscle that results in energy expenditure. Energy expended in physical activity can account for about 30% of the total daily energy expenditure. Physical activity and energy expenditure are not exactly the same. Physical activity is a behavior that can be characterized by intensity (how strenuous the activity is), duration (how long the activity lasts), frequency (how often the activity occurs), and mode (the type of activity being performed). Energy expenditure, on the other hand, is the total metabolic heat that is produced in the performance of any activity involving muscle contraction. One may expend the same amount of energy in a short burst of strenuous exercise as in a less intense but more sustained activity. Although physical activity and energy expenditure appear to be quantitatively related, several factors—including age, body size, weight distribution, fitness level, and mechanical efficiency—can affect this relationship. For example, being heavier will cost more energy during walking, and the difference becomes greater as walking speed increases (figure 6.1).

The energy expended due to regular physical activity can also be accompanied by an increase in fat utilization. Aerobic training has been shown to result in a number of structural and metabolic adaptations that favor fat oxidation. The **capillary density** of muscle tissue increases, which improves the delivery of oxygen and substrates (Gollnick and Saltin 1982). There is an increase in mitochondrial content as well as oxidative enzyme activity in skeletal muscle (Gollnick and Saltin 1982; Hoppeler et al. 1985; Baldwin et al. 1972). Thus, trained individuals are able to oxidize more fat than those who are untrained. Trained muscle cells also store more fat and have greater activity of lipase, which functions to split triglycerides into individual fatty acids. This helps facilitate the entry of fatty acids into mitochondria and thus the utilization of intramuscular triglycerides. In addition, there is evidence that the transport of fatty acids into the muscle cell involves a carrier molecule whose capacity to transport fatty acids can become saturated at high plasma fatty acid concentrations (Saltin and Åstrand 1993). Aerobic training has been shown to increase the capacity to transport fatty acids such that more of them are transported at a given fatty acid plasma concentration in trained individuals (Kiens et al. 1993).

> ### ▶ K E Y P O I N T ◀
>
> Aerobic training has been proven effective in weight control and management. It helps to increase energy expenditure via repeated muscle contractions. It also augments fat utilization, and this can be explained in part by an improved transport of oxygen and fatty acids due to increased capillary density of the muscle tissue being trained.

ENHANCING ENERGY EXPENDITURE THROUGH EXERCISE AND PHYSICAL ACTIVITY

As already mentioned, physical activity is defined as any bodily movement produced by skeletal muscle that results in energy expenditure. Physical activity is frequently categorized by the context in which it occurs. Common categories include occupational,

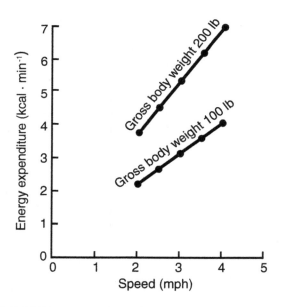

Figure 6.1 Effect of body weight on energy expenditure of walking at various speeds.

Adapted, by permission, from P.O. Astrand and K. Rodahl, 1977, *Textbook of work physiology*, 2nd ed. (New York, NY: McGraw-Hill), 583.

household, leisure time, and transportation. Leisure-time activities can be further divided into competitive sport, recreational activities, and exercise training (United States Department of Health and Human Services 1996). The terms *exercise* and *physical activity* have been used interchangeably in the past; but today, *exercise* is commonly used to refer to physical activity that is planned, structured, repetitive, and purposive in the sense that improvement or maintenance of one or more components of physical activity is the objective (Caspersen et al. 1985). Pursuing exercise and physical activity regularly has long been a part of recommendations made by various organizations such as the American College of Sports Medicine (ACSM), American Heart Association, and U.S. Department of Health and Human Services because exercise and physical activity enable individuals to increase their energy expenditure or create an energy deficit while gaining cardiorespiratory fitness. Exercise in conjunction with dietary modification has been recommended as the most effective behavioral approach for weight loss (National Institutes of Health 1998).

> ▶ **K E Y   P O I N T** ◀
>
> Physical activity differs from exercise in that the former is defined as any bodily movement produced by skeletal muscle that results in energy expenditure, whereas the latter is physical activity that is planned, structured, repetitive, and purposive in the sense that improvement or maintenance of one or more components of physical activity is the objective.

An exercise routine typically used in a weight management program consists of continuous, large-muscle activities with moderate to high caloric cost, such as walking, running, cycling, swimming, rowing, and stair stepping. This approach increases daily energy expenditure and helps to tip the caloric equation so that energy output is greater than energy input. Most training studies that have demonstrated exercise-induced weight loss have adopted exercise programs that elicit a weekly energy expenditure of 1500 to 2000 kcal (Ross and Janssen 2001). This suggests that a minimum energy expenditure of 300 kcal should be achieved during each exercise session and that exercise should be performed no less than five times a week. This amount of energy expenditure generally occurs with 30 min of moderate to vigor-ous running, swimming, or cycling or with 60 min of brisk walking (McArdle et al. 2001).

Energy expenditure during exercise can be influenced by intensity, duration, and mode of exercise. Unlike most fitness programs in which intensity is an important part of the exercise prescription, any program aimed at weight loss should ultimately be guided by the measure of total energy expenditure and its relation to energy consumption. To burn the most calories, exercise **duration** is considered more important, and exercise **intensity** must be adapted for the amount of time one would like to exercise (ACSM 2006). It has been suggested that exercise intensity should be tailored to allow a minimum expenditure of 300 kcal during each exercise session. A study that included dietary modification suggested that overweight women who exercise for a longer duration each week were able to lose more weight during an 18-month intervention (Jakicic et al. 1999). In this study, individuals reporting >200 min of exercise per week also reported >2000 kcal · week^{-1} of leisure-time physical activity as measured by a questionnaire. Table 6.1 illustrates minutes of continuous activity necessary to expend 300 kcal based on body weight as well as the type and intensity of activity.

The higher the exercise intensity, the greater the energy expenditure during exercise per unit of time. In other words, it costs you more energy to move your body weight or a given resistance at a faster pace. However, it is often difficult to maintain vigorous exercise for a period of time sufficient to maximize energy expenditure. This is especially the case for overweight or obese individuals who have low exercise tolerance in general. Some may also have risk factors that contraindicate vigorous exercise. Consequently, those who need to maximize their energy expenditure may have to depend more on the expansion of exercise duration in order to achieve this goal. The recently published ACSM position stand suggests that to be effective in maximizing energy expenditure over the long term, exercise intensity should not exceed 70% of an individual's maximal heart rate, which corresponds to ~60% $\dot{V}O_2$max or **heart rate reserve** (HRR) (ACSM 2001). This intensity allows one to attain the recommended exercise duration of ~45 min. Mild exercise is also considered ideal for people who are beginning an exercise program. This allows adequate time for them to adapt to their exercise routine and to progressively increase exercise intensity over time. Clearly, exercise at moderate to high intensities has proven effective in augmenting

Table 6.1   Minutes of Continuous Activity Necessary to Expend 300 kcal Based on Body Weight and Type and Intensity of Activity

	Body weight						
Pounds	120	140	160	180	200	220	240
Kilograms	55	64	73	82	91	100	109
Conditioning exercise							
Cycling							
Stationary	66	57	50	44	40	36	33
Outdoor	83	71	62	55	50	45	41
Walking (level)							
2.5 mph (4.0 km · h⁻¹)	110	94	83	73	66	60	55
3.0 mph (4.8 km · h⁻¹)	94	81	71	63	57	52	47
3.5 mph (5.6 km · h⁻¹)	83	71	62	55	50	45	41
Water aerobics	83	71	62	55	50	45	41
Lap swimming	41	35	31	28	25	23	21
Yoga	83	71	62	55	50	45	41
Resistance exercise	55	47	41	37	33	30	28
Dancing							
Aerobic dance	55	47	41	37	33	30	28
Low-impact aerobic dance	66	57	50	44	40	36	33
Ballroom dance (fast)	60	52	45	40	36	33	30
Ballroom dance (slow)	110	94	83	73	66	60	55
Lifestyle activities							
Golf (walking)	73	63	55	49	44	40	37
Raking the lawn	83	71	62	55	50	45	41
Lawn mowing							
Power mower	73	63	55	49	44	40	37
Riding mower	132	113	99	88	79	72	66
Vacuuming or sweeping	132	113	99	88	79	72	66

Reprinted, by permission, from American College of Sports Medicine (ACSM), 2001, "Appropriate intervention strategies for weight loss and prevention of weight regain for adults," *Medicine and Science in Sports and Exercise* 33 (12): 2145-2156.

aerobic fitness. However, the level of exercise intensity necessary to improve fitness is generally higher than the level necessary to facilitate energy expenditure and thus weight loss.

KEY POINT

Exercise intensity and duration are the two important components of exercise prescription. Unlike most fitness programs emphasizing intensity, any program aimed at weight loss should be guided by the measure of total energy expenditure and its relation to energy consumption. To burn the most calories, exercise duration is more important, and exercise intensity must be adapted for the amount of time allocated or the number of calories one would like to expend.

Exercise **frequency** complements duration and intensity. Frequency refers to how often each week one performs exercise and is often determined based on how vigorous each exercise session is. According to recent exercise guidelines (ACSM 1998; United States Department of Health and Human Services 1996), a training program consisting of exercise of moderate intensity (i.e., 50-70% V̇O₂max) should be carried out no fewer than three times a week in order to develop and maintain cardiorespiratory fitness and improve body composition. This recommendation in part originated from early studies in which Pollock and colleagues (1975) found that training twice weekly produced no significant changes in body composition whereas training three or four days weekly did. More recent studies have suggested that energy expenditure to bring about weight loss should be no less than 1500 kcal ·

week^{-1}. To achieve this threshold, energy expended during each exercise session should be at least 500 kcal if exercise is performed three times a week, or 300 kcal if exercise is performed five times a week. Obviously, this latter exercise arrangement (i.e., >5 times · week^{-1}) is more tolerable for sedentary and overweight people who have difficulty sustaining vigorous exercise. The more frequent exercise regimen will also help in maximizing fat utilization and minimizing exercise-related injury (Wallace 1997).

For meeting the caloric threshold necessary to bring about weight reduction, ACSM recommends a target range of 150 to 400 kcal of energy expenditure per day in physical activity, exercise, or both (ACSM 2006). The lower end of this range is considered a minimal threshold and an initial goal for previously sedentary or overweight individuals. If this level of light activity can be performed daily, it will help in achieving a total of ~1000 kcal of energy expenditure each week. The caloric expenditure of ~1000 kcal · week^{-1} as suggested by ACSM may be adequate for an overweight individual beginning an exercise intervention. This level of expenditure, however, has been found to be insufficient for long-term weight maintenance and should be increased progressively. It has been reported that those who reported >2000 kcal · week^{-1} of physical activity showed no weight

regain from 6 to 18 months of treatment, whereas a significant weight regain was observed in individuals who had energy expenditures below this threshold (ACSM 2001). Table 6.2 provides a summary of exercise guidelines as well as a sample prescription plan for maximizing energy expenditure and long-term weight control.

> **▶ K E Y P O I N T ◀**
>
> Proper exercise progression is important to the maintenance of weight loss. It appears that as the body adapts to an exercise program, the energy expenditure should increase to maintain a favorable energy balance and thus prevent weight regain.

EXERCISE INTENSITY AND FAT UTILIZATION

Is mild exercise superior to higher-intensity exercise in facilitating fat utilization? Exercise at lower intensities has always been found to be associated with a lower respiratory exchange ratio, which suggests a greater percentage of energy derived from fat oxidation. This

Table 6.2 Exercise Guidelines and Sample Prescription Plan for Maximizing Energy Expenditure and Long-Term Weight Control

Component	Guidelines	Sample prescription
Mode	Exercises include low-impact and non-weight-bearing activities involving large muscle groups. Other nonconventional modes such as yoga, weightlifting, and household activities may also be considered.	Walking, cycling, swimming, low-impact exercise such as water aerobics.
Frequency	Exercise should be performed daily or at least five times per week. If necessary, a single session can be split into two or more mini-sessions.	Daily or 5-7 times a week.
Duration	Exercise duration should be maximized within the limit of tolerance and can be determined by time or caloric expenditure.	40-60 min a day, 20-30 min twice per day, or 150-400 kcal per day.
Intensity	Intensity should generally stay at the lower end of the target range and be determined in accordance with amount of time allocated or number of calories one would like to expend.	40-70% $\dot{V}O_2$max, 40-70% HR reserve, or 60-80% age-predicted HRmax.
Progression	For successful weight loss and maintenance, the volume of exercise should progressively increase as the intervention continues.	Increments in exercise volume should be gradual as one's tolerance increases. Both intensity and duration may eventually reach the upper end of their respective range.

observation has resulted in the current myth that to burn fat, you must exercise at a lower percentage of your maximal oxygen uptake ($\dot{V}O_2$max). It may also explain the growth of "fat burner" classes, which operate on the assumption that low-intensity exercise will lead to more weight loss. It is important to be clear that at higher intensities, though the percentage of total calories derived from fat is lower, the total energy expenditure resulting from the exercise can be higher, especially if the duration is sufficient. Consequently, the absolute quantity of calories burned due to fat oxidation can also be higher. As shown in table 6.3, those who exercised at 50% $\dot{V}O_2$max consumed oxygen at 1 L · min^{-1}, and their respiratory exchange ratio was 0.86. At this ratio, according to table 4.1, a total of 4.875 calories per minute (1 L · min^{-1} × 4.875 kcal · L^{-1}) are expended, and 2.24 or 46% of these are fat calories. On the other hand, those who exercised at 70% $\dot{V}O_2$max had a level of oxygen consumption and a respiratory exchange ratio of 1.5 L · min^{-1} and 0.88, respectively. Thus, a total of 7.349 calories per minute (1.5 L · min^{-1} × 4.899 kcal · L^{-1}) were expended, and 2.87 or 39% of these were fat calories. Although the percentage of fat calories is lower, the total number of fat calories is higher. From this example, it is clear that even though a greater percentage of fat is elicited at a lower level of exercise intensity, this does not necessarily mean that a greater quantity of fat is burned.

It appears that so long as people are physically capable, it is always possible to try to pursue an exercise routine of intensity sufficient to bring about a decent level of total energy expenditure. However, exercise intensity should always be carefully chosen to prevent premature fatigue and to allow the exercise program to continue. Exercise duration could decrease drastically if intensity is set too high or exceeds lactate threshold, the intensity at which blood lactic acid begins to accumulate dramatically.

Based on the current literature and as shown in figure 2.9 (chapter 2), exercise at 60% to 65% $\dot{V}O_2$max helps in eliciting the maximal rate of fat oxidation (Achten et al. 2002). This moderate-level intensity can allow most people to achieve the recommended exercise duration (i.e., ~45 min). Thus this intensity appears to be effective in not only maximizing the total energy expenditure but also facilitating fat utilization.

Performing more vigorous exercise has also been related to greater reduction in fat from the abdominal areas. This is evidenced by a number of epidemiological studies that involved middle-aged and elderly individuals (Buemann and Tremblay 1996; Tremblay et al. 1990, 1994; Visser et al. 1997). For example, Visser and colleagues (1997) found that intense exercise such as playing sports was negatively associated with abdominal fat. Tremblay and colleagues (1990) also observed a preferential reduction in abdominal fat in those who performed more intense exercise. This association favoring the use of more vigorous exercise may be explained by the fact that in contrast to gluteal fat, fat stored in the abdominal area is more responsive to lipolysis, which is intensity dependent. Lipolysis is subject to the influence of epinephrine, which increases with increased exercise intensity (Wahrenberg et al. 1991).

> **▶ K E Y   P O I N T ◀**
>
> Contrary to the common myth, it appears that a relatively more vigorous exercise program is necessary to facilitate fat utilization. The exercise intensity associated with the maximal rate of fat oxidation has been found to be about 60% to 65% $\dot{V}O_2$max. This finding may be explained in part by the fact that the process of lipolysis is intensity dependent.

Table 6.3   Comparisons of Fat and Total Calories Expended During Stationary Cycling at 50% and 70% $\dot{V}O_2$max

	Exercise intensity	
	50% $\dot{V}O_2$max	70% $\dot{V}O_2$max
Oxygen uptake (L · min^{-1})	1.0	1.5
Respiratory exchange ratio	0.86	0.88
Caloric equivalent (kcal · L^{-1})	4.875	4.899
Energy output (kcal · min^{-1})	4.875	7.349
Relative fat contribution (%)	46	39
Fat calories (kcal · min^{-1})	2.24	2.87

Data are unpublished observations obtained from Human Performance Laboratory, The College of New Jersey.

OTHER EXERCISE STRATEGIES

Exercise intensity and duration are the two prescription indices often used to develop an effective yet safe exercise program. An easy way of applying such a prescription is to have steady-state exercise performed at the target intensity and for the duration desired. In reality, however, many exercise sessions are conducted in a more complex fashion. For example, as mentioned earlier, a single exercise session can be divided into two smaller sessions performed at different times of the day so that the target caloric expenditure can be achieved through accumulation. In many cases, despite the provision of target intensity, exercise is performed with a fluctuating intensity—for example Spinning, Treading, and many other forms of group activities involving changing workloads during different stages of the work bout. Additionally, recent evidence suggests that resistance training should be included in a comprehensive weight loss program to maximize its effectiveness. The following sections illustrate the scientific evidence available to support these specific and "nonconventional" exercise strategies.

Intermittent Exercise

A few studies have examined the efficacy of adopting **intermittent** exercise for weight management (Donnelly et al. 2000; Jakicic et al. 1995, 1999). This research direction was driven by the question whether people can achieve the same metabolic and weight loss benefits by exercising in multiple sessions of shorter duration throughout the day rather than in one session. Intermittent exercise is typically defined as the accumulation of 30 to 40 min of exercise each day through participation in multiple 10 to 15 min exercise sessions. This exercise strategy is considered advantageous for those who dislike or are unable to tolerate continuous exercise or have a schedule that prohibits carrying out a typical workout session. Intermittent exercise has long been known to be effective in improving exercise compliance and enhancing cardiorespiratory fitness, as well as improving risk factors for cardiovascular diseases. Its direct impact on energy metabolism and weight loss has also been recognized. Jakicic and colleagues (1995) reported that exercising in multiple short bouts per day was just as effective as a single long exercise session in producing weight loss while increasing cardiorespiratory fitness over a 20-week intervention period that included dietary

modification. In this same study, the program of multiple short bouts of exercise was also found to increase exercise adherence, which implies that this strategy has the potential to facilitate the long-term adoption of a weight loss program and thus to prevent weight regain (Andersen 1999).

Whether performing multiple short sessions of exercise instead of a long exercise bout will elicit more energy expenditure is another intriguing question that has emerged recently but has remained largely unanswered. The total energy expenditure of a single exercise session includes the energy expended during the actual exercise period as well as that expended during the recovery period, or excess postexercise oxygen consumption (EPOC). As discussed in chapter 5, EPOC can be viewed as a compensatory response resulting from a disruption to homeostasis caused by the preceding exercise. In this context, one may speculate that exercise of shorter duration performed more than once daily may be associated with a greater EPOC due to the multiple occurrences of recovery. Almuzaini and colleagues (1998) found that splitting a 30 min session into two 15 min sessions elicited a greater overall postexercise $\dot{V}O_2$. Although the long-term impact of this arrangement on energy metabolism and weight loss remains to be elucidated, exercise for 20 to 30 min twice daily has been adopted by the ACSM as an alternative approach to exercise programming to combat overweight and obesity (Wallace 1997).

Variable-Intensity Protocol

In line with the concept of EPOC, we recently examined whether aerobic exercise performed at variable intensities, such as Spinning, would produce greater energy expenditure than constant-activity exercise (Kang et al. 2005a). This type of exercise has gained popularity in recent years because it replicates the experience of outdoor cycling, during which intensity often varies and which appears to be effective in engaging exercise participants, especially when accompanied by music or visualization or both (Francis et al. 1999). In the study, we observed a greater EPOC following variable-intensity or Spinning exercise as compared to constant-intensity exercise even though the average intensity was kept the same. We attributed the greater EPOC to the fact that intensity fluctuated during the variable-intensity exercise, which may have disturbed homeostasis to a greater extent. This exercise arrangement was also associated with a greater accumulation of blood lactic acid. However, the level of exertion during the entire workout was not perceived as greater than that with the constant-intensity exercise, making the variable-intensity protocol attractive to those who seek to maximize energy expenditure while participating in an exercise program. Note that a variable-intensity exercise regimen differs from an intermittent exercise protocol in that there is no rest period and the intensity fluctuates in a repeating pattern.

Inclusion of Resistance Exercise

Aerobic or endurance exercise is often prescribed to increase daily energy expenditure and fat oxidation and to maintain proper levels of body weight and composition. However, recent evidence suggests that a comprehensive exercise program should also include resistance training, which has its own unique advantages. Resistance training is a potent stimulus to increase **fat-free mass** (FFM), which may help in preserving lean body mass while reducing body weight and body fat. An increase in FFM will also help maintain or augment resting energy rate, which accounts for up to 60% of the total daily energy expenditure (Poehlman and Melby 1998). The resting metabolic rate is primarily related to the amount of FFM. It is known that the resting metabolic rate decreases with advancing age at a rate of 2% to 3% per decade, and this decrease is primarily attributed to the loss of FFM. So incorporating resistance training is especially important as people age and can help counteract the age-related decrease in neuromuscular function and resting energy rate.

Resistance training also improves muscular strength and power. This adaptation is necessary in terms of athletics. It also increases ordinary individuals' capability to perform daily tasks such as carrying, lifting, and changing body posture. The improvement in muscular strength may not affect energy balance directly. However, it will facilitate adoption of a more active lifestyle in sedentary overweight and obese individuals (ACSM 2001). Other metabolic benefits of resistance training may include improvements in **blood lipid profile** (Kokkinos and Hurley 1991; Yki-Jarvinen et al. 1984) and increases in fat oxidation (Van Etten et al. 1995; Treuth et al. 1995b), although these conclusions should be viewed with caution. It has been suggested that some of the studies reporting an improved blood lipid profile did not control for day-to-day variations of **lipoproteins,** concurrent body composition changes, and the impact of dietary intake (Kokkinos and Hurley 1991).

As with other forms of exercise, resistance training increases energy expenditure during both exercise and recovery. However, resistance exercise differs from aerobic exercise in that its energetic contribution to the daily total energy expenditure is very small. As mentioned in chapter 5, energy expenditure during a typical weightlifting workout ranges from 150 to 225 kcal, a figure that is less than half that normally reached during a single session of aerobic exercise. On the other hand, resistance training can trigger a profound EPOC because the exercise is performed intermittently at a very high intensity. In fact, a number of studies showed that the average oxygen consumption following resistance exercise was even greater than that following aerobic exercise when the two types of exercise were equated for total energy expenditure (Gillette et al. 1994; Burleson et al. 1998). The EPOC following resistance exercise has also been shown to last for an extended period of time. For example, by employing a resistance training program that lasted 90 min and included six sets of 10 different lifting exercises, Melby and colleagues (1993) found an average increase in oxygen consumption of 12% during 2 h of recovery and of 9% the morning after exercise. It must be noted that the resistance exercise protocols employed in these studies were quite vigorous and thus may not be applicable in the scenario of obesity treatment. However, the results do suggest that resistance exercise can produce a significant perturbation of homeostasis, which leads to prolonged elevation of the metabolic rate following exercise.

Similar to prescriptions for cardiorespiratory fitness, the resistance training prescription should be made based on the health and fitness status and the specific goals of the individual. For weight management purposes, the major goal of a resistance training program is to develop sufficient muscular strength that an individual is able to sustain a regular training routine and also to live a physically independent lifestyle. To maximize energy expenditure, a **circuit training** program may be introduced to allow people to work on 8 to 10 major muscle groups (e.g., gluteals, quadriceps, hamstrings, pectorals, biceps, triceps, latissimus dorsi, deltoids, and abdominal muscles) in one session that can last for as long as 60 min. The metabolic advantage is that this training format involves multiple sets of low intensity (i.e., 40% 1-repetition maximum [1RM]) and high repetitions (i.e., 10-15 repetitions) for each muscle group, coupled with a relatively shorter rest interval between sets (i.e., ~15 s) (figure 6.2). As

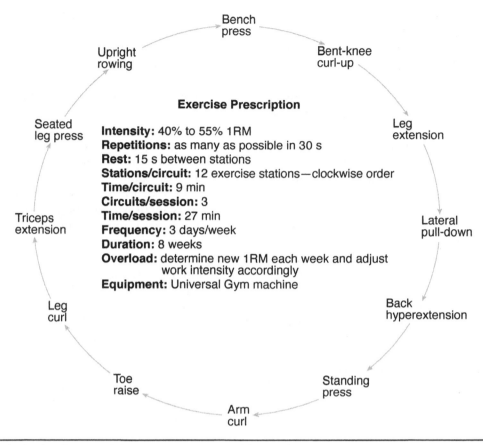

Exercise Prescription

Intensity: 40% to 55% 1RM
Repetitions: as many as possible in 30 s
Rest: 15 s between stations
Stations/circuit: 12 exercise stations—clockwise order
Time/circuit: 9 min
Circuits/session: 3
Time/session: 27 min
Frequency: 3 days/week
Duration: 8 weeks
Overload: determine new 1RM each week and adjust
work intensity accordingly
Equipment: Universal Gym machine

Figure 6.2 Sample circuit resistance training program. 1RM: 1 repetition maximum.

Reprinted, by permission, from V.H. Heyward, 2002, *Advanced fitness assessment and exercise prescription*, 4th ed. (Champaign, IL: Human Kinetics), 141.

mentioned in chapter 5, in addition to the greater energy expended during the workout, this type of training elicits a greater EPOC as compared to regular strength training programs. It is recommended that resistance training be performed at least twice a week, with at least 48 h of rest between sessions to allow proper recuperation (ACSM 2006).

▶ K E Y P O I N T ◀

Circuit weight training allows people to work on 8 to 10 major muscle groups in one session that can last for as long as 60 min. It involves multiple sets of low intensity and high repetitions, coupled with a relatively shorter rest interval between sets. Circuit weight training can effectively stimulate both the muscular and cardiorespiratory systems and thus may be the choice for those interested in losing fat tissue while gaining muscle mass.

LIMITATIONS OF EXERCISE ALONE IN WEIGHT MANAGEMENT

Despite the ability of physical activity and exercise to create a **negative caloric balance,** the actual impact of exercise alone on weight loss has often been found to be minimal (Garrow 1995; Wilmore 1995; Saris 1993). Although there is a negative association between the level of physical activity and the prevalence of obesity, and although those who are physically active tend to be leaner, it remains unclear how much physical activity or exercise can do, even combined with dietary restriction, in treating those who are already overweight or obese. In a meta-analysis of 493 studies over a 25-year period, Miller and colleagues (1997) reported that exercise alone had a relatively minor effect on weight loss and did not add much to the weight loss effect of a reducing diet. As shown in figure 6.3, during a typical 15-week program, the combination of diet and exercise caused a 24.2 lb (11 kg) weight loss, which is virtually no different from the 23.5 lb (10.7 kg) loss induced by diet alone. The same study also showed that exercise alone had a relatively minor effect on weight loss. Some studies have demonstrated a positive effect of exercise on weight loss (Ross et al. 2000). However, the exercise intervention adopted was rather vigorous and resulted in an energy deficit of 700 kcal per day—twice the caloric value normally recommended for an exercise session. It appears that in order for physical activity and exercise to have a major impact on body weight reduction, the exercise prescription should entail daily exercise, at moderate to high intensity, for more than an hour. Obviously, most obese individuals cannot tolerate or sustain this exercise dosage.

The weak effect of exercise alone on weight loss may be explained by the fact that when people begin exercise training, they tend to rest more after each exercise session, which negates the calories expended during exercise. Total energy expenditure for any given day can be partitioned into energy

expended via (1) the **resting metabolic rate,** (2) the thermic effect of food, and (3) physical activities. Normally, no more than 30% of the total daily energy expenditure can be attributed to physical activity, whereas the resting metabolic rate—the energy required by the body in a resting state—can often account for up to 60% of the daily energy expenditure. One should realize that the amount of energy expended during exercise that is suitable to sedentary or obese individuals is actually rather small. For example, the net energy cost during 3-mile (4.8 km) brisk walking for a 70 kg (154 lb) obese woman is only about 150 to 160 kcal (215 total calories minus 50 calories for the basal metabolic rate) (Nieman 1999). Given that 1 lb (0.5 kg) of fat contains 3500 calories, it would take nearly a month of daily walking 3 miles to lose 1 lb of fat if all else stayed the same. That exercise in and of itself is ineffective in weight loss should not be attributed to a tendency to eat more when one becomes more active. In fact, in those studies that failed to demonstrate an effect of added exercise, the exercise intervention was implemented under controlled dietary conditions in which all subjects, whether treated with exercise or not, were fed similarly.

It appears that in terms of exercise alone, the key question is, What is the minimal exercise dose necessary to achieve desirable weight loss? While this issue remains to be investigated, future researchers should also be able to adequately control energy intake as well as all other components of daily energy expenditure when introducing a planned exercise intervention. From a public health standpoint, it seems prudent to conclude that any exercise is better than none and that, within the range of tolerance, more is probably better (Volek et al. 2005). Still, irrespective of body weight control, regular physical activity has proven beneficial in many ways, for example in improving cardio-respiratory fitness, augmenting overall feelings of well-being, and reducing risk factors for developing chronic conditions such as coronary heart disease, hypertension, diabetes, and osteoporosis.

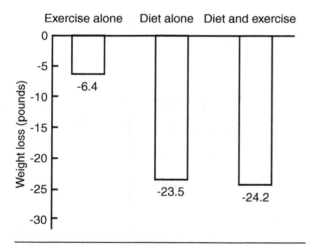

Figure 6.3 Average weight loss during a 15-week intervention with diet, exercise, or both. Results were derived from a meta-analysis of 493 studies over a 25-year period.

Reprinted from D.C. Nieman, 2003, *Exercise testing and prescription*, 4th ed. (Mountain View, CA: Mayfield Publishing Company), 491, by permission of The McGraw-Hill Companies.

SUMMARY

Both leisure and planned physical activities are effective means of elevating energy expenditure via repeated muscular contraction. This increased energy expenditure helps create the negative caloric balance necessary for weight loss. An exercise intervention aimed at weight loss should be designed so as to provide the maximal caloric expenditure possible. Caloric expenditure of an exercise session is the product of intensity, duration, and frequency. For burning the most calories, exercise duration is considered the most important; and exercise intensity and frequency must be adapted for the amount of time one would like to devote to exercise and the number of calories one would like to expend on a weekly basis. Though mild-intensity exercise appears to be adequate for most overweight people, especially at the beginning of the program, a gradual increase in exercise intensity should take place as people adapt to the program. Exercise performed at a relatively higher intensity can augment fat utilization, and this together with increased overall energy expenditure may be more effective in the long run than an exercise of mild intensity.

To increase variety, an exercise intervention may incorporate nonconventional approaches such as intermittent exercise, exercise with variable intensities, and resistance exercise, which have all proven beneficial in assisting or maximizing energy expenditure. Increasing physical activity level alone has often been found to be ineffective for quick weight loss. Possible reasons are that the exercise dosage did not reach the threshold needed to bring about changes in body fat or that there might have been a compensatory decrease in other components of energy expenditure such as resting metabolic rate. Weight management should be viewed as a concerted effort and a commitment to engage in changes in all aspects of one's lifestyle, including physical activity.

KEY TERMS

blood lipid profile
capillary density
circuit training
duration
fat-free mass

frequency
heart rate reserve
intensity
intermittent

lipoproteins
negative caloric balance
positive energy balance
resting metabolic rate

REVIEW QUESTIONS

1. Explain the differences between physical activity and exercise.

2. What are the major components of daily energy expenditure? What does *negative energy balance* mean? Why is it important to achieve this caloric balance, and how can a person achieve it?

3. Define *intensity, duration, frequency,* and *progression*. How would you use these terms in establishing an exercise program for people who want to lose excess body fat?

4. How is circuit weight training carried out? What are the advantages of using this type of resistance training program?

III

BIOENERGETICS IN SPECIAL CASES

Energy transformation responds rapidly to a metabolic stress in an effort to maintain energy homeostasis. However, it is important to know that such acute metabolic responses can be altered following exercise training. In general, repeated exposure to a heightened metabolic demand makes energy transfer more efficient. Our ability to store and use energy fuels and to utilize metabolic pathways can also be influenced by changes in structure and function associated with age and gender, as well as many metabolic diseases.

Part III stresses the uniqueness and specificity of energy transformation. This understanding enables us to recognize distinctions in energy metabolism in individuals differing in training status, age, gender, and medical conditions. Chapter 7 discusses training-induced cellular adaptations and their impact on fuel storage and utilization. Chapter 8 illustrates differences in bioenergetics in particular subpopulations—children, women, and persons who are elderly. Chapter 9 presents evidence concerning altered energy metabolism due to diabetes and obesity.

CHAPTER

7

METABOLIC ADAPTATIONS TO EXERCISE TRAINING

This chapter reviews chronic metabolic changes in response to aerobic and anaerobic training. In particular, it covers training-induced alterations in the storage of energy substrate, enzymatic content and activities, patterns of substrate utilization, muscle respiratory capacity, and hormonal responses. The chapter also includes a discussion of how training affects exercise economy and metabolic efficiency.

CELLULAR ADAPTATIONS TO AEROBIC TRAINING

Exercise training can be viewed as repeated bouts of physical activities aimed at improving one's functional capacity. The chronic stimulus provided by training is able to produce a series of metabolic and morphological changes that will lead to improvement in fitness and sport performance. This increase in exercise capacity is manifested in the ability to perform more exercise during a given period of time or to perform exercise of a given intensity for a longer period of time. Underlying this training-induced increase in exercise capacity is a unique combination of physiological and morphological adaptations in the skeletal muscles, cardiorespiratory system, and **autonomic** nervous and endocrine systems. However, the nature and magnitude of these changes vary depending on the type of training performed. For example, aerobic training improves central and peripheral blood flow and enhances the capacity

of muscle fibers to generate a greater amount of adenosine triphosphate (ATP) via the aerobic energy-yielding pathway. Anaerobic training, on the other hand, increases the muscles' ability to generate ATP independent of oxygen and thus augments force production and promotes a greater tolerance for acid–base imbalances during highly intense effort. The fact that the kinds of physiological adaptations are specific to the type of training used is referred to as **specificity,** which has been an important guiding principle for conditioning and testing.

> ### ▷ K E Y P O I N T ◁
>
> Aerobic training improves capacity for energy transformation that requires oxygen, whereas anaerobic training enhances muscle's ability to generate energy independent of oxygen. Such differential responses resulting from specific training protocols reflect the principle of training specificity—that is, the idea that selecting a proper training program is important for producing the expected physiological effect.

Mitochondrial and Capillary Density

As mentioned in chapter 1, the function of mitochondria plays a key role in bioenergetics. During exercise, energy turnover in skeletal muscle may increase by 400 times compared with that in muscle at rest (Tonkonogi and Sahlin 2002). It is in the mitochondria that the aerobic energy system operates. The ability to use oxygen and produce ATP via the oxidative pathway depends on the number, size, and efficiency of a muscle's mitochondria. Endurance training that comprises repeated bouts of submaximal exercise has been shown to increase the number and size of mitochondria within the muscles being conditioned (Holloszy and Coyle 1984). For example, Holloszy and colleagues (1971) reported that in endurance-trained rats, the actual number of mitochondria increased by 15% during 27 weeks of exercise. The average size of mitochondria was also found to increase by 35%. The increase in mitochondria is also associated with increases in the activity of enzymes involved in oxidative pathways such as the Krebs cycle, β-oxidation, and the electron transport chain. Changes also occur in the process that moves NADH (the reduced form of nicotinamide adenine dinucleotide) produced in cytoplasm via glycolysis to the mitochondria, where it is used to produce ATP via the electron transport chain. Recall from chapter 1 that using energy fuels to produce ATP depends on the action of enzymes. The activity of these enzymes improves in response to the stimulus of training. Table 7.1 presents data on selected muscle enzyme activities of each energy system for untrained, anaerobically trained, and aerobically trained men.

Of equal importance to energy provision are oxygen delivery and fuel supply via capillaries within the working muscle. During aerobic energy transformation, oxygen serves to accept electrons that are passed through the electron transport chain and combines with hydrogen to form water. Several early studies showed greater capillary density (capillaries per muscle fiber) in trained as compared to untrained individuals (Hermansen and Wachtlova 1971; Ingjer 1979). Recent longitudinal studies also demonstrated a more than 15% increase in the number of capillaries surrounding each muscle fiber following prolonged periods of aerobic training (Rico-Sanz et al. 2003). Having more highly proliferated capillaries facilitates diffusion of oxygen from the capillary to the mitochondria. This, together with increased number and size of mitochondria, helps maintain an internal environment conducive to energy production via oxidative metabolism. In addition to augmented oxygen diffusion, the increase in capillary density allows greater exchange of metabolic waste products such as carbon dioxide, as well as heat and nutrients, between the blood and the working muscle fibers.

Myoglobin

When oxygen enters the muscle cell, it binds to **myoglobin,** a compound that shuttles the oxygen molecules from the cell membrane to the mitochondria. Myoglobin is a protein that turns red when bound to oxygen. Slow-twitch muscle fibers contain a large quantity of oxygen-carrying molecules. This gives these fibers their red appearance. Fast-twitch muscle

Table 7.1   Selected Muscle Enzyme Activities (mmol · g^{-1} · min^{-1}) for Untrained, Anaerobically Trained, and Aerobically Trained Men

	Untrained	Anaerobically trained	Aerobically trained
ATP-PCr system			
Creatine kinase	609.0	702.0*	589.0
Myokinase	309.0	350.0*	297.0
Glycolytic system			
Phosphorylase	5.3	5.8	3.7*
Phosphofructokinase	19.9	29.2*	18.9
Lactate dehydrogenase	766.0	811.0	621.0
Aerobic system			
Succinate dehydrogenase	8.1	8.0	20.8*
Malate dehydrogenase	45.5	46.0	65.5*
Carnitine palmitoyl transferase	1.5	1.5	2.3*

*Significantly different from untrained values.

Adapted, by permission, from J.H. Wilmore and D.L. Costill, 2004, *Physiology of sport and exercise*, 3rd ed. (Champaign, IL: Human Kinetics), 201.

fibers, on the other hand, are more specialized for glycolysis and therefore require and have less myoglobin. This gives these fibers their lighter or whiter appearance. Myoglobin may be viewed as an oxygen reservoir. It is saturated with oxygen, but releases the oxygen to the mitochondria when oxygen levels become deficient during muscle action. For example, during the transition from rest to exercise, when there is usually a lag in oxygen delivery, the myoglobin-oxygen compound stored in the exercising muscle can release its oxygen to fill the void so that the body will experience a smaller oxygen deficit. Whether or not the myoglobin level increases in response to aerobic training is less certain, because an increase in myoglobin content associated with the trained state was reported only in early studies that used animals (Lawrie 1953; Pattengale and Holloszy 1967). It remains largely unknown whether or how regular training affects myoglobin content in humans.

Muscle Fiber Types

Competitive endurance athletes tend to have a higher percentage of type I (or slow-twitch) muscle fibers, while sprinters have a higher percentage of type II (or fast-twitch) fibers, although the majority of people have equal percentages of these two types of muscle fibers (Gollnick et al. 1972). **Type I muscle fibers** differ from **type II muscle fibers** in that type I fibers contain a greater volume of mitochondria, a higher concentration of myoglobin, and a larger number of oxidative enzymes. They are also surrounded by a more highly proliferated capillary system. Consequently, type I muscle fibers have a large capacity for aerobic metabolism and a high resistance to fatigue. Thus they are primarily recruited during endurance events such as long-distance running, as well as during most daily activities in which the muscle force requirements are low. Type II muscle fibers, on the other hand, have comparatively lower aerobic capacity. However, they are better suited to generating energy anaerobically. Type II muscle fibers can be further divided into type IIa (fast oxidative) and IIb (or fast glycolytic) muscle fibers. While type IIa fibers are used mainly during shorter, high-intensity endurance events such as the mile run, type IIb fibers are predominantly used in highly explosive events such as the 60 m dash. Table 7.2 provides a summary of structural and functional characteristics of various muscle fiber types.

Training appears to have a minor impact on the composition of muscle fibers (Baldwin et al. 1972;

Winder et al. 1974). In other words, the difference in fiber types seen in athletes who perform in different events should be considered primarily the result of selection; it is believed that this can be changed via training, but only to a small extent. Training may change the size of the corresponding muscle fibers. For example, in response to aerobic training, the cross-sectional area of type I muscle fibers was shown to have increased by as much as 25%, whereas such a change did not occur in type II fibers (Wilmore and Costill 2004). Training can also help in maximizing already existing aerobic or anaerobic potential. As mentioned earlier, many adaptive changes result from aerobic training, including increases in the size and number of mitochondria and the activities of oxidative enzymes in type I muscle fibers. Studies have also shown that endurance training can lead to a conversion of type IIb to type IIa fibers, as well as an increase in the mitochondrial content in type II muscle fibers (Chi et al. 1983; Jansson and Kaijser 1977). Of particular note, despite a long-standing belief that the makeup of muscle fiber types is a natural endowment, a recent study by Rico-Sanz and colleagues (2003) demonstrated an increase in type I muscle fibers, from 43.2% pretraining to 46.7% posttraining, following 20 weeks of aerobic training. This finding suggests that it is still uncertain whether, or to what degree, training influences the composition of muscle fiber types.

> **KEY POINT**
>
> Among the major cellular changes resulting from aerobic conditioning are increases in capillary density and mitochondrial content, as well as in a series of enzymes that are important during the aerobic production of energy. Improved oxidative capacity of fast-twitch muscle fibers has also been reported, although it is debated whether fast-twitch muscle fibers can undergo a complete conversion into slow-twitch muscle fibers.

CHANGES IN FUEL UTILIZATION

Carbohydrate, such as muscle and liver glycogen, is the primary energy fuel used during exercise. Depletion of muscle glycogen has been regarded as the major cause of fatigue especially during strenuous exercise

Table 7.2   Metabolic and Performance Characteristics of Various Muscle Fiber Types

	Slow fibers	Fast fibers	
Characteristics	**Type I**	**Type IIa**	**Type IIb**
Number of mitochondria	High	High/Moderate	Low
ATPase activity	Low	Moderate	Highest
Oxidative capacity	High	Moderate	Low
Glycolytic capacity	Low	Moderate	High
Contractile speed	Low	Moderate	High
Force production	Low	High	High
Fatigue resistance	High	High/Moderate	Low

Adapted from S.K. Powers and E.T. Howley, 2004, *Exercise physiology: Theory and applications to fitness and performance*, 5th ed. (New York, NY: McGraw-Hill), 149, by permission of The McGraw-Hill Companies.

of prolonged duration. One important consequence of the cellular adaptations to aerobic training is modification of the rates at which various fuels are used during exercise. For example, for exercise performed at the same intensity, trained individuals are able to generate energy by oxidizing less carbohydrate and more fat (figure 7.1). This modification of fuel utilization is considered desirable in that it allows the body to postpone depletion of limited carbohydrate reserves and therefore delay the onset of fatigue. This adaptation was first demonstrated in 1939 by Christensen and Hansen, who observed a lower respiratory exchange ratio (RER) in endurance-trained individuals. This finding was later confirmed by numerous studies using the respiratory quotient (RQ) measured directly across exercising limbs or tracer-dilution methodology that quantified CO_2 produced directly from substrate oxidation. As discussed in chapter 4, a lower RER or RQ value represents a greater contribution of fat to energy provision. The following sections provide more detail on the effect of aerobic training on carbohydrate, fat, and protein utilization.

Carbohydrate Utilization

With use of the **muscle biopsy** procedure, it is now widely recognized that endurance training spares muscle glycogen utilization. Classic studies in which subjects were trained with one leg (Henriksson 1977; Saltin et al. 1976) showed that when subjects were tested during two-legged exercise, glycogenolysis was slower in the trained leg as compared to the untrained leg. Liver glycogen is another depot of endogenous carbohydrate fuels. Although utilization of this source of energy is less well elucidated, it also has been found to decrease as a result of endurance training. More recent studies employing tracer-dilution technique have demonstrated a reduced rate of liver glycogenolysis and thus glucose output from the liver (Coggan et al. 1990, 1993). The tracer-dilution technique is designed to track how much glucose is utilized by the muscle and how much is released by the liver. The reduced reliance on liver glycogen following training was attributed to a slow rate of glucose oxidation during exercise (Brooks and Donovan 1983). It was also found that

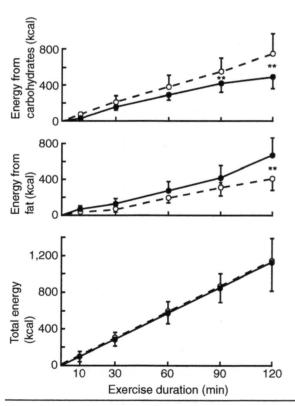

Figure 7.1 Total energy utilized, energy from fat oxidation, and energy from carbohydrate oxidation during prolonged exercise testing before (open circles) and after (filled circles) aerobic training.

Reprinted, by permission, from B.F. Hurley et al., 1986, "Muscle triglyceride utilization during exercise: Effect of training," *Journal of Applied Physiology* 60(2): 566.

such a reduction in liver glycogenolysis accounted for two-thirds of the total training-induced reduction in carbohydrate utilization (Fitts et al. 1975). As liver glycogen is the sole provider of circulating glucose in the blood, the sparing of liver glycogen would enable trained individuals to better maintain plasma glucose homeostasis during prolonged exercise.

Indeed, endurance training has also been linked to an enhanced ability to maintain plasma glucose homeostasis. Plasma glucose is commonly observed to fall during prolonged exercise. While the decrement in blood glucose level during exercise appears to be relatively mild for most individuals, severe hypoglycemia, that is, a level less than 2.5 mM, has been reported to occur during prolonged exercise (Felig et al. 1982). Plasma glucose serves as the sole fuel for the central nervous system, and the onset of hypoglycemia has often been associated with the induction of fatigue with sustained exercise. As already noted, training reduces the utilization of muscle glycogen. This reduction can decrease the reliance on plasma glucose by working muscle. A number of studies have demonstrated that maintenance of plasma glucose is accomplished by a diminished utilization of plasma glucose as an energy source by the working muscle (Coggan et al. 1990, 1995; Phillips et al. 1995; Jansson and Kaijser 1987; Turcotte et al. 1992). The reduced plasma glucose uptake and utilization not only help to maintain plasma glucose homeostasis, but also promote the sparing of liver glycogen. It has been suggested that the liver glycogen saved can be important during the later stages of prolonged exercise (Donovan and Sumida 1997). Several animal studies have also demonstrated a training-induced increase in glucose production via liver gluconeogenesis (Donovan and Pagliassotti 1990a, 1990b). This improvement is thought to be particularly responsible for the ability of trained individuals to maintain plasma glucose during the later portion of exercise lasting more than 2 h (Donovan and Sumida 1997).

Fat Utilization

Fat is regarded as a desirable fuel during exercise because for every gram of fat used, 2 g of carbohydrate may be spared. Fat serves as the primary energy fuel during light-intensity exercise of prolonged duration. However, in most competitive sports in which intensity is vigorous, fat remains a secondary fuel of energy. With the cellular adaptations discussed earlier in this chapter, endurance-trained individuals are capable of deriving proportionally more of their total energy from fat sources during exercise. Turcotte and colleagues (1992) reported that uptake of plasma fatty acids in the thigh was significantly higher in trained than in untrained men during prolonged exercise. Hurley and colleagues (1986) also revealed a greater utilization of **intramuscular triglycerides** following a 12-week endurance exercise program. In this latter study, the authors estimated that the energy expenditure derived from fat during exercise rose from 35% before training to 57% after training. Training has also been reported to increase the breakdown of triglycerides circulating in the blood (Kiens and Lithell 1989). Interestingly, although fatty acids derived from adipose tissue and circulating triglycerides are the two important sources of energy, the training-induced increase in fat oxidation has been found to result primarily from increased utilization of triglycerides stored in the muscle, rather than those in adipose tissue and blood.

> ### ▶ K E Y   P O I N T ◀
>
> Cellular changes in response to aerobic conditioning make it possible for an individual to be more effective in using stored fat. This improved fat utilization in turns helps to spare the use of glycogen in muscle and liver and to maintain blood glucose homeostasis. Competitive endurance performance is largely dependent on carbohydrate availability and utilization.

Protein Utilization

Considerably less information is available on how exercise training affects the utilization of protein or amino acids as energy substrates. The reason may be that the oxidation of amino acids contributes less than 5% to the total energy cost during endurance exercise. Much of the current literature on this issue has dealt with the oxidation of leucine, one of the three branched-chain amino acids. Via cross-sectional comparisons, Lamont and colleagues (1999) found an increase in leucine oxidation during an acute bout of exercise, but no difference in this parameter between trained and untrained individuals either at rest or during exercise. However, in a longitudinal study, McKenzie and colleagues (2000) observed that before training, leucine oxidation increased during exercise in both men and women. Nevertheless, after eight weeks of aerobic training, this increase was not seen in either men or women. The authors speculated that the training-induced attenuation in leucine oxidation during exercise was due to an increase in the ability to use more fat so that utilization of carbohydrate was reduced.

RESPONSES OF OXYGEN UPTAKE AND ENDURANCE PERFORMANCE

Aerobic training also improves the body's ability to transport oxygen. For example, it is well established that aerobic training can augment one's maximal oxygen uptake ($\dot{V}O_2$max). As $\dot{V}O_2$max increases, any given workload represents a relatively lower metabolic strain on the body. Increased $\dot{V}O_2$max allows a better and more sustained oxygen supply during submaximal exercise so that disruption to homeostasis can be minimized or endurance can be increased. The improved oxygen transport can also be reflected in more efficient use of oxygen by working muscles. As discussed previously, compared to the energetic process that occurs without oxygen, energy transformation using oxygen is always more productive in that more of the energy stored in food substrates can be used to synthesize ATP. The following sections deal more specifically with the impact of aerobic training on various physiological indices related to oxygen utilization.

$\dot{V}O_2$max

Maximal oxygen uptake, or $\dot{V}O_2$max, is the upper limit of energy transfer through oxidative pathways and is often used as a measure of cardiorespiratory fitness. Endurance training has been shown to improve $\dot{V}O_2$max in previously sedentary individuals if the training entails some vigorous exercise sessions. The improvement in $\dot{V}O_2$max has been linked to training-induced increases in mitochondrial content and activity of the enzymes involved in aerobic pathways (Green et al. 1995; Costill et al. 1976). The adaptive increase in $\dot{V}O_2$max is also caused by an enhanced ability of working muscle to extract oxygen from blood (Ekblom 1969). This notion is supported by evidence that mitochondrial content and capillary density increase following endurance training, as mentioned earlier. The amount of oxygen taken in by the body is also influenced by the function of the circulatory system, which is responsible for oxygen

transport. It has been documented that enhanced $\dot{V}O_2$max is attributable to the enhanced capacity of **myocardial** contraction, which is determined by maximal **stroke volume** and thus **cardiac output** (Powers and Howley 2001). The fact that the amount of oxygen taken in by the body is subject to changes in both oxygen transport and oxygen utilization is described by the Fick equation:

$$\dot{V}O_2 = \text{cardiac output} \times \text{arterial-mixed venous oxygen difference.}$$

Metabolic Responses During Submaximal Exercise

Typically, people who adapt to endurance training are able to exercise longer at submaximal intensities that require less than $\dot{V}O_2$max. However, unlike the situation with $\dot{V}O_2$max, this improved submaximal performance appears to be independent of increased capacity for oxygen delivery and extraction. A number of studies showed that oxygen uptake, as well as blood flow to the working muscle at a given submaximal work rate, remained unchanged from before to after training (Hagberg et al. 1980; Hickson et al. 1978). In fact, the improved endurance has been ascribed to a slower utilization of carbohydrate and an increased reliance on fat oxidation as a source of energy. In addition, due to an improved oxidative energy system concomitant with increased utilization of fat, both muscle and blood lactate levels were found to be lower at the same relative intensities in the trained as compared to the untrained state (Hurley et al. 1984; Karlsson et al. 1972). Lactate accumulation plays a key role in the development of fatigue. Lower lactate production allows people to exercise longer at a given submaximal intensity or to exercise at a higher intensity for a given time period.

Endurance training also reduces the magnitude of the $\dot{V}O_2$ slow component for a given work rate (Carter et al. 1999; Poole et al. 1990; Casaburi et al. 1987; Womack et al. 1995). Recall from chapter 5 that the $\dot{V}O_2$ slow component is defined as an additional increment in oxygen consumption despite the fact that exercise is performed at a constant work rate. Carter and colleagues (1999) reported that six weeks of running training resulted in a significant reduction in the $\dot{V}O_2$ slow component, from 0.321 L · min^{-1} to 0.217 L · min^{-1}, for the same treadmill running speed. This reduced oxygen cost following training can be explained by the reductions in blood lactate levels, ventilation, heart rate, and plasma catecholamine levels that accompany endurance training (Jones and Carter 2000). It is also attributed to an increased recruitment of type I muscle fibers, which are metabolically more efficient (Barstow et al. 1996; Poole et al. 1991). In fact, being able to recruit more type I muscle fibers has been considered the most plausible explanation for the training-induced reduction in the $\dot{V}O_2$ slow component. The relative contribution of the $\dot{V}O_2$ slow component to the total oxygen consumption during exercise is negatively related to aerobic fitness, or the proportion of type I muscle fibers in the working muscle, or both (Barstow et al. 1996); that is, the greater one's fitness, the smaller the ratio of the $\dot{V}O_2$ slow component to total oxygen consumption.

Oxygen Deficit

Endurance training is also associated with a reduced oxygen deficit (figure 7.2). Oxygen deficit occurs when the aerobic system is unable to match the energy demand at the onset of exercise. In this context, a reduction in oxygen deficit means that once exercise begins, the body is able to mobilize aerobic energy system more quickly. This improvement with respect to oxygen deficit can be explained by a training-induced increase in mitochondrial content (Powers and Howley 2004). At the onset of steady-state exercise, ATP provides the most immediate energy. The

resulting increase in adenosine diphosphate (ADP) concentration then activates energy-yielding systems to meet the energy demand of exercise. The ATP-PCr system responds immediately to this demand, followed by glycolysis and the aerobic Krebs cycle. The aerobic pathway provides essentially all the ATP needed during the subsequent steady-state phase of exercise. During aerobic metabolism, mitochondria increase oxygen consumption in response to an increased concentration of cellular ADP.

With the training-induced increase in mitochondrial content, the task of producing ATP aerobically can be shared by more mitochondria within the working muscle. For example, if a muscle cell has only one mitochondrion, an increase of 100 units in the ADP concentration is needed in order for the muscle to consume 2.0 L of oxygen per minute. After training, when the number of mitochondria has doubled, the ADP concentration increases only half as much because each mitochondrion requires only half the stimulation to consume 1 L of oxygen per minute. Consequently, individuals who adapt to endurance training are expected to experience a lesser change in ADP concentration. Since less of a change in ADP concentration is needed to stimulate the mitochondria, it is possible that in trained individuals, the rising ADP concentration causes the oxidative system to be activated more spontaneously. This means that as exercise begins, the rise in oxygen uptake is faster and therefore the oxygen deficit is smaller.

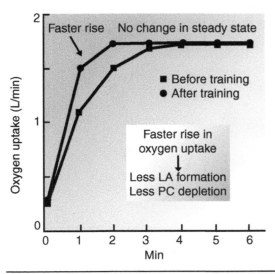

Figure 7.2 O_2 deficit at the onset of exercise before and after aerobic training.

Adapted from S.K. Powers and E.T. Howley, 2004, *Exercise physiology: Theory and applications to fitness and performance*, 5th ed. (New York, NY: McGraw-Hill), 260, by permission of The McGraw-Hill Companies.

> ## ▷ K E Y   P O I N T ◁
>
> Training-induced improvements in aerobic metabolism can be reflected by an augmented oxygen transport system, which can be manifested by an increase in $\dot{V}O_2$max and a reduction in the oxygen deficit and oxygen slow component. An improved oxygen transport system allows the body to derive proportionally more energy from aerobic metabolism. Being able to rely on oxidative energy transformation is desirable in most circumstances.

Excess Postexercise Oxygen Consumption

As some of the oxygen consumed during recovery is used to restore the level of ATP and PCr used during the preceding exercise, one can reasonably expect that due to a smaller increase in ADP at the onset of exercise, the level of excess postexercise oxygen consumption (EPOC) will be lower in trained than in untrained individuals. This contention has been proven by a number of studies demonstrating an inverse relationship between the level of fitness and the length of EPOC (Hagberg et al. 1980; Frey et al. 1993; Short and Sedlock 1997). Hamilton and colleagues (1991) also found that endurance-trained athletes were able to consume more oxygen during the early phase of recovery than those who trained anaerobically. Speedier oxygen consumption during recovery can be viewed as an advantage for those who engage in repeated exercise. In the same study, Hamilton and colleagues observed that endurance-trained athletes were more successful in maintaining their initial power output throughout 10 repeated bouts of sprinting. In contrast, those with an anaerobic training background generated higher power output during the first four sprints, but failed to maintain their initial power outputs during the final six sprints. Some studies have failed to demonstrate a relationship between aerobic fitness and EPOC (Bell et al. 1997). However, these studies appear to have involved a relatively homogenous subject group or an exercise protocol that was too mild relative to the fitness of the subjects. Both these factors can diminish the ability to detect significant correlations.

Lactate Threshold

Another physiological marker of the status of aerobic conditioning is the lactate threshold—the intensity or work rate at which the blood concentration of lactic acid begins to increase sharply. The lactate threshold is also viewed as the maximal steady-state intensity that can be sustained for a long period of time. Exercise above the lactate threshold is associated with more rapid fatigue, either through the effect of metabolic acidosis on muscle contraction (Sahlin 1992) or through an accelerated depletion of muscle glycogen (Boyd et al. 1974). Although lactate is the product of anaerobic metabolism, this physiological marker has been commonly used by endurance athletes to determine the maximal steady-state intensity they can tolerate. As shown in figure 7.3a, the lactate threshold is higher in trained than in untrained individuals. Studies have demonstrated that it can be improved with 6 to 12 months of endurance training. As seen in figure 7.3b, following endurance training the lactate threshold shows a rightward shift with higher work rates or running speeds. This adaptive response allows one to exercise at a higher percentage of one's $\dot{V}O_2max$ before lactate begins to accumulate in the blood. Because race pace in endurance events such as running and cycling is closely related to the lactate threshold (Conley and Krahenbuhl 1980), this means that one would be able to maintain a much

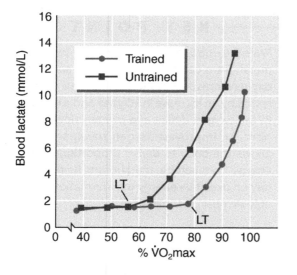

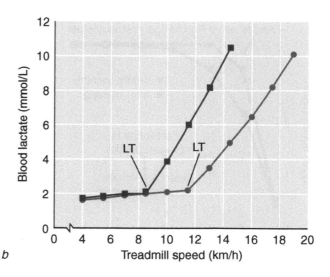

a   *b*

Figure 7.3 Comparisons of lactate threshold (LT) between the trained and the untrained state. Lactate thresholds are expressed as *(a)* percentage of $\dot{V}O_2max$ and *(b)* treadmill running speed.

faster race pace via energy derived from the oxidative pathway. In general, the improved lactate threshold is caused by a reduction in lactate production and an increase in lactate removal within the working muscle. Whereas the reduced production is considered a consequence of enhanced aerobic respiration, the increased removal may be viewed as the outcome of improved blood circulation.

Exercise Efficiency

Exercise efficiency is defined as the amount of energy required for performance at a given exercise intensity and is often expressed as the ratio of work accomplished to oxygen consumed. Being more efficient is considered advantageous because better preservation or budgeting of energy sources prolongs endurance performance. In a classic study, Conley and Krahenbuhl (1980) reported that performance in a 10K race was closely related to running economy in a group of well-trained athletes who had similarly high $\dot{V}O_2$max. Although it is generally believed that trained individuals have better exercise efficiency than untrained individuals, it is still unclear whether this metabolic parameter responds to endurance training similarly to other physiological variables such as mitochondrial content and $\dot{V}O_2$max. Whereas some studies failed to show improvement in running economy following 6 to 12 weeks of training (Wilcox and Bulbulian 1984; Overend et al. 1992; Lake and Cavanagh 1996), others demonstrated that six weeks of training caused a significant improvement in running economy (Franch et al. 1998; Billat et al. 1999; Jones 1998). Improved exercise efficiency can be attributed in part to a training-induced enhancement of aerobic metabolism, since the oxidative pathway is the most efficient energy transfer process. Increased efficiency can also be induced by improved exercise technique and **motor unit** recruitment patterns.

Exercise efficiency adapts to training much less rapidly than other physiological parameters. This is especially the case in individuals who are already trained. Since the best economy values are often found in older or more experienced athletes or in those who do a large volume of running, it is possible that a longer training period is necessary to improve exercise economy. It should be noted that exercise efficiency is also associated with factors that may not be altered with training, such as percentage of type I muscle fibers, muscle elasticity, and body mass and body mass distribution (Bailey and Pate 1991).

ADAPTATIONS TO ANAEROBIC AND RESISTANCE TRAINING

Much less attention has been given to the impact that anaerobic and resistance exercise have on energy metabolism. Part of the reason is that aerobic exercise has been traditionally associated with many health benefits such as improving fitness and reducing weight. This type of exercise is also easier to study experimentally, as subjects generally exercise in a controlled manner (i.e., exercise in the same body position and at a constant intensity for an extended period of time). In contrast, with resistance exercise, it is often difficult to quantify physiological responses accurately. Also, if done improperly, this type of exercise can cause muscular and skeletal injuries. Anaerobic training activities such as repeated 40 yd dashes may be limited to the athletic population. However, resistance training has gained tremendous popularity in the fitness industry in recent years. This form of exercise has unique advantages in that it increases muscle mass and thus resting metabolic rate. With the use of machines, the exercise can be performed quite safely. Resistance exercise also helps to increase the muscular strength and endurance essential to performance of many daily tasks such as carrying, lifting, and changing body posture.

Adaptations to Short-Term Exercise

Anaerobic exercises may be further divided into ultra-short-term and short-term. The category of ultra-short-term includes the shot put, high jump, long jump, and 60 m dash as well as a majority of the plays and movements in team sports such as football and baseball. These events last less than 10 s and are powered primarily by energy produced from the ATP-phosphocreatine (PCr) system. Glycogen may be used to a certain extent, but mainly for the replenishment of PCr, especially during rest intervals between exercise bouts. Success in these events also depends on the proportion and recruitment of type II muscle fibers. Costill and colleagues (1979) examined training adaptations to short, maximal exercise bouts aimed at developing the ATP-PCr system. The design of the study was unique in that subjects trained one leg using repeated 6 s maximal work bouts and the other leg using repeated 30 s work bouts. The two forms of training produced the same improvement in muscle strength and resistance to fatigue. However, the activity of an enzyme involved in ATP production, creatine kinase, increased in the leg trained with 30 s exercise bouts but remained unchanged in the leg trained with 6 s exercise bouts. These findings suggested that training emphasizing the ATP-PCr system minimally affects energy release from this system. Training-induced gains in strength may therefore be mediated by other factors such **muscle hypertrophy** and motor unit recruitment.

The short-term category of anaerobic exercise includes events that last longer than 10 s but less than 3 min, such as the 400 and 800 m run and 100 and 200 m swimming, as well as wrestling and boxing. These events are still predominantly anaerobic, using the high-force, fast-twitch muscle fibers; but when maximal performance is extended beyond 2 min, more than 50% of the energy comes from the slower, aerobic ATP-generating process. This category differs from the ultra-short-term category in that the sporting events are primarily powered by anaerobic glycolysis using glycogen as the energy fuel. Anaerobic training that emphasizes the use of glycolysis for energy transfer has been reported to increase the activity of several key enzymes involved in energy production through the glycolytic pathway. These enzymes include **phosphorylase,** phosphofructokinase (PFK), and **lactate dehydrogenase.** With repeated 30 s training bouts, Costill and colleagues (1979) revealed that the activities of all three of these increased by 10% to 25%. Nevertheless, the activities remained unchanged with a 6 s training program that stressed the ATP-PCr system. More recently, MacDougall and colleagues (1998) also observed a 49% increase in PFK activity following training consisting of 30 s maximal sprint efforts. Despite the augmented activities of glycolytic enzymes, it continues to be debated whether these biochemical changes would actually translate into an enhancement of anaerobic performance. In the same study in which Costill and colleagues (1979) observed an improvement in glycolytic enzymes, the peak power output achieved and the rate of decline in performance remained unaffected from pre- to posttraining.

> **▷ K E Y P O I N T ◁**
>
> Training-induced improvements in anaerobic performance may not necessarily result from enhanced bioenergetic pathways. Training with ultra-short-term activities such as 6 s sprints improves performance, but not because of increased energy release through the ATP-PCr system. On the other hand, training with short-term activities such as 30 s maximal sprints increases the activity of glycolytic enzymes such as PFK, but this favorable biochemical change is not accompanied by enhanced performance.

Interestingly, anaerobic training can also lead to some aerobic adaptations. Using high-intensity interval training involving repeated 30 s maximal sprints, MacDougall and colleagues (1998) observed an increase in $\dot{V}O_2$max, which reflects an individual's aerobic capacity. They also found a significant increase in the activity of a host of oxidative enzymes such as **succinate dehydrogenase** and **citrate synthase** as a result of anaerobic training. Recently, using a similar training format, Burgomaster and colleagues (2005) demonstrated a nearly 100% increase in endurance as well as a 38% increase in the activity of citrate synthase. In this study, endurance, measured by having subjects cycle at 80% $\dot{V}O_2$max until exhaustion, improved from ~26 min pretraining to ~51 min posttraining despite no change in $\dot{V}O_2$max. These findings challenge the concept that aerobic performance is improved only by aerobic training. Such a "cross" training effect may be due in part to the fact that in high-intensity anaerobic training, though to a smaller extent than in aerobic training, the slow-twitch muscle fibers are also recruited during muscle contraction. An enhancement in the oxidative potential of muscle will help anaerobic performance in a number of ways. Increased oxidative potential can speed up replenishment of ATP and PCr during rest intervals between exercise bouts or during recovery following exercise. It can also reduce anaerobic ATP generation and levels of the related by-product, lactic acid, thereby increasing resistance to fatigue.

Adaptations to Resistance Training

The major adaptation that occurs in resistance training is an increase in the cross-sectional area of trained muscle, or muscle hypertrophy. Muscle hypertrophy appears to result primarily from increased size of individual muscle fibers, which can be observed within two months following the onset of training (Hakkinen et al. 1981; Thorstensson et al. 1976). Muscle hypertrophy is further brought about by increased synthesis of contractile proteins. Protein synthesis occurs in both fast- and slow-twitch muscle fibers, but is greater in the fast-twitch fibers (Hakkinen et al. 1981; Thorstensson et al. 1976). Although resistance training has no effect on the absolute quantity of mitochondria and capillaries of exercising muscle, the densities of these two elements have been reported to decrease in parallel with muscle hypertrophy (MacDougall 1986). Obviously, the decrease is the direct result of a dilution of these cellular elements due to an increase in cell volume with training. From a practical standpoint, reductions in mitochondrial and capillary density may explain why endurance capacity often decreases in those who undergo resistance training. It was proposed years ago that some form of endurance work should be incorporated into a regular resistance training routine to enable hypertrophied muscle fibers to maintain their metabolic properties associated with endurance training (Hickson 1980). Nevertheless, the dosage of added endurance training should be carefully selected and monitored so as not to interfere with the gain in muscle hypertrophy.

Resistance training may be viewed as similar to ultra-short- and short-term anaerobic training in that it depends primarily on energy from the ATP-PCr system and, to a lesser extent, glycolysis. Studies have failed to demonstrate concurrent increases in activities of the enzymes of these energy systems such as adenosine triphosphatase (ATPase), creatine kinase, and PFK. However, resistance training seems to be able to augment the storage of a variety of energy fuels necessary for muscle performance. For example, MacDougall and colleagues (1977) reported increases in concentrations of ATP, PCr, and glycogen in exercising muscle in subjects who had undergone a five-month strength training program.

There is speculation that resistance training can alter fat utilization during recovery. Treuth and colleagues (1995b) found that daily fat oxidation toward the end of a 16-week strength training program was double that during pretraining. Van Etten and colleagues (1995) also observed increased fat oxidation during sleep following a 12-week weight training program. In these studies, the level of fat utilization was determined from the RER measured in a respiratory chamber. The mechanism underlying this increased fat utilization is unclear. It may be that training-induced physiological and biochemical adaptations increase the body's ability to mobilize adipose tissue stores or oxidize fatty acids during nonexercise periods (or both). Despite this favorable metabolic effect following resistance training, it appears that the actual amount of fat oxidized is too small to effect changes in body weight and composition (Binzen et al. 2001; Van Etten 1995).

SUMMARY

The chronic stimulus provided by exercise training has been shown to be capable of producing a series of cellular changes that can augment energy transfer and therefore lead to improvement in fitness and sport performance. Among the major cellular changes resulting from aerobic conditioning are increases in capillary density and mitochondrial content and in the activity of a series of enzymes important during the aerobic production of energy. Improved oxidative capacity in fast-twitch muscle fibers has also been reported, although it is debated whether fast-twitch muscle fibers can undergo complete conversion into slow-twitch muscle fibers. Aerobic training also improves the oxygen transport system, and this can be manifested in an increase in $\dot{V}O_2$max and a reduction in the oxygen deficit and the oxygen slow component.

Improvement in the oxygen transport system allows the body to derive proportionally more of its energy from aerobic metabolism, which is desirable in most circumstances.

Cellular adaptations also occur in response to anaerobic and resistance training; these adaptations include increases in muscle size, storage of ATP, PCr, and glycogen, and the activity of glycolytic enzymes such as PFK. While these changes are considered favorable, whether they translate into an improvement in performance remains to be substantiated. It appears that training-induced improvements in anaerobic performance can also be attributed to factors that are not metabolic in origin, such as muscle hypertrophy and improved motor unit recruitment.

KEY TERMS

autonomic

cardiac output

citrate synthase

intramuscular triglycerides

lactate dehydrogenase

motor unit

muscle biopsy

muscle hypertrophy

myocardial

myoglobin

phosphorylase

specificity

stroke volume

succinate dehydrogenase

type I muscle fibers

type II muscle fibers

REVIEW QUESTIONS

1. Define the concept of training specificity.

2. How does aerobic training affect carbohydrate and fat utilization? Why are these training-induced changes considered favorable?

3. Describe how increases in capillary density and mitochondrial content are related to a lower oxygen deficit.

4. How does an improvement in one's maximal aerobic power or $\dot{V}O_2$max help exercise metabolism?

5. Describe the differences between ultra-short-term and short-term activities.

6. What are the major metabolic adaptations associated with anaerobic training? Why is there often a decrease in capillary density and mitochondrial content following resistance training?

CHAPTER
8

INFLUENCE OF GENDER AND AGE ON METABOLISM

This chapter provides information on a more contemporary topic—how energy metabolism may be influenced by age and gender. It presents research literature on energy expenditure and fuel utilization in women, people who are elderly, and children. The chapter also explores mechanisms that may explain the differences in energy metabolism seen in these subpopulations.

GENDER DIFFERENCES IN SUBSTRATE METABOLISM

In the not-so-distant past, many societies were influenced by the notion that boys were meant to be active and athletic whereas girls were weaker and thus less well suited to physical activity. In fact, until the 1960s, women were prohibited from running any race longer than 800 m (Wilmore and Costill 2004). This notion is no longer held to be true, and girls and women are given equal access to most athletic activities. Because of the increased involvement of girls and women in physical activity and training, there has been a tremendous decrease in the gender gap in relation to athletic performance. As shown in table 8.1, in events other than those requiring muscular strength and power, performance differences between the genders are no more than 15%.

Current knowledge of metabolic responses to exercise and training is based largely on the responses of young adult males. This is the case because in the

past, much of the research in the area of exercise metabolism was conducted using primarily male subjects. Due to the ever-increasing involvement of women in sport as well as in leisure-time and occupational physical activities, there has been a steady increase in research aimed at comparing exercise-induced metabolic responses and adaptations between genders. The proliferation of this research will serve to improve our understanding of gender differences in exercise metabolism. It will also help to establish evidence-based guidelines to be used for designing gender-specific training regimens and exercise prescriptions.

Fat Oxidation

Though there is some disagreement, perhaps the most repeatedly evidenced metabolic difference between genders is that compared to men, women are able to derive proportionally more of the total energy expended during aerobic exercise from fat oxidation. This conclusion was drawn primarily from studies using indirect calorimetry (Blachford et al. 1985; Froberg and Pedersen 1984; Friedlander et al. 1998; Horton et al. 1998; Phillips et al. 1993; Tarnopolsky et al. 1995, 1997). In these studies, a lower respiratory exchange ratio (RER) during submaximal endurance exercise was found in females as compared to males. This translated into increased fat oxidation, which was shown to be 70% higher in females than in males during exercise (i.e., ~39% vs. ~22% of the

Table 8.1   Comparisons of Recent Men's and Women's World Records in Track and Field, Swimming, and Weightlifting

Event	Men	Women	% difference
Track and field			
100 m	9.78 s	10.49 s	6.8
200 m	19.32 s	21.34 s	9.5
400 m	43.8 s	47.60 s	9.3
800 m	1:41.11 min:s	1:53.28 min:s	10.7
1500 m	3:26.00 min:s	3:50.46 min:s	10.6
3000 m	7:20.67 min:s	8:06.11 min:s	9.3
5000 m	12:39.36 min:s	14:28.09 min:s	12.5
10,000 m	26:22.52 min:s	29:31.78 min:s	10.7
Marathon	2.04:55 h:min:s	2:15.25 h:min:s	7.8
High jump	2.45 m	2.09 m	14.7
Long jump	8.95 m	7.52 m	16.0
Swimming			
50 freestyle	21.64 s	24.13 s	11.0
100 freestyle	47.84 s	53.77 s	10.7
200 freestyle	1:44.06 min:s	1:56.64 min:s	8.6
400 freestyle	3:40.17 min:s	4:03.85 min:s	10.8
800 freestyle	7:36.19 min:s	8:16.22 min:s	6.1
1500 freestyle	14:34.56 min:s	15:52.10 min:s	7.4
Weightlifting (68 kg class)			
Snatch	165.0 kg	117.5 kg	28.8
Clean and jerk	197.5 kg	148.5 kg	24.8
Total	357.5 kg	265.0 kg	25.9

Adapted from J.H. Wilmore and D.L. Costill, 2004, *Physiology of sport and exercise,* 3rd ed. (Champaign, IL: Human Kinetics), and G.A. Brooks, T.D. Fahey and K.M. Baldwin, 2005, *Exercise physiology: Human bioenergetics and its applications,* 4th ed. (New York, NY: McGraw-Hill).

total energy). In these studies, an effort was made to match male and female subjects for their $\dot{V}O_2$max or training status. This approach was used to preclude the potential confounding effect of fitness on any gender-related difference in exercise metabolism. Matching is a time-consuming process because investigators have to recruit sedentary males and females, expose them to the same training program, and then test them at the same relative and absolute exercise intensity following training. In these studies, oxygen uptake was also normalized relative to lean body mass in order to minimize the gender differences in energy metabolism attributable to percent fat. Controlling for these confounding factors is critical because otherwise it may be counterargued that a lower RER in females during aerobic exercise is due to a systematic error owing to the selection of better-trained females.

The evidence with indirect calorimetry that women oxidize less carbohydrate and more fat during exercise is consistent with recent investigations that have involved more sophisticated laboratory techniques, that is, muscle biopsy and isotopic tracer methods.

For example, Tarnopolsky and colleagues (1990) found that **vastus lateralis** glycogen concentration was less depleted in women than in men following exercise. The authors had six males and six equally trained females run for more than 90 min at 65% $\dot{V}O_2$max following three days on a controlled diet. Muscle glycogen utilization was calculated from pre- and postexercise needle biopsies of vastus lateralis. As mentioned in chapter 4, the level of muscle glycogen utilization can also be determined when indirect calorimetry is used in conjunction with isotopic tracer method. In this case, the researcher computes muscle glycogen utilization by subtracting the rate of blood glucose disappearance from the rate of total carbohydrate oxidation. Using this approach, Carter and colleagues (2001b) and Friedlander and colleagues (1998) observed a lower utilization of muscle glycogen in women than men during endurance exercise both before and after endurance training. In these studies, lipid oxidation as determined by indirect calorimetry was found to be uniformly higher in women than in men during exercise at the same relative intensity.

Not all studies have yielded evidence supporting gender differences in substrate metabolism. For example, Tarnopolsky and colleagues (1997), after observing lower glycogen utilization in women during running, failed to confirm this finding when exercise was conducted on a stationary cycle ergometer. Thus, it remains to be determined whether the difference can be attributed to differences in muscle recruitment between the sexes during running as compared to cycling. With indirect calorimetry in conjunction with isotopic tracer method, Romijn and colleagues (2000) found that RER, as well as rates of fat and carbohydrate oxidation, was similar between trained men and women during exercise at either low or high intensity. These findings are inconsistent with those reported by Friedlander and colleagues (1998), who used the same experimental approach. It must be noted that Romijn and colleagues did not control for the effect of menstrual phases, whereas most studies have tested exercising women in the midfollicular phase. There are indications that the menstrual cycle can affect the metabolic responses to exercise (Hackney et al. 1994). However, the impact of menstrual phases on exercise metabolism is abolished when the exercise is performed at higher intensities. In light of these discrepancies, it appears that studies that carefully control for potential confounding factors are needed in order to allow further evaluation of the relation between gender and substrate utilization.

Effects of Sex Hormones

Very little investigation has addressed potential mechanisms for the higher reliance in women on lipid oxidation versus the higher reliance on carbohydrate oxidation in men during exercise. Part of the reason is that the concept of gender differences in exercise metabolism has not been universally accepted. According to the currently available research, gender differences in exercise metabolism seem to be mediated primarily by sex hormones such as **estrogen** and **progesterone,** which are present in small quantities in men as well. Progesterone, released from the **corpus luteum, placenta,** and adrenal glands, is considered a precursor to the male and female sex hormones, testosterone and estrogen, respectively. Estrogen is a collective term for a group of 18-carbon steroid hormones. The most biologically active estrogen is 17β-estradiol (E_2), and there are other less potent estrogens such as estrone (E_1) and estriol (E_3) (Vander et al. 2001). Estrogens are secreted mainly by the ovaries and, to a lesser extent,

the adrenal glands. Estrogens are also synthesized from androgens such as testosterone in blood or other organs such as adipose and muscle tissues.

Animal Studies of Estrogen and Progesterone

A number of animal studies have been undertaken to examine the impact of estrogen on the utilization of glycogen (Ellis et al. 1994; Kendrick and Ellis 1991; Kendrick et al. 1987; Rooney et al. 1993). In these studies, the experimental approach was to alter the hormonal environment by injecting estrogen and then to evaluate the metabolic consequences. For example, Kendrick and colleagues (1987), who administered E_2 to rats in doses sufficient to achieve blood levels of estrogen in the physiological range, showed decreased utilization of glycogen stored in skeletal muscle as well as in the heart and liver. The role of progesterone in exercise metabolism is less clear. It has been reported that this hormone increases liver glycogen content and suppresses hepatic gluconeogenesis, and that these effects can be enhanced by concurrent administration of E_2 (Kalkhoff 1982; Matute and Kalkhoff 1973). In this context, it appears that the two female hormones may work additively or synergistically in reducing carbohydrate utilization during exercise.

High levels of E_2 have also been found to increase the availability of free fatty acids (FFA) during exercise in rats. For example, Ellis and colleagues (1994) observed that during exercise, E_2 increased lipolysis in adipose tissue and enhanced distribution of FFA to the muscle. This increased availability of FFA can be further attributed to alterations in the activity of **lipoprotein lipase** (LPL), which regulates fat metabolism. In the same study, Ellis and colleagues demonstrated a decreased activity of adipocyte LPL, which promotes fat synthesis, and an increased activity of muscle LPL, which promotes fat utilization. Of particular interest is that estrogen and progesterone play opposing roles in regulating fat metabolism, which is not the case with respect to their action on carbohydrate metabolism. Hansen and colleagues (1980) found that the rate of fatty acid synthesis was lower in E_2-treated rats compared with progesterone-treated rats. More recently, Campbell and Febbraio (2001) also observed increased activity in several key enzymes involved in fatty acid oxidation as a result of estrogen supplementation, an effect that was reversed with concurrent administration of progesterone. The roles that estrogen and progesterone play in regulating fat and carbohydrate metabolism are illustrated in table 8.2.

Table 8.2 The Actions of Estrogen and Progesterone on Carbohydrate and Fat Metabolism

Action	Estrogen	Progesterone
Carbohydrate metabolism		
Muscle glycogenolysis	Inhibiting	Inhibiting
Liver glycogenolysis	Inhibiting	Inhibiting
Glucose transport into muscle	Inhibiting	Inhibiting
Fat metabolism		
Adipose tissue lipolysis	Stimulating	Inhibiting
Fatty acid transport into mitochondria	Stimulating	Inhibiting

Adapted from T. D'Eon and B. Braun, 2002, "The roles of estrogen and progesterone in regulating carbohydrate and fat utilization at rest and during exercise," *Journal of Women's Health and Gender-Based Medicine* 11(3): 225-237.

Observations With Human Subjects

The effects of sex hormones on energy metabolism have also been examined in humans. Hackney (1990) performed muscle biopsies on the vastus lateralis of 10 healthy women in both the **follicular** and **luteal** phases of the menstrual cycle. The findings showed that under resting conditions, muscle glycogen content was higher in the luteal than in the follicular phase. The same research group later reported lower carbohydrate oxidation in the luteal phase during exercise at 35% and 60% $\dot{V}O_2$max (Hackney et al. 1994). As the luteal phase is a time when the production of both estrogen and progesterone is higher, this finding is consistent with the conclusion from animal studies that estrogen attenuates the utilization of carbohydrate. The inhibitive role of estrogen on carbohydrate utilization was also evidenced in studies in which subjects were supplemented with **exogenous** estrogen. After providing 17β-estradiol or E_2 to a group of amenorrheic females, Ruby and colleagues (1997) observed altered carbohydrate metabolism. In that study, with use of an isotopic tracer method, the investigators were able to determine muscle glucose utilization and hepatic glucose production. They found that during exercise the release of glucose from the liver was reduced as a result of the increased E_2 level, while glucose utilization by muscle remained similar between the E_2 and the placebo groups. A reduction in hepatic glucose output due to E_2 supplementation was also observed by Carter and colleagues (2001a), who administered E_2 to a group of men, although this study showed a decrease in muscle glucose utilization as well. In both studies, no differences in whole-body substrate oxidation were found between the experimental and placebo groups. It appears that despite the indications from animal studies that estrogen may mitigate muscle glycogen utilization, such a role for estrogen in humans is less clear.

KEY POINT

During aerobic exercise, it is generally accepted that in comparison with men, women are able to derive proportionally more of the total energy expended from fat oxidation. This gender difference may be explained by experimental observations that estrogen stimulates lypolysis and also inhibits carbohydrate utilization.

Ruby and colleagues (1997), as well as Carter and colleagues (2001a), reported a decrease in blood concentrations of epinephrine during exercise in subjects with estrogen treatment. This observation has led to the hypothesis that observed gender differences in substrate metabolism may be mediated by changes in the circulating level of, or tissue sensitivity to, epinephrine. The level of epinephrine during submaximal exercise is generally lower in women than in men, although it has been shown that this gender difference disappears during strenuous exercise (i.e., >88% $\dot{V}O_2$max) (Marliss et al. 2000).

Epinephrine plays a key role in regulating carbohydrate metabolism during exercise. In muscle, epinephrine activates **adenylate cyclase,** a second messenger, and thus stimulates glycogenolysis and glycolysis. In the liver, epinephrine also stimulates glycogenolysis and thus glucose output. Additionally, epinephrine stimulates lipolysis in adipose tissue, thereby increasing circulating levels of FFA. Given that women normally produce less epinephrine than men, their fat utilization should be lower.

However, as shown in the preceding discussion, this is not the case. A number of studies suggest that estrogen potentiates the lipolytic effect of epinephrine. For instance, Flechtner-Mors and colleagues (1999) found a greater release of glycerol in response to infused epinephrine in women than in men at rest. In addition, Horton and colleagues (1998) observed no gender differences in blood glycerol concentration during exercise at 40% $\dot{V}O_2$max, although the men had higher levels of epinephrine than the women. It seems possible that women are able to use more fat primarily because their higher level of estrogen functions to augment the lipolytic effect of epinephrine. If this is indeed the case, then the question remains: Why does this potentiation effect not appear to occur with respect to carbohydrate utilization?

Estrogen has also been shown to cause dilation of blood vessels (Braun and Horton 2001). However, it is unknown whether or not this vasodilation is specific to adipose tissue. If women do indeed maintain a higher blood flow to adipose tissue, one would expect an increased interaction between epinephrine and β receptors in the adipose tissue of women. This could ultimately lead to an increase in fat utilization.

PREGNANCY

Pregnancy places unique demands on women's metabolism. It affects the metabolic cost and physiological strain imposed by exercise. An early investigation involved 13 women from six months of pregnancy to six weeks after gestation (Knuttgen and Emerson 1974). The results showed that during walking, heart rate and $\dot{V}O_2$ increased progressively despite an unchanging exercise intensity. However, these two parameters remained constant throughout steady-state cycle exercise. These findings suggest that an increase in body mass, including fetal tissue, leads to an increase in energy cost during weight-bearing activities like walking, jogging, and running.

In addition to this added energy cost during exercise, it has been demonstrated that pregnant women experience an increase in resting metabolism especially during the later stages of the pregnancy. Table 8.3 provides a comparison of caloric costs of common household activities between pregnant and nonpregnant women. It has been estimated that throughout an entire pregnancy, an additional 80,000 kcal are required to build new tissues and to meet the demand imposed by higher energy costs of daily activities (Brooks et al. 2005). This figure represents an extra expenditure of 250 kcal per day during a 40-week pregnancy. It has been assumed that the increased energy cost of weight-bearing activities is offset by a decrease in the amount of time pregnant women spend engaged in these activities, as well as by the relaxed and economical fashion in which they tend to move. Consequently, the net increase in energy expenditure associated with pregnancy may only reflect an increase in resting metabolism as a consequence of the growth of both maternal and fetal tissues.

Table 8.3 Comparisons of Energy Cost of Household Activities in Pregnant and Nonpregnant Women

	Energy Cost (kcal · min⁻¹)	
Activity	**Pregnant**	**Nonpregnant**
Lying quietly	1.11	0.95
Sitting	1.32	1.02
Sitting, combing hair	1.36	1.22
Sitting, knitting	1.55	1.47
Standing	1.41	1.12
Standing, washing dishes	1.63	1.33
Standing, cooking	1.66	1.41
Sweeping with broom	2.90	2.50
Making a bed	2.98	2.66

Adapted from G.A. Brooks, T.D. Fahey and K. Baldwin, 2005, *Exercise physiology: Human bioenergetics and its applications*, 4th ed. (New York, NY: McGraw-Hill), 799, by permission of The McGraw-Hill Companies.

> ▶ **K E Y   P O I N T** ◀
>
> Due to an increase in body mass, including fetal tissue, pregnant women incur an increase in energy cost during weight-bearing activities such as walking, jogging, and running. Their energy cost during nonexercise periods also increases, primarily due to increased resting metabolism. It has been estimated that throughout an entire pregnancy, an additional 80,000 kcal are required to build new tissues and to meet the higher energy cost of daily activities.

Substrate Metabolism During Pregnancy

Although pregnancy consists of a series of small, continuous physiological adjustments, the alterations in substrate metabolism appear to occur primarily during the later phase of a pregnancy. From a metabolic standpoint, pregnancy may be divided into two halves. The first half is primarily a time of preparation to meet the demands of the rapid fetal growth that will occur later. This period is characterized by a continuous increase in the production of estrogen and progesterone. The presence of these hormones can help not only to mobilize fat for energy, but also to stabilize plasma glucose at relatively high levels in order to meet the needs of the fetus. There is evidence that perhaps as a means of protecting the fetus from hypoglycemia, pregnancy reduces the ability of the mother to metabolize carbohydrate (Clapp et al. 1987). This metabolic alteration could limit pregnant women from performing anaerobic or strenuous aerobic exercises in which carbohydrate is a primary fuel.

For pregnant women, measurement of insulin sensitivity is often used to detect the possibility of **gestational diabetes.** The reason is that estrogen and **placental lactogen** have been considered diabetogenic hormones due to their inhibitive effects on insulin-mediated glucose uptake by various tissues. A number of studies have shown that during the early phase of pregnancy, the sensitivity of peripheral tissues to insulin is either normal or slightly increased (Buch et al. 1986; Gatalano et al. 1991, 1992). However, longitudinal studies of glucose tolerance have shown that as gestation continues, there is a progressive increase in the insulin response to a given dose of glucose challenge (Sivan et al. 1997). This greater than normal insulin response is consistent with the phenomenon of **insulin resistance** and suggests that pregnancy can potentially diminish a woman's ability to handle glucose via insulin. This is especially the case in obese women, who have a high risk of developing diabetes even without pregnancy (Sivan et al. 1997). The reduced insulin sensitivity is thought to be secondary to gestation-induced changes in hormones including estrogen, progesterone, cortisol, and **prolactin,** although the precise mechanism remains unclear. From a fetal standpoint, a certain degree of insulin resistance is considered desirable in that it can serve to shunt ingested nutrients to the fetus (Butte 2000).

Exercise During Pregnancy

During pregnancy, the metabolic reserve available for performing exercise is diminished owing to increased resting metabolism and blunted sympathetic response to physical activity. However, during the early stages of pregnancy, light to moderate activity can be pursued safely, provided that blood glucose is carefully monitored to prevent hypoglycemia. Regular exercise during pregnancy counteracts the effects of deconditioning. It attenuates pregnancy-related fatigue. It helps maintain muscular strength, which may speed delivery. Regular exercise can also prevent excessive weight gain, insulin resistance, and type 2 diabetes. In general, the quantity of exercise performed should not be increased over that at the prepregnancy level, especially during the first 14 and the final 12 weeks of pregnancy (Shephard 2000). Based on the literature, it appears that pregnant women ordinarily are able to tolerate light- to moderate-intensity exercise sessions of up to 30 min in duration, four times per week (Ohtake and Wolfe 1998; Webb et al. 1994; Wolfe et al. 1994), although exercise tolerance can be affected by environmental conditions as well as the fitness level of the mother. Caution is warranted in selection of the appropriate exercise modality. With the advancement of pregnancy, the capacity for exercise, especially for activities that are performed against gravity, decreases. Therefore, during the later stages of pregnancy, it is helpful to introduce weight-supported activities such as cycling, swimming, and water aerobics. Exercise can be dangerous if excessive. Important contraindications to vigorous exercise include hypoglycemia, intrauterine growth retardation, premature labor or ruptured membrane, placental injury or dysfunction, incompetent cervix, pregnancy-induced hypertension, and blood poisoning (American College of Obstetricians and Gynecologists 1994).

▶ K E Y   P O I N T ◀

The inhibitive effect of estrogen as well as other placental hormones on carbohydrate metabolism can place pregnant women at high risk of developing insulin resistance, which can lead to gestational diabetes. Being physically active during pregnancy is important in helping to prevent these metabolic disorders. Pregnant women should choose primarily non-weight-bearing activities and always exercise at low to moderate intensity—an intensity level that depends comparatively less on carbohydrate utilization.

EXERCISE METABOLISM IN ELDERLY PEOPLE

The elderly population—those who reach and pass the age of 65—make up the fastest-growing segment of many countries today. For example, currently in the United States approximately 35 million, or nearly 12% of Americans, are older than 65. It is predicted that this figure will climb to 70 million or 22% of the population by 2030. The trend seen in the United States also applies in many Western nations and may soon emerge in developing countries. Aging refers to the normal yet irreversible biological changes that occur throughout an individual's life span. It involves a diminished capacity to regulate the internal environment in order to meet external challenges. As shown in table 8.4, this reduced ability can be further attributed to a series of attenuated or impaired metabolic functions, which ultimately reduce one's ability to generate needed energy. Aging is influenced by genetics. This is attested to by observations that the life spans of twins are remarkably similar. It has been reported that identical twins usually die with two to four years of each other, and nonidentical twins within seven to nine years (Brooks et al. 2005). Aging is also affected by lifestyle factors. It is considered a process associated with an accumulation of wear and tear that leads to gradual loss of the ability to respond to stress. On the other hand, although regular physical activity and a healthy diet may not halt the aging process, they can help improve quality of life and prolong life expectancy.

Reduced $\dot{V}O_2$max and Energy Expenditure

Reductions in physical capacity with age can be characterized by a decrease in aerobic power or $\dot{V}O_2$max, which was observed more than a half-century ago. With cross-sectional comparisons, Robinson (1938) demonstrated that in men, $\dot{V}O_2$max declines an average of 0.44 ml $\cdot$ kg^{-1} $\cdot$ min^{-1} per year up to age 75. This translates to about 1% per year or 10% per decade. For women between the ages of 25 and 65 years, Åstrand (1960) showed a decline of 0.38 ml $\cdot$ kg^{-1} $\cdot$ min^{-1}, or 0.9% per year. Since these early observations, numerous cross-sectional and, to a lesser extent, longitudinal studies have been conducted with the aim of further characterizing the age-related decline in $\dot{V}O_2$max and its metabolic consequences. The rate of decline in

Table 8.4 Aging-Related Metabolic Changes and Their Physiological Consequences

Metabolic change	Physiological consequences
Myosin-ATPase↓	Reduced muscle contractility
Lactate dehydrogenase↓	Reduced oxidative capacity
Succinic dehydrogenase↓	Reduced oxidative capacity
Malic dehydrogenase↓	Reduced oxidative capacity
Cytochrome c oxidase↓	Reduced oxidative capacity
Mitochondrial size and number↓	Reduced oxidative capacity
Type II muscle fibers↓	Reduced muscle strength and power
Capillary density↓	Reduced blood flow and oxygen delivery
Glucose tolerance↓	Increased risk of diabetes and heart disease
Blood insulin↑	Increased risk of diabetes and heart disease
Insulin sensitivity↓	Increased risk of diabetes and heart disease
Sympathetic stimulation↓	Reduced maximal heart rate and lipolysis
Muscle mass↓	Reduced basal metabolism and fat oxidation

$\dot{V}O_2$max found in these studies in general agrees with that reported initially by Robinson in 1938. A reduction in $\dot{V}O_2$max due to aging has been further ascribed to a decrease in maximal heart rate, maximal cardiac output, and maximal ability of working muscle to utilize oxygen for energy transfer.

Goran and Poehlman (1992) observed a significant correlation between $\dot{V}O_2$max and the total daily energy expenditure. They also found a modest relationship between the total daily energy expenditure and the level of physical activity. These findings suggest that people with greater aerobic fitness tend to be more physically active and therefore have greater daily energy expenditure. However, such a linkage between fitness level and energy expenditure provides no information on cause and effect. In other words, it is unclear whether the increased total energy expenditure associated with a physically active lifestyle leads to a higher $\dot{V}O_2$max, or alternatively, whether individuals with a higher $\dot{V}O_2$max engage in physical activities more frequently and intensely because of their higher work capacity.

A reduction in total energy expenditure has been well evidenced in elderly persons (Margaret-Mary and Morley 2003; Elia et al. 2000). Interestingly, many normal-weight healthy older men and women decrease their energy intake well below their energy expenditure and thus lose weight. In addition to a lack of physical activity, the age-related decrease in energy expenditure has been linked to reductions in basal metabolic rate and diet-induced thermogenesis. Chapters 10 and 11 provide more detailed discussion regarding the impact of aging on these two energy components.

Changes in Enzymes in the Bioenergetic Pathway

As discussed earlier in this book, mitochondria serve to allow biologically usable energy to be generated via oxidative pathways. Therefore, the activities of selected oxidative enzymes and the rate of adenosine triphosphate (ATP) production have been frequently examined as markers of mitochondrial function in skeletal muscle. Papa (1996), Rooyackers and colleagues (1996), and Houmard and colleagues (1998) all reported age-related decline in enzymes such as citrate synthase, succinate dehydrogenase, and **cytochrome c oxidase.** Furthermore, from data provided by Holloszy and colleagues (1991), it appears that this reduction in activity of oxidative enzymes occurs primarily in red, predominantly oxidative muscle rather than in the whiter glycolytic muscle. As a result of these enzymatic changes,

mitochondrial oxidative capacity is impaired. With the use of muscle samples, Papa (1996) observed a decreased ability of mitochondria to consume oxygen for generating energy. This in vitro observation was later confirmed by an in vivo study in which Conley and colleagues (2000) used nuclear magnetic resonance technique and found that the average rate of ATP formation in the quadriceps muscles of subjects between the ages of 65 and 80 was approximately half that of subjects between the ages of 25 and 48. Not all reports agree with these findings, however. Discrepancies in the literature may in part stem from the fact that different anatomical sites have been used for muscle sampling in various studies.

Alterations in Carbohydrate and Fat Metabolism

Another metabolic hallmark of the aging process is the impairment of carbohydrate and fat metabolism. Substantial evidence has shown that increasing age is associated with decreased **glucose tolerance.** The glucose tolerance test measures the body's ability to metabolize glucose. It is performed after an overnight fast. During the test, the patient drinks a solution containing a known amount of glucose. Blood is obtained before the patient drinks the glucose solution and again every 30 to 60 min afterward, for 2 or 3 h. Blood glucose levels above normal limits at the times measured can be used to diagnose type 2 diabetes or gestational diabetes. It has been estimated that the 2 h plasma glucose level during an oral glucose tolerance test rises on average 5.3 mg · dL^{-1} per decade, and that the fasting plasma glucose rises on average 1 mg · dL^{-1} per decade (Davidson 1979). Levels of insulin, a hormone produced by the pancreas that moves glucose from the bloodstream into cells, have also been found to be higher in many older individuals (Brooks et al. 2005). This observation suggests that as people age, they may lose the ability to respond to insulin effectively and therefore require extra insulin to maintain normal blood glucose levels.

Impairment in fat metabolism is another age-related metabolic disorder. Aging has been associated with reduced fat oxidation at rest (Nagy et al. 1996) and following a meal (Roberts et al. 1996). Sial and colleagues (1998) also demonstrated an age-related reduction in fat oxidation during aerobic exercise. It is thought that these reductions in fat utilization play an important role in mediating the age-related increase in adiposity, especially in the abdominal compartment. Sial and colleagues also reported greater carbohydrate oxidation, a finding that was thought to result from the impaired fat utilization. All these age-related metabolic changes gradually deprive elderly people of the ability to use their energy fuels efficiently during exercise as compared to younger individuals.

Fat oxidation is mainly a function of two processes: the release of fatty acids from adipose tissue and the capacity of respiring tissue to oxidize fatty acids. Studies using aging rats and humans have demonstrated a diminished sympathetic stimulation of lipolysis (James et al. 1971; Lonnqvist et al. 1990). However, when examined in relation to the needs of the metabolically active tissue, the release of FFA was found to be greater in older compared to younger individuals (Toth et al. 1996). The age-related reduction in fat utilization is therefore considered to be due primarily to the loss of the size or the oxidative capacity, or both, of metabolically active tissues such as skeletal muscle. Fatty acids that are released but not metabolized could have adverse metabolic effects such as **hyperlipidemia** and insulin resistance. In this regard, aerobic training becomes particularly important because it can maintain or increase the size and function of skeletal muscle, thereby improving the health status of elderly individuals.

▶ K E Y   P O I N T ◀

Both carbohydrate and fat utilization decrease as one ages, and these declines can impair the ability of elderly people to tolerate strenuous physical activity. The age-related reduction in substrate utilization is due to the loss of size or oxidative capacity (or both) of metabolically active tissues such as skeletal muscle, as well as to a decrease in tissue sensitivity to insulin, which moves glucose from the bloodstream into cells.

ENERGY METABOLISM IN CHILDREN AND ADOLESCENTS

The period of life from birth to the start of adulthood may be divided into three phases: **infancy, childhood,** and **adolescence.** Infancy is defined as the first year of life. The span of childhood is from the first birthday to the beginning of adolescence. The period of adolescence is more difficult to define, but is often considered to begin at the onset of **puberty**

and to terminate as growth and development are completed. Research on metabolism with regard to these early stages of the life spectrum is rather limited. The main reasons are ethical considerations and methodological constraints in studying children and adolescents. For example, very few investigators would puncture a child's artery or take a needle biopsy of a child's muscle. In addition, efforts are still being made to develop instruments and protocols that are age and size appropriate. Consequently, our understanding of children's metabolic response to exercise has been based on a limited number of investigations. Many conclusions regarding exercise metabolism in children and adolescents are derived primarily from measurement of cardiorespiratory parameters such as oxygen uptake and RER.

Children and adolescents should not be regarded as miniature adults (Bar-Or and Rowland 2004). In other words, age-related functional deficiency in children and adolescents is not always attributable to the fact that they are smaller. It is generally true that children are less capable of performing a given task than adults. Their physiological function increases as they grow older and bigger, but only some gains in physiological function are proportional to changes in size. For example, muscle strength increases in direct proportion to muscle cross-sectional area. Many changes in function have been found to be either partially related to or completely independent of changes in size. For example, anaerobic capacity depends on the activity of certain key anaerobic enzymes in addition to muscle size. Additionally, it has been found that some physiological parameters, such as the blood concentrations of oxygen and glucose, remain unchanged despite a gain in body size. It is important to understand the patterns of function–size relationship in growing individuals. This helps us to interpret age-related physiological differences properly and to determine whether there is a need to normalize a physiological parameter for body size before making age-related comparisons.

▶ K E Y   P O I N T ◀

Children and adolescents should not be regarded as miniature adults, because age-related functional deficiency in children and adolescents is not always attributable to the fact that they are smaller in size. Many differences in function have been found to be either partially related to or completely independent of changes in size.

Aerobic Capacity

As mentioned earlier, $\dot{V}O_2$max reflects the highest metabolic rate made available by aerobic energy transfer, and this parameter can be expressed in both L · min⁻¹ and ml · kg⁻¹ · min⁻¹. As shown in figure 8.1a, which depicts the chronological changes in $\dot{V}O_2$max in L · min⁻¹ reported in studies involving boys ($n = 2180$) and girls ($n = 1730$), $\dot{V}O_2$max increases continuously until the age of 17 to 18 in boys, but increases minimally beyond the age of 14 to 15 in girls. The gender difference in $\dot{V}O_2$max can be ascribed in part to the differences in muscle mass between boys and girls (Davies et al. 1972). With use of ratio scaling, however, in which $\dot{V}O_2$max in L · min⁻¹ is divided by body mass, the average $\dot{V}O_2$max in ml · kg⁻¹ · min⁻¹ was still somewhat higher in boys than girls, especially during the later periods of adolescence (figure 8.1b). This finding suggests that the increase in $\dot{V}O_2$max can also be explained by other factors, such

as those involved in oxygen transport and utilization, that are gender specific. Over the years, $\dot{V}O_2$max in ml · kg⁻¹ · min⁻¹ remains essentially unchanged in boys and slightly declines among girls. This finding suggests that when comparing aerobic capacity using a $\dot{V}O_2$max already adjusted for body mass, one should expect no differences between children and adults. The decline in the relative $\dot{V}O_2$max seen in girls is due to a progressive increase in body fat in girls during adolescence (Bar-Or and Rowland 2004).

Oxygen Deficit and Respiratory Exchange Ratio

Among other gas exchange parameters that have received a great deal of attention in children are $\dot{V}O_2$ kinetics and the RER. $\dot{V}O_2$max kinetics assesses the integrated responses of oxygen requirement and supply at the onset of exercise and during exercise of varying intensity. As mentioned in chapter 5, $\dot{V}O_2$ kinetics at the onset of exercise can be characterized by the phenomenon of oxygen deficit, which is defined as a lag of oxygen supply in relation to oxygen demand. A number of research groups have examined $\dot{V}O_2$ kinetics at the onset of aerobic exercise in children (Armon et al. 1991; Hebestreit et al. 1998; Sady 1981). In general, these studies agreed in finding that children demonstrated a faster increase in oxygen uptake at the onset of exercise than adults. It appears that children have the ability to adapt their oxidative metabolism faster to meet the energy demand imposed by exercise. Aside from being able to activate the aerobic system more rapidly, children are also able to derive proportionally more of the total energy from fat oxidation. A number of studies have shown a lower RER in children than in adults (Martinez and Haymes 1992; Morse et al. 1949; Rowland et al. 1987). This contention, however, needs to be further evaluated, as Rowland and Rimmy (1995) and Macek and colleagues (1976) failed to observe this age-related difference.

Metabolic Efficiency

When performing aerobic exercise, children are less efficient than adults. This is manifested by a greater mass-specific oxygen uptake or $\dot{V}O_2$ expressed in ml · kg⁻¹ · min⁻¹ in children. This metabolic feature has been observed particularly during weight-bearing activities such as walking and running (Fawkner and Armstrong 2003; Bar-Or and Rowland 2004). The aim of a study by Sallis and colleagues (1991) was to quantify the excessive metabolic cost of walking and running in children

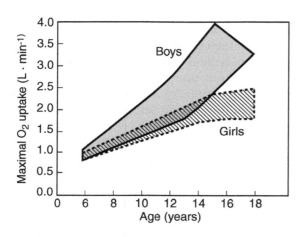

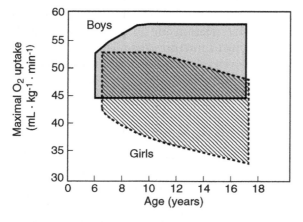

Figure 8.1 Relationship between age and maximal aerobic power expressed in L · min⁻¹ (top) and in ml · kg⁻¹ · min⁻¹ (bottom).

Reprinted, by permission, from O. Bar-Or and T.W. Rowland, 2004, *Pediatric exercise medicine* (Champaign, IL: Human Kinetics), 6-7.

by compiling data from various studies. As shown in table 8.5, on average, a 5-year-old child would expend 37% more oxygen than an adult performing the same task. This excess, however, decreases with age. It has been suggested that the low economy of locomotion in children is caused by multiple factors, including high resting metabolic rate, high stride frequency, mechanically wasteful locomotion style, and excessive cocontraction of antagonist muscles (Bar-Or and Rowland 2004).

Carbohydrate Storage and Utilization

Much less is known about carbohydrate utilization during exercise in children. This lack of information is obviously attributable to ethical concerns about taking muscle biopsies from children. Based on very limited data from muscle biopsies performed on children (Eriksson et al. 1973), it appears that children have a lower glycogen content at rest and a reduced rate of glycogenolysis during exercise as compared to adults. Glycogenolysis is regulated by the activity of such enzymes as phosphofructokinase, and this enzyme has been found to be less active in muscle cells of boys as compared to adults (Eriksson et al. 1971, 1973; Bell et al. 1980). These findings suggest that it would be a disadvantage for children to compete in prolonged strenuous events that are glycogen dependent. Endurance capacity may not be limited by the volume of mitochondria in children, as Bell and colleagues (1980) reported a similar mitochondrion to muscle fiber ratio in prepubertal and adult muscle tissues.

Anaerobic Capacity

Despite the fact that children often perform activities in an intermittent fashion, their anaerobic capacity is lower than that of adults. This lower anaerobic capacity may be manifested particularly in short-term events that last 1 to 2 min, such as the 400 to 800 m run or the 100 to 200 m swim. The reason is that children are less able to store glycogen and also to extract energy from glycogen via glycolysis, as mentioned earlier. The lower glycolytic ability is consistent with evidence that peak lactate concentration is lower in children than in adults (Eriksson et al. 1971). The reduced anaerobic capacity in children may also be explained by a lower level of sympathetic activity, which functions to simulate glycogenolysis and glycolysis. Pullinen and colleagues (1998) showed that adolescent males aged 15 ± 1 years had lower levels of blood catecholamines during resistance exercise as compared to adult males aged 25 ± 6 years. Interestingly, unlike what happens with glycogen, both storage and utilization of ATP and PCr were found to be comparatively similar between children and adults (Zanconato et al. 1993; Eriksson et al. 1971). In these comparisons between adults and children, the level of energy substrates was expressed relative to muscle mass in order to account for differences in body size.

KEY POINT

In comparison with adults, children are inefficient metabolically because they are less coordinated in performing physical activities. They are also less able to store and use carbohydrate as an energy source, and this can limit their tolerance to strenuous exercise of extended duration. However, they have a lower oxygen deficit, suggesting that they can activate their aerobic system more spontaneously at the start of an activity, and are also able to derive proportionally more of the total energy from fat oxidation.

Table 8.5   Excess Oxygen Cost of Walking and Running per Kilogram Body Mass in Children of Various Ages Compared With Young Adults

Age (years)	Excess cost (%)
5	37
7	26
9	19
11	13
13	9
15	5
17	3

Reprinted, by permission, from O. Bar-Or and T.W. Rowland, 2004, *Pediatric exercise medicine* (Champaign, IL: Human Kinetics), 9; adapted from J.F. Sallis, M.J. Buono and P.S. Freedson, 1991, "Bias in estimating caloric expenditure from physical activity in children: Implications for epidemiological studies," *Sports Med* 11: 203-209.

SUMMARY

Gender and age are two major biological factors that can affect energy metabolism and substrate utilization. Of particular note with regard to gender differences is that women are generally able to derive proportionally more energy from fat during exercise than men. This observation is underscored by repeated evidence in both animals and human subjects that the hormone estrogen is able to facilitate lipolysis and thus attenuate glycogen utilization. Such gender differences, however, are less clear for exercise performed at high intensities.

As with aerobic capacity, the ability to use carbohydrate and fat appears to diminish in older individuals, and this can impair their ability to tolerate strenuous physical activity. The age-related reduction in substrate utilization can be attributed to loss in the size of metabolically active tissues such as skeletal muscle, or their oxidative capacity, or both. It can also be a consequence of a decrease in tissue sensitivity to insulin, which functions to move glucose from the bloodstream into cells.

Although $\dot{V}O_2$max expressed in ml $\cdot$ kg^{-1} $\cdot$ min^{-1} does not differ much between children and adults, anaerobic capacity as measured by short-term performance (1-2 min) is lower in children. Children have a limited ability to use carbohydrate due to their lesser glycogen storage and utilization, and they are generally less efficient than adults in performing physical activities. On the other hand, children are able to oxidize more fat and have a smaller oxygen deficit at the onset of exercise. Understanding these gender- and age-related differences is essential in guiding our efforts to develop effective yet safe training programs for people in different subpopulations.

KEY TERMS

adenylate cyclase

adolescence

childhood

corpus luteum

cytochrome c oxidase

estrogen

exogenous

follicular

gestational diabetes

glucose tolerance

hyperlipidemia

infancy

insulin resistance

lipoprotein lipase

luteal

placenta

placental lactogen

progesterone

prolactin

puberty

vastus lateralis

REVIEW QUESTIONS

1. What is the experimental evidence supporting the observation that women are able to oxidize proportionally more fat than men during exercise?

2. Why isn't the gender difference in substrate utilization always observed?

3. Provide a specific explanation for the impaired utilization of carbohydrate and fat seen in older individuals.

4. How are children different from adults in terms of their aerobic and anaerobic capacity?

5. How would you explain the notion that children should not be viewed as miniature adults?

6. How would you apply current knowledge to develop a training regimen for children?

9

ENERGY METABOLISM IN PEOPLE WITH OBESITY AND DIABETES

This chapter examines the impact of selected metabolic disorders on energy metabolism and substrate utilization. Special attention is given to obesity and diabetes, as energy metabolism and substrate utilization have been found to be altered by physiological and biochemical changes related to these diseases. Central to these changes is tissue resistance to insulin, a disorder that can diminish the body's ability to use carbohydrate and fat as energy. Thus the chapter also discusses the efficacy of using exercise as a therapeutic modality to improve insulin sensitivity.

GENERAL DESCRIPTION OF THE DISEASES

Obesity and diabetes are both associated with a diminished ability to maintain blood glucose homeostasis, although the etiologies of the metabolic defect are different from a pathological standpoint. As blood glucose is one of the carbohydrate energy substrates, an inability to regulate blood glucose concentration suggests that utilization of carbohydrate and fat will be altered. This section describes each of these diseases with particular regard to their development, diagnosis, and metabolic consequences.

Obesity

Obesity is defined as an excess accumulation of body fat (i.e., body mass index ≥30 kg · m⁻²). It refers to an overfat condition that is associated with a number of **comorbidities,** including glucose intolerance, insulin resistance, dyslipidemia, non-insulin-dependent (or type 2) diabetes, hypertension, and increased risk of coronary heart disease and cancer. Obesity can be attributed simply to an energy imbalance in which energy intake chronically exceeds energy expenditure. Disruption in energy balance often begins in childhood, and those who are overweight during childhood have a significantly greater chance of becoming obese adults as well. Childhood obesity has been ascribed in part to parental obesity. For example, if parental obesity also exists, the child's risk of obesity in adulthood is two to three times that of normal-weight children without obese parents (Whitaker et al. 1997). The ages of 25 to 45 years are another dangerous period in which there is a progressive weight gain over time (Crawford et al. 2000). Reports indicate that despite a progressive decrease in food consumption, a 35-year-old male will gain an average of 0.5 kg (or 1 lb) of fat each year until the sixth decade of life. It remains unclear whether this "creeping" obesity results from alterations in lifestyle or reflects a normal biological pattern.

Do people who are obese eat more than lean individuals? This question remains unresolved at present, partly because the techniques used to determine food intake are not sufficiently accurate in quantifying how much energy obese individuals consume compared to their lean counterparts. The general techniques used to obtain information about energy consumption in humans include the use of nutritional histories and records of food eaten and observation of the types and amounts of food chosen. Most surveys of nutritional histories suggest that energy intake is significantly lower in people who are overweight than in those of normal weight (Baecke et al. 1983). However, a major concern is the validity of studies using data based on self-reports. By repeating dietary surveys on obese individuals during a three-month period, Bray and colleagues (1978) demonstrated an apparent underreporting of food intake in the initial interview. With direct observations, on the other hand, it was found that obese people tended to choose larger meals than did lean persons (Stunkard and Kaplan 1977). Nevertheless, there is evidence suggesting that this overeating behavior appears to occur primarily in public places. Stunkard and Waxman (1981) found that obese adolescent boys ate more than their siblings at school but not at home. In general, energy intake declines with age, with the peak values occurring in the second decade of life in both sexes (Bray 1983). In this context, obesity must also be attributed to a greater reduction in energy expenditure.

As mentioned in chapter 6, total energy expenditure can be partitioned into energy expended via (1) the resting metabolic rate, (2) the thermal effect of food, and (3) physical activities. Of these three energy components, resting metabolic rate and energy cost due to physical activity appear to receive the most attention in research on the etiology of obesity. This may be the case because thermogenesis associated with food consumption constitutes a very small portion of daily energy expenditure. Resting metabolism is the energy required by the body in a resting state. It can be influenced by age, gender, drugs, climate, body weight, and body composition and accounts for a majority of daily energy expenditure. There is a gradual decline in resting metabolism as people age. Those with greater lean body mass have greater resting metabolism. However, in relation to obesity, several studies have failed to prove that resting metabolic rate is responsible for obesity. For example, Seidell and colleagues (1992) and Weinsier and colleagues (1995) reported that a gain in body weight occurred independent of changes in resting metabolism over 10 and four years, respectively. In fact, obese individuals were found to have an expanded lean body mass and thus a greater resting metabolic rate (Bray 1983). Please refer to chapters 10 and 11 for further details on resting metabolism and the thermic effect of food and their impact on weight gain and obesity.

Energy output associated with physical activity can vary tremendously. Thus this energy component has been the center of a majority of the research dealing with obesity and its prevention and treatment. Despite some controversy, it has been a popular belief that a reduced level of physical activity leads to the development of obesity. Indeed, Andersen and colleagues (1998) and Gortmaker and colleagues (1996) have demonstrated a positive relationship between time spent viewing television and the incidence of obesity in children. It is important to note that those who are obese tend to expend relatively more energy than leaner people for any given movement (Bray 1983). Consequently, despite reduced physical activity, energy expenditure may not necessarily be less in obese as compared with lean individuals. This raises the question whether, or to what extent, a decrease in physical activity actually contributes to the occurrence of obesity. It has been recently suggested that the impact of energy expenditure on the cause of obesity can vary from individual to individual and can also differ within individuals at different stages of development. Perhaps future studies on the etiology of obesity should examine the impact of energy balance over time using relatively homogeneous groups in terms of age, gender, fitness, and severity of obesity.

> **KEY POINT**
>
> Overweight or obesity can be attributed simply to an energy intake greater than the energy expenditure. However, the etiology is rather complex, as either an increase in energy intake, a decrease in energy expenditure, or a combination of the two can contribute to this positive energy balance. In addition, the impact of the energy imbalance can vary from individual to individual and can differ at different stages of development within the same individual.

Diabetes Mellitus

Diabetes mellitus is defined as abnormally high levels of blood glucose due to the body's inability to manufacture or respond to insulin. Worldwide, 100 to 120 million people have this chronic condition. In the United States, the prevalence currently stands at about 16 million people, nearly half of whom do not yet know they have the disease. The major types of diabetes are insulin-dependent diabetes mellitus (IDDM) and non-insulin-dependent diabetes mellitus (NIDDM). Insulin-dependent diabetes mellitus, also called type 1 diabetes, usually emerges before age 30 and tends to come on suddenly. Non-insulin-dependent diabetes mellitus, also referred to as type 2 diabetes, is far more common than IDDM. It usually starts after age 30, and the majority of those who have the disease are obese. Recently there has been a steady increase in diagnoses of NIDDM in younger people (<30 years of age), particularly among those who are overweight. The onset of NIDDM tends to be more gradual than that of IDDM, and blood glucose levels remain more stable.

Diabetes is regarded as a metabolic disorder in that it impairs the way the body utilizes glucose due to a deficiency of insulin. Each cell needs a regular supply of glucose. The cells absorb glucose from the blood and use some of it immediately for various metabolic functions. The rest of the glucose is converted to glycogen in the liver and muscles and stored there for future use. However, the body's ability to store glycogen is limited, and glucose that is not used immediately or stored as glycogen will be converted to triglycerides stored in adipose tissue. Insulin is the key regulator of glucose in the body. As blood glucose levels rise, for example after a meal, the pancreas produces insulin, which is then transported via the circulation to target organs such as muscles and the liver. The insulin then attaches to sites on the surface of cells called receptors. The binding of insulin to these receptors causes carrier proteins or **glucose transporters** to move from inside the cell to the cell's surface. The glucose transporters travel back and forth across the cell membrane, picking up glucose from the blood and dropping it off inside the cell. In diabetes, insufficient insulin production or a tissue's insensitivity to insulin results in elevated blood glucose, which, if remained uncontrolled, can cause many chronic complications, including cardiovascular diseases, kidney damage, neuropathy, and foot problems.

Glucose transporters (GLUT) are a family of membrane proteins found in most mammalian cells that are responsible for transporting glucose across the cell membrane. Multiple **isoforms** of glucose transporter proteins (i.e., GLUT 1, GLUT 2, GLUT 3, GLUT 4) have been identified; these are distributed differently throughout different body tissues. Much of the research in this area has involved GLUT 4, in part because this isoform is found primarily in skeletal muscle, which is considered a major depot for storing carbohydrate. In addition, unlike the situation with the other glucose transporters, the function of GLUT 4 can be affected by insulin. The working

mechanism of glucose transporters remains hypothetical at this point. It is thought that the binding of insulin to its receptors on the cell membrane triggers a series of events involving second messengers that lead to translocation of GLUT 4 from inside the cell to the cell surface (Brooks et al. 2005). The GLUT 4 proteins then bind with glucose molecules, and the binding provokes a **conformational change** associated with transport and thus releases glucose to the other side of the membrane (Hebert and Carruthers 1992; Cloherty et al. 1995).

Insulin-dependent diabetes mellitus is an autoimmune disease in which the body produces antibodies that attack and damage the pancreatic β-cells. At first, the ability of the β-cells to secrete insulin is merely impaired, but usually within a year or so these cells stop producing or produce little insulin. People with IDDM must inject insulin daily, because the function of their body tissues in responding to insulin remains normal. Although heredity plays some role in IDDM, there is no known family history of diabetes in most cases. Non-insulin-dependent diabetes mellitus, on the other hand, begins with impairment of the body's tissues in responding to insulin. As a consequence, to get the cells the glucose they need, the β-cells must increase their production of insulin. Diabetes results when the β-cells are unable to secrete enough extra insulin to overcome the tissue resistance to insulin. Most people with NIDDM can be treated with oral drugs aimed at improving insulin sensitivity or simply with a lifestyle intervention that promotes weight loss. About 30% to 40% of NIDDM patients need insulin to achieve adequate control of their blood glucose. Heredity plays an important role in NIDDM, and those with NIDDM are highly likely to have at least one relative with diabetes.

> **▶ K E Y P O I N T ◀**
>
> Both IDDM and NIDDM are metabolic disorders in that the body's ability to regulate blood glucose concentration is impaired, and both are accompanied by hyperglycemia. However, the etiology of the impairment differs in IDDM and NIDDM. Insulin-dependent diabetes is associated with a lack of insulin, a hormone required to move glucose from the blood into the cell. Non-insulin-dependent diabetes involves a reduced ability of tissues to respond to insulin; that is, despite the presence of insulin, little blood glucose is being transported into cells.

INSULIN RESISTANCE

Many individuals with obesity and diabetes mellitus can be characterized as having insulin resistance. As discussed in the preceding chapters, insulin is an anabolic hormone that promotes the synthesis of glycogen and triglycerides. Insulin resistance is a decreased ability of insulin to stimulate cellular glucose uptake and storage and to suppress hepatic glucose production. This condition is also associated with a reduced ability of insulin to suppress fat mobilization and thus with increased levels of circulating fatty acids. Insulin resistance can occur in individuals without NIDDM, but in most cases these people are obese. It has been suggested that insulin resistance can be influenced by many factors including obesity, physical activity, and dietary composition (Walker 1995).

Testing for Insulin Resistance

Insulin resistance can be assessed with the use of an oral glucose tolerance test. In this test, 75 to 100 g of glucose in water is given orally to a fasting subject. Blood levels of glucose and insulin are subsequently measured at intervals for 2 to 3 h. During the measurement period, the blood level of glucose rises initially and then falls due to the action of insulin (figure 9.1). The response curve for the blood level of insulin generally follows a similar but lagging time course. Insulin resistance is therefore judged from the insulin response compared to the glucose response (Reaven and Miller 1968). Those who demonstrate a high insulin response in the face of a normal or high glucose response are considered insensitive to insulin. The simplicity of the technique is an advantage, but a weakness is that the results are difficult to interpret because blood glucose concentration depends not only on the insulin sensitivity of the liver and peripheral tissues but also on many other factors such as glucose absorption, insulin secretion, and insulin clearance. In this test, concentrations of glucose and insulin are interrelated, and changes in one variable can result in changes in the other. Thus, an effect of insulin on glucose metabolism cannot be determined, at least in a sequential manner.

The shortcomings associated with the oral glucose tolerance test have promoted the development of a more precise but invasive technique called the **euglycemic, hyperinsulinemic** glucose clamp (DeFronzo et al. 1979). With this technique, the blood glucose concentration is kept constant by

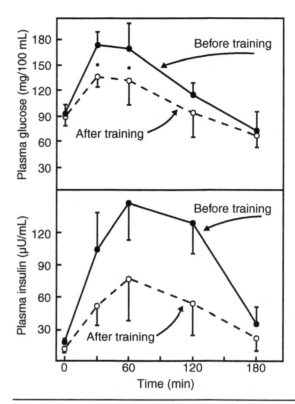

Figure 9.1 Sample plasma glucose and insulin responses during a 3 h oral glucose tolerance test before and after aerobic training.

Reprinted, by permission, from J.O. Holloszy et al., 1986, "Effects of exercise on glucose tolerance and insulin resistance," *Acta Medica Scandinavica* 711(Suppl.): 55-65.

glucose infusion that is regulated according to repeated, rapid blood glucose measurements. The blood insulin concentration is initially raised and then held constant via a prime-continuous infusion of insulin. Under these steady-state conditions of euglycemia and hyperinsulinemia, the glucose infusion rate equals glucose disposal or uptake by the cell, which can then be used to determine the severity of insulin resistance. This technique, if conducted in conjunction with indirect calorimetry or isotopic tracing technique, can also allow one to partition the amount of glucose taken up by tissues into the amount being oxidized and the amount being stored. Such information is valuable for exploring the cellular mechanisms that account for insulin resistance. The glucose clamp technique has been widely used in most clinical studies aimed at investigating glucose metabolism at a given insulin concentration. With the use of multiple levels of insulin concentration, it becomes possible to generate a dose–response relation between the insulin concentration and its effect on glucose disposal, which allows for more complete examination of insulin action (figure 9.2). Two terms have been derived as a result of this analytic approach: **insulin sensitivity** and **insulin responsiveness.** Increased insulin sensitivity is defined as a reduction in the insulin concentration that produces half of the maximal response, whereas increased insulin responsiveness is defined as an increase in the maximal response to insulin.

Insulin Resistance and Body Fat Distribution

Insulin resistance is associated not only with the overall accumulation of fat in the body, but also with how the body fat is distributed. Considerable evidence suggests that excess accumulation of fat in the upper body, or truncal region, is a strong predictor of insulin resistance. For example, Banerji (1995) observed that variance in **visceral** adiposity accounted for much of the interindividual variation in insulin resistance among individuals with type 2 diabetes. In addition, a weight loss intervention study conducted by Goodpaster and colleagues (1999) revealed that among nondiabetic obese subjects, a decrease in visceral adiposity was the body composition change that best predicted improvement in insulin sensitivity after weight loss.

Distinctive roles of different patterns of fat distribution are also supported by several in vitro studies

on the metabolic heterogeneity of adipose tissue (Richelsen et al. 1991; Jansson et al. 1990; Lafontan et al. 1979). The general experimental approach of these studies was to isolate adipose tissue from abdominal and lower body **subcutaneous** regions so that the lipolytic activity of adipose tissue could be compared between the two regions. Collectively, these studies revealed that adipose tissue from the abdominal region, particularly within visceral compartments, is metabolically more active and has a greater tendency to be broken down into free fatty acids. As free fatty acids formed due to lipolysis in this central region are directly released into the **portal circulation,** it is thought that in persons with central obesity, the liver may have been exposed to high concentrations of free fatty acids, which can ultimately decrease hepatic insulin sensitivity (Ferrannini et al. 1983). An excess of fatty acids in the systemic circulation derived from the abdominal region can also inhibit skeletal muscle glucose metabolism, according to Randle and colleagues (1963), and this has been considered a cause of insulin resistance manifested in the peripheral region such as in skeletal muscle.

> **▶ K E Y   P O I N T ◀**
>
> Many people with obesity and NIDDM have insulin resistance. Insulin resistance is defined as a decreased ability of insulin to stimulate cellular glucose uptake and storage and to suppress hepatic glucose production. Insulin resistance is associated not only with an overall level of body fat but also with body fat distribution. Obese individuals whose body fat is distributed primarily in the abdominal region are most prone to insulin resistance, which then leads to NIDDM.

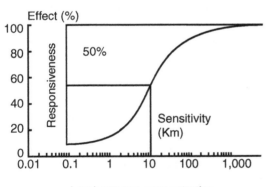

Dose-response relation
Sensitivity and responsiveness

Figure 9.2 Hypothetical dose–response relation between hormone concentration and its biological effect. Insulin responsiveness is the maximal response of glucose disposal. Insulin sensitivity is the hormone concentration eliciting half of the maximal response.

Adapted, by permission, from K.J. Milkines, 1992, "Influence of physical activity and inactivity on insulin action and secretion in man," *Acta Physiologica Scandinavica* 609(Suppl.): 1-43.

Subcutaneous adipose tissue in the legs is generally regarded as a relatively weak marker of insulin resistance, although considerable adipose tissue is found in this region. However, a growing number of researchers have examined the impact of peripheral adiposity on insulin resistance. Goodpaster and colleagues (2000) used computed tomography imaging to measure the quantity and distribution of adipose tissue in the thigh. Via the novel approach of subdividing adipose tissue into that present above the fascia lata (termed subcutaneous adipose tissue) and that present below the fascia lata (termed subfascial adipose tissue), the authors observed that variance in the amount of adipose tissue beneath muscle fascia correlated with insulin resistance, whereas no correlation was found between insulin sensitivity and the subcutaneous adiposity of the legs. These findings suggest that the amount of fat beneath the fascia, as well as within the muscle tissue in the lower extremities, is a key determinant of insulin resistance.

Glucose and Fat Utilization

Numerous researchers have examined the mechanisms underlying the impairment in insulin-mediated glucose utilization seen in people who are obese and people with NIDDM. These studies generally used the glucose clamp technique in conjunction with muscle sampling, indirect calorimetry, or the isotope tracer method (or some combination of these) so that the metabolic fate of glucose taken in by skeletal muscle could be further divided into glucose oxidation, glucose storage, and non-oxidized glycolysis. This experimental approach allows examination of the mechanisms responsible for insulin resistance at the cellular level. DeFronzo and colleagues (1985) found an approximately 45% reduction in insulin-stimulated leg glucose uptake in nonobese diabetic subjects. Using obese diabetic subjects, Kelley and colleagues (1992) observed a 60% decrement in insulin-stimulated leg glucose uptake. They further estimated that of this deficit in glucose uptake, 66% was due to decreased leg glucose storage, whereas 33% was due to decreased leg glucose oxidation. This finding, together with the fact that these patients had lower than normal glycogen synthase activity (Kelley et al. 1992), suggests that reduced insulin-mediated glucose uptake can be attributed mainly to decreased leg glucose storage. It is now widely believed that glycogen synthesis is the metabolic pathway in skeletal muscle most severely affected by insulin resistance and is primarily responsible for decreased rates of glucose utilization. From a practical standpoint, individuals with obesity, NIDDM, or both are at a disadvantage with regard to physical activity due to insulin resistance and the associated reduction in glycogen storage. These individuals should avoid performing sustained strenuous exercise that depends heavily on carbohydrate as a source of energy.

Reduced muscle glycogen content leads to an increase in fat utilization for the maintenance of an adequate energy supply. This homeostatic adaptation has been well documented in healthy individuals (Henriksson 1995). However, recent studies have suggested that skeletal muscle in people with insulin resistance is unable to switch easily from carbohydrate to fat utilization (Kelley 2005). In other words, as a result of insulin resistance, in addition to impaired glucose metabolism, fat utilization is lower. This conclusion was derived from observations that during fasting conditions, respiratory quotient (RQ) across the tissue bed of the leg was comparatively higher in people who were obese and insulin resistant than in metabolically healthy individuals. As discussed in chapter 4, a high RQ indicates a greater reliance on carbohydrate oxidation. Recently, Ukropcova and colleagues (2005) found that fat oxidation of skeletal muscle increased in subjects with improved insulin sensitivity, leanness, and aerobic fitness. An inability to increase the reliance on fat oxidation in the face of reduced muscle glycogen has been termed **metabolic inflexibility.** As shown in figure 9.3, healthy individuals use fat as the predominant source of energy during fasting but are able to shift efficiently to glucose oxidation upon insulin stimulation. However, those with obesity, NIDDM, or both demonstrate a constrained adjustment to the transition between fasting and insulin stimulation conditions. Their metabolic responses are blunted in terms of fat utilization during fasting and of carbohydrate utilization upon insulin stimulation. The phenomenon of metabolic inflexibility emphasizes the importance of physical activity as a part of interventions for treating obesity and NIDDM. As discussed in chapter 7, regular aerobic exercise training increases the ability of skeletal muscle to oxidize fat and thus delay the consumption of glycogen.

> ### ▶ KEY POINT ◀
>
> Recently, insulin resistance has been associated with impaired fat utilization. In healthy individuals, reduced muscle glycogen content results in an increase in fat utilization. However, the skeletal muscle in those with insulin resistance is unable to switch easily from carbohydrate to fat utilization. This condition is also known as metabolic inflexibility.

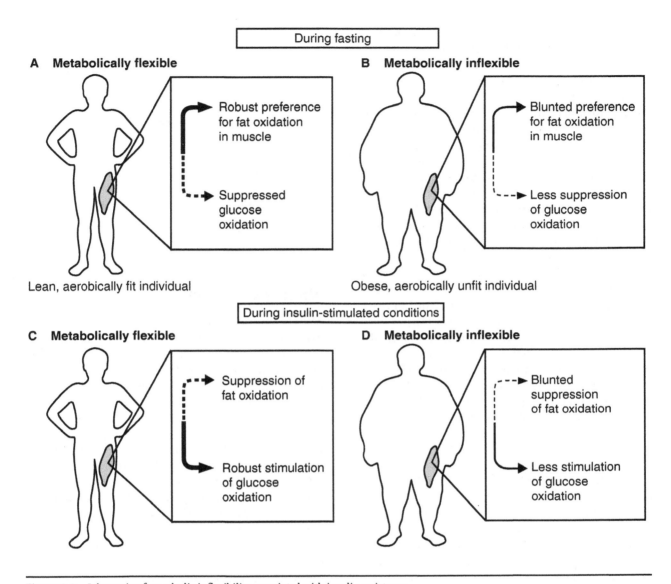

Figure 9.3 Schematic of metabolic inflexibility associated with insulin resistance.

Reprinted, by permission, from D.E. Kelley, 2005, "Skeletal muscle fat oxidation: Timing and flexibility are everything," *Journal of Clinical Investigation* 115(7): 1699-1702.

ALTERATIONS IN METABOLISM DURING EXERCISE IN OBESITY AND DIABETES

Exercise has long been regarded as beneficial in the treatment of obesity and diabetes. However, not until the last several decades has the interaction between exercise and these metabolic disorders been extensively studied. Research in this area has provided insight into the unique characteristics of exercise responses related to these metabolic disorders. This information will ultimately help in developing effective exercise programs aimed at treating or preventing obesity and diabetes.

Blood Glucose

In early investigations that used patients with IDDM, it was generally found that a bout of aerobic exercise caused a fall in blood glucose to normal levels if the patients had been treated regularly with insulin and had mild hyperglycemia (Lawrence 1926; Barringer 1916). This observation suggests that regular exercise training may reduce the exogenous insulin requirement. Plasma glucose concentration reflects a balance between glucose uptake by peripheral tissues, mostly muscles, and glucose production by the liver. Wahren and colleagues (1975) found that in insulin-withdrawn diabetic subjects, muscle glucose uptake was greater during exercise while hepatic glucose production was similar to that in nondiabetic controls. According to the authors, the greater glucose uptake seen in IDDM was primarily driven by the mass action of hyperglycemia: More glucose can rush into tissue due to a greater concentration gradient. In these patients, it is the lack of insulin that prevents their tissues from successfully drawing glucose from the blood. However, muscle contraction can allow this favorable response to occur. The glucose-lowering effect of exercise, however, was not always observed, especially in patients with more severe IDDM. Patients with more marked hyperglycemia may respond to exercise with a further rise in blood glucose levels (Wahren et al. 1978).

Within the last several decades, considerable effort has also been devoted to examining patterns of glucose turnover and its impact on blood glucose levels during exercise in patients with NIDDM. The favorable glycemic response seen in IDDM was also observed in NIDDM. Additionally, the mechanism underlying this response appears to be similar. Researchers have observed that in NIDDM, a fall in blood glucose during exercise is accompanied by a greater increased peripheral glucose uptake, whereas hepatic glucose production remains the same between diabetic subjects and nondiabetic controls (Colberg et al. 1996; Kang et al. 1999; Martin et al. 1995). As already discussed, exercise enhances glucose uptake, thereby reducing the level of hyperglycemia in IDDM. However, whereas exercise can augment glucose uptake to a greater extent in those with NIDDM than in nondiabetic controls, this has not been observed for IDDM. Non-insulin-dependent diabetes is characterized by marked insulin resistance in skeletal muscle. The greater increase in glucose uptake seen in NIDDM suggests that insulin resistance does not substantially impede the cellular uptake of blood glucose during exercise. In fact, it seems possible that muscle contraction abetted by hyperglycemia and hyperinsulinemia can have an additive or synergistic effect on glucose uptake. This contention is underscored by the findings of DeFronzo and colleagues (1981), who reported that exercise in combination with experimentally induced hyperinsulinemia produced glucose uptake greater than that following either treatment alone in nondiabetic individuals.

> ### ▶ KEY POINT ◀
>
> Aerobic exercise helps reduce blood glucose levels in people with diabetes. This effect of exercise can occur with or without insulin. In NIDDM associated with hyperinsulinemia and hyperglycemia, the exercise-induced reduction in blood glucose can be greater than in nondiabetics. For both IDDM and NIDDM, an exercise program should be accompanied by modifications in the diet as well as medication or insulin to prevent an overreduction in blood glucose concentration or hypoglycemia.

Muscle Glycogen

Muscle glycogen is a major energy depot during exercise, and its utilization increases as exercise intensity increases. Earlier studies using subjects with IDDM showed that the rates of glycogen utilization during exercise were no different in people with diabetes than in nondiabetic controls (Saltin et al. 1979; Maehlum et al. 1977; Roch-Norland 1972). Furthermore, the glycogen depletion pattern in the different fiber types during exercise was also similar in people with diabetes and nondiabetic controls (Saltin et al. 1979). However, there is some indirect evidence to suggest that patients with IDDM may use less muscle glycogen. This contention was derived from the observation that people with diabetes had reduced rates of total carbohydrate oxidation concomitant with increased rates of muscle glucose uptake, and was based on the assumption that all glucose molecules taken up by muscle are oxidized. Resynthesis of muscle glycogen during postexercise recovery is an insulin-dependent process. Maehlum and colleagues (1977) found that in the absence of insulin injection, muscle glycogen repletion during recovery from exercise was minimal in diabetic patients, while with insulin, the rate of repletion was the same as in nondiabetic subjects.

The reduced ability of skeletal muscle to use glycogen during exercise has been more uniformly reported in patients with NIDDM or obesity (Colberg et al. 1996; Kang et al. 1999; Goodpaster et al. 2002). By having three groups (healthy, obese, and NIDDM) exercise at a mild intensity, Colberg and colleagues (1996) found that while utilization of glycogen was lower in both the obese and the NIDDM groups, it was only half as much in the NIDDM group as in the healthy controls. This finding was confirmed by Kang and colleagues (1999) and Goodpaster and colleagues (2002) using patients with NIDDM and obesity, respectively. To date, the cause of this reduced glycogen utilization remains unclear. In these studies, the researchers determined muscle glycogen indirectly by subtracting the rate of glucose uptake from the rate of total carbohydrate oxidation, and a decrease in muscle glycogen utilization was accompanied by an increase in plasma glucose uptake. In this context, it is possible that the reduced utilization of muscle glycogen may be secondary to compensatory responses resulting from greater glucose utilization. Both NIDDM and obesity have been associated with lower muscle glycogen content. Therefore, it is also likely that a reduced utilization of muscle glycogen during exercise is brought about by lower muscle glycogen content prior to exercise (Sherman et al. 1981).

Fatty Acids and Triglycerides

Circulating fatty acids and intramuscular triglycerides are the two major sources of fat energy utilized during exercise. Studies in exercising humans have provided estimates that intramuscular triglycerides generally contribute more to the total fat oxidation than circulating fatty acids. However, circulating fatty acids become more important oxidative fuels during prolonged exercise. The relationship between intramuscular triglycerides and circulating fatty acids appears to resemble that between muscle glycogen and plasma glucose; that is, while the intramuscular substrate stores are relatively more important at the start of work, fuels supplied via the blood are the predominant substrates when work is prolonged. In patients with IDDM in whom insulin is completely absent, Wahren and colleagues (1984) found a greater increase in fat oxidation during exercise as compared to that in healthy controls. The authors also observed a more exaggerated exercise-induced rise in plasma norepinephrine. Norepinephrine is a lipolytic hormone that helps mobilize fatty acids from adipose tissue. It is thought that the increased fat oxidation seen in patients with IDDM may result from increased lipolysis. The utilization of intramuscular sources of triglycerides was also found to be higher in the diabetic state. This finding was reported in studies using IDDM patients and depancreatized dogs (Standl et al. 1980; Issekutz and Paul 1968).

More recently, greater fat oxidation was also reported during exercise in patients with NIDDM or obesity (Goodpaster et al. 2002; Horowitz et al. 2000; Blaak et al. 2000). However, this increased fat oxidation can be explained only by an increased oxidation of intramuscular triglycerides, because oxidation of blood-borne fatty acids was found to be either the same or lower in patients with NIDDM or obesity as compared to healthy controls. The greater utilization of intramuscular sources of fat has been attributed to an increased accumulation of intramuscular triglycerides frequently found in these patients during the **postabsorptive state** and considered a cause of insulin resistance. The postabsorptive state refers to the period during which the gastrointestinal tract (e.g., stomach and small intestine) is empty of nutrients and body stores must be used to supply required energy. It appears that patients with NIDDM or obesity are generally not limited in their ability to oxidize fatty acids during exercise, although they tend to accumulate excessive intramuscular triglycerides. Given that intramuscular triglycerides have been related to insulin resistance, exercise is considered necessary, and ideal, for these patients in that it can help stimulate greater fat utilization and may help alleviate or prevent insulin resistance. Table 9.1 provides a summary of altered carbohydrate and fat utilization in patients with IDDM and NIDDM.

 KEY POINT

Insulin resistance is associated with an increased accumulation of intramuscular triglycerides, suggesting impaired fat utilization at rest. However, people with insulin resistance are not limited in their ability to use fat as an energy source during exercise. Therefore, exercise is considered ideal in that it can help stimulate greater fat utilization, which may alleviate insulin resistance.

ROLE OF EXERCISE IN IMPROVING INSULIN SENSITIVITY

Exercise can produce many favorable responses with respect to carbohydrate and fat metabolism. For example, glucose uptake by peripheral tissues, such as muscle, increases during exercise. Such an increase in glucose uptake helps improve hyperglycemia. Exercise also allows a greater utilization

Table 9.1   Substrate Utilization During Aerobic Exercise in Patients With IDDM and NIDDM as Compared to Nondiabetic Controls

Substrate	IDDM	NIDDM
Carbohydrate		
Muscle glycogen	Same	Decreased
Plasma glucose	Increased	Increased
Fat		
Muscle triglycerides	Increased	Increased
Plasma fatty acids	Increased	Decreased/same

of fatty acids and thus is considered a quite viable option for reversing insulin resistance. Because of these beneficial patterns of metabolism, which clearly suggest the potential of exercise in treating or preventing metabolic diseases, another growing line of research has focused on the role of exercise in reversing diabetes- and obesity-related symptoms such as insulin resistance. Unlike the studies discussed in the preceding sections, the scientific quest in this regard has generally involved examination of insulin sensitivity or insulin-mediated glucose disposal sometime after the cessation of a single bout of exercise or a period of regular exercise training. Much of the pertinent literature is related to NIDDM and obesity because insulin resistance is the hallmark of these metabolic disorders. Part of the reason for this emphasis may be that previous studies generally failed to demonstrate significant improvements in glycemic control assessed by glycosylated hemoglobin (HbA_{1c}) and fasting glucose concentration in IDDM after physical training (Wallberg-Henriksson et al. 1982; Zinman et al. 1984). For patients with IDDM, the emphasis should be on devising regimens that will allow them to exercise regularly and safely to avoid hypoglycemia and hyperglycemia, and not on promoting exercise as a means of improved glycemic control.

The impact of aerobic exercise on subsequent insulin sensitivity or insulin-mediated glucose disposal has been typically examined using the euglycemic, hyperinsulinemic glucose clamp. This procedure is often performed during the recovery period so that the acute effect of prior exercise can be examined, although in some studies it was performed one or two days later. With the selection of multiple levels of insulin to be infused, a dose–response curve can be generated between insulin concentrations and the resultant rates of glucose disposal in order to assess insulin sensitivity. It has been well documented that in healthy humans, insulin-mediated whole-body glucose disposal is

increased following an acute bout of exercise and that this improved insulin action continues for as long as 48 h (Mikines et al. 1988; Richter et al. 1989). The improved insulin action following an acute bout of exercise has also been demonstrated in patients with NIDDM or obesity (Devlin and Horton 1985; Devlin et al. 1987; Burstein et al. 1990). These findings clearly support the use of aerobic exercise in treating insulin resistance.

In some cases these studies used the clamp technique combined with indirect calorimetry and the isotopic tracer technique in an effort to delineate the metabolic fate of enhanced glucose uptake (Devlin and Horton 1985; Devlin et al. 1987). Results showed that the exercise-induced improvement in insulin-mediated glucose disposal can be largely explained by an increase in glucose storage as glycogen. A handful of studies of animals and humans also revealed an increase in glycogen **synthase** and hexokinase during exercise recovery. Overall, a plausible conclusion is that glycogen formation is an ultimate destination for the augmented glucose taken up by muscle after exercise.

The improved insulin action following exercise can also be ascribed to augmented glucose transport across the cell membrane. Glucose transport into tissues is achieved by the action of protein molecules called glucose transporters. As mentioned earlier, a number of different glucose transport proteins have been identified and are manifested in a variety of tissues. GLUT 4 is the form of transporter found in skeletal muscle and adipose tissues. In a series of experiments using the hindquarter of diabetic rats, Richter and colleagues (1982, 1984, 1985) found an acute increase in glucose transport following exercise. These investigators determined glucose transport by quantifying how much of the specially marked glucose (i.e., 3-O-methylglucose) entered the muscle cell. Ren and colleagues (1994) also observed an increase in GLUT 4 expression, along with improved insulin-stimulated glycogen storage, in rat muscle following prolonged swimming. It appears that the improvement in insulin action occurs primarily in exercising muscle. Since exercise is also accompanied by major cardiovascular and hormonal changes, it is possible that insulin action is also affected in extramuscular tissues. It has been recently shown that the liver, like muscle, becomes more insulin sensitive after exercise (Pencek et al. 2003). Similarly to what occurs with muscle, a major portion of glucose taken up by the liver is channeled into glycogen synthesis (Hamilton et al. 1996).

As glycogen synthesis is a major metabolic fate for enhanced glucose uptake, one may speculate that the greater the depletion of glycogen during prior exercise, the greater the improvement in insulin sensitivity following exercise. However, this hypothesis remains questionable. It has been found that an improved insulin effect on glucose uptake can persist even when preexercise glycogen levels have been restored (Hamilton et al. 1996; Richter 1996). This finding suggests that the exercise benefit is not necessarily dependent on glycogen depletion. On the other hand, several lines of research appear to support the hypothesis. For example, Bogardus and colleagues (1983) found an inverse relation between insulin sensitivity and muscle glycogen content after a single bout of exercise. Ivy and colleagues (1985) observed a greater reduction in plasma insulin response to oral glucose challenge at lower concentrations of muscle glycogen. Using a glucose tolerance test, we also reported an improved insulin action following exercise at 70% but not 50% $\dot{V}O_2$max (Kang et al. 1996). It may be safe to suggest that if an exercise program is prescribed with the goal of improving insulin sensitivity, the program should entail at least some component of vigorous exercise.

Physical training consists of repeated bouts of acute exercise. With the use of an oral glucose tolerance test, studies comparing trained and untrained individuals have shown that trained individuals are better able to tolerate glucose and are more sensitive to insulin (King et al. 1987; Heath et al. 1983; Seals et al. 1984). This benefit associated with training was also shown in studies using the euglycemic, hyperinsulinemic glucose clamp technique, although it is of interest that these studies demonstrated an improved insulin responsiveness, but not improved insulin sensitivity (Mikines et al. 1989a, 1989b). Remember that insulin responsiveness is the maximal ability of the whole body to handle glucose; thus an improvement in this measure may result from wider changes that occur not only in skeletal muscle, but also in liver, adipose tissue, and the pancreas. Exercise training has been found to increase muscle GLUT 4 expression (Lee et al. 2002; Terada et al. 2001). This increase in GLUT 4 may have contributed to the enhanced capacity for insulin-stimulated glucose disposal in trained subjects. The training-induced improvement in insulin action may also be attributable to the fact that trained subjects are able to utilize more of their fat energy sources. This augmented fat utilization may be in part explained by increased lipolysis. It has been found that training can make adipose tissue more sensitive to adrenergic stimulation (Izawa et al. 1991). As discussed earlier, insulin resistance is linked to abdominal obesity as well as to an excess accumulation of intramuscular fatty acids. Training has also been shown to result in a decrease in the mRNA (messenger ribonucleic acid) for **proinsulin** and **glucokinase** in the pancreas (Koranyi et al. 1991). A reduction in these protein molecules suggests decreased insulin secretion at a given level of blood glucose concentration.

> **KEY POINT** ◄

Research has shown that a greater reduction in muscle glycogen following an acute bout of exercise can produce a greater improvement in insulin sensitivity. In this context, it may be safe to suggest that if an exercise program is prescribed with the goal of improving insulin sensitivity, it should entail at least some component of vigorous exercise so that utilization of muscle glycogen will be greater.

> **KEY POINT** ◄

Regular physical activity has proven beneficial in improving insulin sensitivity. This can be explained by cellular changes including increased glucose transport and activity of enzymes that are responsible for glycogen synthesis. The improved insulin sensitivity is observed not only in skeletal muscle, but also in the liver and adipose tissue. These positive changes justify the use of exercise as part of therapy in treating insulin resistance associated with obesity and NIDDM.

SUMMARY

Obesity and diabetes are two major metabolic disorders that cause serious health concerns in today's society. These disorders are commonly associated with an impaired ability of the body to regulate blood glucose concentration. The failure to modulate blood glucose seen in individuals with obesity or NIDDM is due primarily to a reduced tissue response to insulin. On the other hand, hyperglycemia seen in IDDM is related to a complete absence of pancreatic secretion of insulin. Being overweight or obese is often accompanied by insulin resistance, which can then lead to NIDDM. In this regard, obesity and NIDDM share many of the same cellular changes associated with insulin resistance. These cellular changes include reductions in glucose transport, glycogen synthesis, and lipid utilization in response to a given level of insulin.

Muscle contraction has insulin-like effect in facilitating glucose uptake, thereby lowering blood glucose concentration. As such, regular physical activity is recommended as a therapeutic modality in treating these metabolic disorders. Indeed, both short- and long-term exercise have proven beneficial in improving insulin sensitivity. Among the exercise-induced adaptations that may contribute to improved insulin-mediated glucose disposal are increases in glucose transporter proteins as well in enzymes responsible for glycogen synthesis. Though regular exercise cannot change pancreatic function so that the pancreas releases insulin in patients with IDDM, regular physical activity should still be considered a part of their daily routine in order to provide other exercise-related benefits such as maintaining cardiorespiratory fitness and muscular strength and endurance.

KEY TERMS

comorbidities

conformational change

euglycemic

glucokinase

glucose transporters

hyperinsulinemic

insulin responsiveness

insulin sensitivity

isoforms

metabolic inflexibility

portal circulation

postabsorptive state

proinsulin

subcutaneous

synthase

visceral

REVIEW QUESTIONS

1. What are the differences between insulin-dependent diabetes mellitus (IDDM) and non-insulin-dependent mellitus (NIDDM)?

2. Define the terms *insulin sensitivity* and *insulin responsiveness*. How can these parameters be determined?

3. Insulin resistance can be assessed by the glucose tolerance test or the euglycemic, hyperinsulinemic glucose clamp test. Provide a brief explanation of these procedures as well as the advantages and disadvantages associated with each test.

4. Those with central obesity often manifest insulin resistance. Why is there a close relationship between insulin resistance and visceral adiposity?

5. Explain how a bout of exercise may cause a fall in blood glucose concentration in patients with IDDM or NIDDM. Why is this glucose-lowering effect of exercise often more marked in NIDDM than IDDM?

6. How does an improvement in insulin sensitivity come about following training?

IV

INFLUENCES ON BIOENERGETICS APART FROM PHYSICAL ACTIVITY

Energy expenditure results from cellular oxidation of stored energy such as fat. Hence, maximizing energy expenditure is an integral part of efforts to lose weight. Energy expenditure can be further ascribed to resting metabolism, thermogenesis of food, and physical activities; among these, resting metabolism is the greatest contributor. Energy expenditure can also be influenced by pharmacological and nutritional substances, and this has given rise to a great deal of public interest in these types of substances as an alternative resource for weight loss.

Part IV deals with overweight and obesity and strategies for maximizing energy expenditure to combat these metabolic disorders, although the focus is on energy components other than physical activity. Chapters 10 and 11 discuss the resting metabolic rate and the thermic effect of food, respectively, and their roles in the pathogenesis of obesity. Chapter 12 offers an objective evaluation of selected nutritional and pharmacologic substances and the roles they play in modulating energy expenditure.

RESTING METABOLIC RATE

This chapter provides an in-depth discussion on the resting metabolic rate and its relation to total daily energy expenditure. We focus particularly on (a) what the resting metabolic rate is, (b) how it is determined, (c) factors that can affect its magnitude, and (d) how it may respond to interventions such as diet and exercise. Given that resting metabolic rate constitutes a majority of the total daily energy expenditure, its role in the pathogenesis of obesity is also discussed.

GENERAL DESCRIPTION OF RESTING METABOLIC RATE

Human metabolism is the total collection of chemical reactions that occur within the organism. Some of these reactions result in the breakdown of organic molecules like carbohydrate, fat, and protein; these reactions are referred to as catabolism. Most catabolic reactions are accompanied by the release of chemical energy, some of which is used by cells to perform various daily functions. Consequently, energy stored in the body decreases. In contrast, other reactions combine small molecules to form larger ones that a cell uses as structural elements or to perform specific functions. This overall synthesis of organic molecules by cells is known as anabolism. Anabolic reactions are associated with a gain in potential energy. However, energy is often a prerequisite for these reactions to occur. The total daily energy expenditure is viewed as the amount of energy used during both anabolic and catabolic processes throughout a 24 h period. This energy index represents the overall utilization of the potential energy that the body possesses. It is often used for comparison against the total energy intake, so that a net gain or loss of energy can be identified over a given period of time. The total daily energy expenditure

can be further divided into (1) resting or basal energy expenditure (resting metabolic rate, RMR), (2) thermic effect of food (TEF), and (3) energy expenditure during physical activities. The preceding chapters presented extensive information on energy expenditure during physical activities. Hence, this chapter and chapter 11 focus specifically on issues related RMR and TEF, respectively.

French chemist Antoine Lavoisier may be considered a founder of the concept of RMR. As mentioned in chapter 4, the studies of respiratory combustion that Lavoisier performed more than 200 years ago served as the foundation for the methodology of indirect calorimetry that has been used to determine human energy expenditure ever since. Lavoisier showed that animal respiration is the combination of oxygen from air with carbon and hydrogen from the animal's body (which he recognized came from food) to produce water, carbon dioxide, and heat (Lutz 2002). He also demonstrated that the rate of oxygen consumption was influenced by (1) the consumption of food, (2) environmental temperature, and (3) the performance of muscular work. Of particular pertinence to RMR is that Lavoisier attempted to measure the minimal metabolic rate in a resting and postabsorptive state. This measurement may be the first scientific effort to determine resting metabolic rate, although the metabolic rate determined by Lavoisier has been referred to as the basal metabolic rate (BMR) rather than resting metabolic rate, or RMR (Blaxter 1989).

Basal metabolic rate may be viewed as a minimal rate of metabolism necessary to sustain life. It represents the energy requirements of a variety of cellular events that are essential to the life of an organism. It is important to recognize that in most instances, so-called basal values measured under controlled laboratory settings always fall slightly below values for RMR typically measured 3 to 4 h after a light meal without prior physical activity. In this context, RMR can be viewed as BMR plus energy expenditure associated with prior eating, muscular activity, or both. It is well acknowledged that the two parameters differ by less than 10% (Williams 2005). For this reason, RMR is often used in lieu of BMR. The determination of BMR requires the subject to fast for 12 h and to remain in a reclined position. In addition, the measurement should take place in a well-controlled environment that produces minimal or no stresses or disturbances. The RMR is generally greater than the BMR and is considered the sum of the BMR and the additional energy expenditure associated with whatever cellular events occur in a resting and postabsorptive state.

▶ KEY POINT ◀

The total energy expenditure on any given day can be divided into resting energy expenditure (RMR), the thermic effect of food, and the energy expenditure during physical activities. The RMR accounts for about 60% to 75% of the total daily energy expenditure. It is because of this large fraction that RMR has drawn a great deal of attention with regard to its role in mediating weight gain and obesity.

MEASUREMENT OF RESTING METABOLIC RATE

The measurement of RMR is an important part of the examination of human bioenergetics. Quantitatively, it accounts for about 60% to 75% of the total daily energy expenditure. For this reason, this energy component has drawn a great deal of attention and has often been treated as a major player in contributing to metabolic disorders associated with energy imbalance. Knowledge of RMR has helped in establishing an important energy baseline for the construction of sound programs of weight control through dietary restriction, physical activity, or a combination of the two. In many exercise intervention studies, RMR has been used as a major dependent variable expected to increase due to an exercise-induced increase in lean body mass. On average, RMR has been estimated to be about 1680 $kcal \cdot day^{-1}$ for men and 1340 $kcal \cdot day^{-1}$ for women. This gender difference can be largely attributed to the **thermogenesis** that takes place in metabolically active organs such as the brain, liver, muscle, kidney, and heart (Elia 1992). Table 10.1 illustrates the contribution of various organs and tissues to body weight and BMR. It is interesting to note that approximately 60% of the resting energy expenditure of the body arises from organs such as the liver, kidney, heart, and brain, although these organs together account for only about

Table 10.1 Relative Contribution of Organs and Tissues to Body Weight and BMR in an Average Man and Woman

Organs	Weight (kg)	Weight (% body weight)	Metabolic rate (kcal · kg⁻¹ · day⁻¹)	Metabolic rate (% total kcal)
		Man		
Liver	1.8	2.57	200	21
Brain	1.4	2.00	240	20
Heart	0.33	0.47	440	9
Kidneys	0.31	0.44	440	8
Muscle	28.00	40.00	13	22
Adipose tissue	15.00	21.43	4.5	4
Other (bones, skin, intestines, glands)	23.16	33.09	12	16
		Woman		
Liver	1.4	2.41	200	21
Brain	1.2	2.07	240	21
Heart	0.24	0.41	440	8
Kidneys	0.28	0.47	440	9
Muscle	17.00	29.31	13	16
Adipose tissue	19.00	32.75	4.5	6
Other (bones, skin, intestines, glands)	18.89	32.58	12	19

Adapted, by permission, from M. Elia, 1992, Organ and tissue contribution to metabolic rate. In *Energy metabolism- Tissue determinants and cellular corollaries*, edited by J.M. Kinney and H.N. Tucker (New York, NY: Raven Press), 61-77.

5% to 6% of body weight. Muscle is the largest tissue component in the body, accounting for about 40% of adult body weight. However, its metabolic rate relative to tissue mass is rather low. The relative contribution of muscle tissue to total energy expenditure ranges from 20% to 25%.

Resting metabolic rate is not measured very often, partly because RMR can be easily influenced by many factors, such as age, gender, body size, body composition, hormonal level, and prior exercise, as well as environmental temperature. In addition, RMR is only a fraction of the total energy expenditure. A change in RMR could be offset by an opposite change in another energy component, so that it may not necessarily be associated with a corresponding change in the total energy expenditure. Accurate determination of RMR may be important in a clinical setting with patients who are obese because such an effort can at least identify whether a lower than normal basal metabolism contributes to the obesity.

A simple way to estimate resting metabolic rate is to use the factor of 1 kcal per kilogram body weight per hour. For example, for a 70 kg (154 lb) male, his daily RMR will be 1680 kcal (i.e., 1 kcal × 70 kg × 24 h). This factor may be reduced from 1 to 0.9 kcal · kg^{-1} · h^{-1} for use in women given that the average BMR is about 10% lower in women than in men. This method is convenient, but it does not discriminate differences related to age or body composition. To eliminate this drawback, a revised equation was developed that allows estimation of RMR from fat-free mass (FFM) in kilograms (McArdle et al. 2001):

$$RMR = 370 + 21.6(FFM).$$

For example, a male who weighs 70 kg (154 lb) and has 20% body fat has an FFM of 56 kg (123 lb). With use of the equation, his estimated daily RMR is 1580 kcal (i.e., 370 + 21.6 × 56 = 370 + 1209.6 = 1579.6). This equation was developed based on studies of a mixed sample of males and females and therefore should apply uniformly to both genders. The major advantage of this equation is that it takes into account the impact of body composition on RMR and in so doing can more accurately reflect gender-related differences in metabolism. Indeed, the reduction in RMR in women has been found to be proportional to their smaller FFM (Cunningham 1982). By including FFM, this equation appears also to be adequate in assessing BMR in individuals of different ages, because a reduction in BMR with advancing age can be largely explained by a loss of FFM. This estimation approach, however, requires measurement of body composition in the first place, which could be problematic for those who do not have access to body composition equipment. Another relatively simple method of quantifying RMR, which does not require the measurement of body composition, is to use multiple age- and gender-specific equations that relate RMR directly to body mass (Williams 2005). These equations are listed in table 10.2. For example, for a male who is 20 years of age and weighs 70 kg, his RMR would be 1750 kcal (i.e., 15.3 × 70 + 679 = 1750).

> **KEY POINT**
>
> Resting metabolic rate can be influenced by many factors including age, gender, body composition, and body size. Thus, estimating RMR using fat-free mass is a convenient approach, because FFM is the location of most energy expenditure and discriminates among individuals of different gender or age. This estimation uses the formula RMR = 370 + 21.6 (FFM).

Table 10.2   Age- and Gender-Specific Equations for Estimating Daily RMR in Kilocalories

Age (years)	Equations Males	Females
3-9	(22.7 × body weight*) + 495	(22.5 × body weight) + 499
10-17	(17.5 × body weight) + 651	(12.2 × body weight) + 746
18-29	(15.3 × body weight) + 679	(14.7 × body weight) + 496
30-60	(11.6 × body weight) + 879	(8.7 × body weight) + 829
>60	(13.5 × body weight) + 487	(10.5 × body weight) + 596

*Body weight is expressed in kilograms (kg).

Adapted from M.H. William, 2005, *Nutrition for health, fitness, and sports*, 7th ed. (New York, NY: McGraw-Hill), 95, by permission of The McGraw-Hill Companies.

FACTORS INFLUENCING RESTING METABOLIC RATE

From the preceding discussion, it is clear that RMR can be affected by body composition; that is, the greater the FFM, the greater the RMR. As FFM differs between men and women and generally decreases with advancing age, the relation between FFM and RMR has enabled us to estimate RMR directly from FFM for various populations. Resting metabolic rate is also influenced by many other factors such as body size, growth and development, climate conditions, and hormonal concentrations, as discussed next.

Body Size

In general, the greater a person's size, the greater the level of resting metabolism. Research examining the relation between RMR and body size can be traced back to the 1800s, when a number of scientists proposed what was called the surface law of metabolism as an attempt to account for individual differences in energy metabolism. For example, Regnault and Reiset in 1850 showed that the resting oxygen consumption of **endothermic** mammals and birds was 10 to 100 times that of **ectothermic** reptiles and amphibians. They also noted that the mass-specific rate of oxygen uptake was smaller in larger species (Lutz 2002). Stated alternatively, larger species have a smaller energy expenditure per unit of body mass. In 1883, Max Rubner measured the heat production of dogs ranging in size from 3 to 31 kg (6.6 to 68 lb) and showed that values were much more similar when energy was expressed relative to body surface area than to body mass (Elia 1992). This finding suggests that energy expenditure of a living organism correlates well with its body surface area.

The surface law of metabolism provides the basis for the common practice of expressing metabolic rate or energy expenditure by body surface per hour (i.e., $kcal \cdot m^{-2} \cdot h^{-1}$). Nevertheless, research in the 1900s showed that the formulation resulting from the surface law of metabolism did not apply universally to different species of temperature-regulating animals. For example, experiments conducted by Kleiber in 1932 showed that BMRs of mammals and birds of different sizes were proportional to the 0.73 power of body mass. This evidence was then confirmed by Benedict, who measured RMR in a wide variety of animals that differed considerably in size (White and Seymour 2005). Benedict's extensive work in this area resulted in the establishment of a benchmark curve that relates BMR to body mass (figure 10.1). This curve illustrated a **logarithmic plot** of body mass ranging from 0.01 to 10,000 kg and a metabolic rate ranging from 1 to 1000 W, and reflected essentially the same allometric exponent as previously reported, which was later approximated to 0.75 mainly for mathematical ease.

Growth and Development

From the preceding discussion, it is clear that RMR can be affected by body size and body composition. As shown in table 10.3, this energy component can also be influenced by many other factors. As compared to adults, infants have a large proportion of metabolically active tissue. Hence, their mass-specific RMR is higher than that of adults. It has been reported that the metabolic rate per kilogram body weight in young children may be two times greater than in adults (Elia 1992). However, RMR declines through childhood, adolescence, and adulthood as full growth and maturation are achieved. In theory, this decline with age can be ascribed to either a reduction in the metabolic rate per gram of individual tissue or a change in the proportion of different tissues, each of which has a relatively fixed metabolic rate. The available evidence appears to support the latter explanation, suggesting that the metabolic rate per gram of each organ changes little during growth and development. For instance, Kennedy and Sokoloff (1957) found no change in oxygen consumption in brain tissue between children aged 3 to 11. The metabolic values they found in children were also similar to those of adults. The energy expenditure of the brain, which accounts for almost one-half of the total energy expenditure in infants, decreases in proportion to body weight during growth and development so that in the adult it is responsible for only 20% of the total energy expenditure (table 10.1). Mass-specific RMR shows a continued decline in those who reach or pass the age of 60. Indeed, as mentioned earlier, aging has been associated with the loss of lean body mass, which includes muscle as well as other metabolically active organs (Going et al. 1995). This loss of fat-free tissues is largely responsible for the reduced RMR frequently seen among those who are elderly (Tzankoff and Norris 1978).

Climate Conditions

Climate conditions, especially temperature changes, can also raise resting energy expenditure. For example, exposure to cold may stimulate

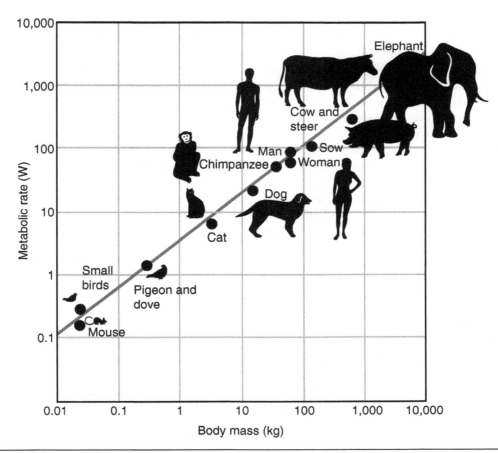

Figure 10.1 Logarithmic relationship of body mass and metabolic rate for a variety of animals and humans differing in body size.

Adapted, by permission, from W.D. McArdle, F.I. Katch and V.L. Katch, 2001, *Exercise physiology*, 5th ed. (Philadelphia, PA: Lippincott, Williams & Wilkins), 189.

Table 10.3 Factors That Affect RMR

Factor	Effect
Body size	An increase in body size increases RMR.
Body composition	An increase in lean body mass increases RMR.
Growth	Mass-specific RMR declines through childhood, adolescence, and adulthood.
Age	Mass-specific RMR continues to decline from adulthood into older age.
Ambient temperature	Both cold and warm exposure stimulate RMR.
Hormones	Both thyroid hormone and epinephrine stimulate RMR.
Aerobic fitness	A high level of aerobic fitness is linked to an increased RMR.
Resistance training	Resistance training increases RMR or prevents a decline in RMR as one ages.
Smoking	Nicotine increases RMR.
Caffeine	Caffeine stimulates RMR (refer to chapter 12).
Sleeping	RMR decreases during sleep.
Nutritional status	Underfeeding tends to reduce RMR, while overfeeding tends to increase RMR.

muscle shivering, as well as the secretion of several thermogenic hormones, such as epinephrine and thyroid hormone. As a result of these physiological responses, metabolic rate can increase four- to fivefold at rest. The heat produced from this augmented metabolism is vital for the maintenance of a stable core temperature. The increment in energy used to maintain body temperature appears to be highest when the body is exposed to cold water. It has been reported that thermogenesis resulting from cold-water immersion is often insufficient to prevent a drop in body temperature even with exercise (Toner and McArdle 1996). Exposure to a warm environment also provokes an increase in energy metabolism, although this increase may not be as great in magnitude as that induced in a cold environment. The resting metabolism of people living in a tropical climate is generally 5% to 20% higher than for those living in a more thermoneutral region (McArdle et al. 2001). The increased thermogenesis due to a warm environment is accomplished primarily by cardiovascular adjustments such as skin **vasodilation** and increased sweat secretion, both of which help to dissipate excess heat and thus maintain a stable core temperature.

Hormonal Concentration

Resting metabolic rate is also subject to changes in hormonal concentration. The two major hormones linked to RMR are epinephrine and thyroid hormone. Epinephrine is secreted by the medullar region of the adrenal glands. A marked increase in plasma concentration of epinephrine can promptly increase heat production by more than 30%. This calorigenic or heat-producing effect of epinephrine may be related to its stimulation of glycogen and triglyceride catabolism, as adenosine triphosphate (ATP) splitting and energy liberation occur in both the breakdown and the subsequent resynthesis of these molecules. The metabolic effect of epinephrine accounts at least in part for the greater heat production associated with emotional stress. Thyroid hormone, which is secreted by the thyroid gland, also increases oxygen consumption and heat production of most body tissues except the brain. Thyroid hormone is important for growth and development, and this effect may be explained by its role in potentiating the effect of growth hormone on protein synthesis. The anabolic role that thyroid hormone plays during growth and development is quite distinct from its calorigenic effect. Individuals with **hyperthyroidism** have been linked to the net catabolism of endogenous protein and fat stores, which then leads to the loss of body weight. One interesting hypothesis for why thyroid hormone augments metabolism is that this hormone may act by reducing the efficiency of oxidative energy transformation (Dauncey 1990). This hypothesis suggests that with an increase in thyroid hormone, a greater amount of oxygen is necessary for a given level of ATP production.

> ### ▶ K E Y   P O I N T ◀
>
> In addition to age and gender, RMR can be influenced by factors such as body size, environmental temperature, and hormonal concentration. Thus, in comparing RMRs across individuals with different body size, it is important to express RMR relative to body size or body surface area. Because RMR is subject to changes in environmental temperature and hormonal concentrations, measurements of RMR should be made in a thermoneutral and nonstressful environment.

INFLUENCE OF EXERCISE ON RESTING METABOLIC RATE

The impact of physical activity on RMR has received a great deal of attention recently, although the answers on this issue remain unclear. The energy expenditure due to physical activity per se is relatively small, accounting for no more than 30% of the total daily energy expenditure. Thus it is thought that regular exercise produces energetic benefits in other components of the daily energy expenditure, thereby generating a net effect on energy balance that can be much greater than the direct energy cost of the exercise alone. As RMR is the largest component of the total daily energy expenditure, in theory even small changes in RMR may significantly affect the regulation of body weight and body composition over an extended period of time. Two experimental approaches have been employed to investigate the effect of exercise training on metabolic rate in humans: (1) cross-sectional comparisons of RMR between trained and untrained individuals and (2) longitudinal follow-up studies of RMR in the same individual before and after an exercise intervention. Both study designs have their unique strengths. Directly comparing groups of different training status is an efficient way of gathering data that reflect information at one point in time. On the other hand, having subjects participate in longitudinal intervention programs during which outcomes are carefully monitored and recorded is a more robust design that allows examination of cause-and-effect relationships.

Effects of Aerobic Training

One would expect training to augment RMR. What the current literature shows, however, is that both cross-sectional and longitudinal studies have yielded quite inconsistent results on this issue. By classifying subjects into high- and low-fit categories according to their $\dot{V}O_2max$, Poehlman and colleagues (1988, 1989) found that the high-fit group had a greater RMR than the low-fit group. These authors also demonstrated a linear relationship ($r = 0.77$, $p < 0.05$) between $\dot{V}O_2max$ and RMR. Using the same experimental approach, Tremblay and colleagues (1985) also noted a higher mass-specific RMR in a group of highly trained athletes. On the basis of their findings, these authors suggested that a high level of maximal aerobic power is needed in order for people to manifest a high RMR. Of particular note is that the group differences in RMR occurred even though RMR was normalized to lean body mass. This observation suggests a role for high levels of physical conditioning separate from the influence of differences in adiposity as a determinant of RMR.

A handful of studies have failed to demonstrate such a training effect on RMR (LeBlanc et al. 1984b; Tremblay et al. 1983; Hammer et al. 1988; Poehlman et al. 1986). It appears that some of these studies failed to recruit trained subjects who had $\dot{V}O_2max$ above 60 ml $\cdot$ kg^{-1} $\cdot$ min^{-1}, as used by Poehlman and colleagues (1988, 1989). Others used a group of obese individuals rather low in fitness. The discrepancies among these studies may also be attributable to smaller sample sizes. Resting metabolic rate can vary a great deal because, as already noted, it is subject to the influence of many factors such as age, body composition, nutrition, hormones, and environment. In this context, a large sample size is of particular importance for the purpose of revealing a statistical difference. There are also limitations specifically associated with the

design of cross-sectional comparisons. With this experimental design, one can argue that differences in RMR between trained and untrained individuals are due to factors such as genetic predisposition rather than physical training.

Longitudinal studies are harder and more expensive to conduct. However, this type of study design is helpful in controlling for some of the confounding variables associated with cross-sectional comparisons. It can also allow examination of changes in fitness level and body composition concomitant with changes in RMR. In longitudinal studies on RMR, a training regimen is given only to the experimental group while both the experimental and placebo groups are instructed to follow a similar dietary plan and similar leisure activity patterns. Outcomes such as RMR, body composition, and $\dot{V}O_2max$ are determined before and after the training intervention. Several longitudinal studies support an increase in RMR following exercise training. For example, Tremblay and colleagues (1986) observed an increase in relative RMR (expressed relative to FFM) following 11 weeks of training, which suggests that being physically active may increase energy expenditure during nonactive times. A training-induced increase in RMR was also noted in a study by Lennon and colleagues (1984), but these authors found an increase in RMR only in the group that was trained at a moderate to high intensity and no difference in the group performing lower-intensity exercise. This finding implies that exercise training should be performed at moderate intensity or higher (i.e., >70% $\dot{V}O_2max$) to produce an increase in RMR.

A number of longitudinal studies have failed to demonstrate a training-induced increase in RMR. For instance, Broeder and colleagues (1992) conducted 12-week aerobic and resistance training programs using two groups of healthy young males. They found that despite a decline in percent body fat and an increase in FFM, RMR adjusted for lean body mass did not change following either training regimen. Westerterp and colleagues (1992) had subjects participate in a 44-week training program for a half-marathon, which was quite intensive and increased energy demand by approximately 30%. As a result of the program, subjects lost fat mass and gained FFM. However, the authors failed to observe an increase in RMR. It is difficult to discern what might have caused the discrepancies among these studies. According to Broeder and colleagues (1992), possible factors that could contribute to such divergent findings on training-induced changes in RMR include (1) the method used to normalize the RMR data when differences in body composition variables are found between comparison groups or between two measurement periods; (2) initial fitness level of subjects; (3) sample size and statistical power; (4) the timing of the RMR measurement in relation to the last bout of exercise; and (5) training dosage factors such as mode, intensity, duration, and frequency. It is of interest to note that in most studies that failed to show an increase in RMR, metabolic measurements were made at least 24 h after the last bout of exercise. This prolonged time delay is important for a true exploration of the impact of chronic exercise on RMR. An increase in RMR measured within 24 h of the last exercise session may be only an acute effect. Such an increase in RMR can also be viewed as excess postexercise oxygen consumption (EPOC). Indeed, EPOC has been found to last as long as 24 h.

Effects of Resistance Training

Not until relatively recently have some longitudinal studies incorporated a regimen of resistance training into an exercise intervention in an effort to examine its impact on RMR. This approach makes sense in that heavy resistance training promotes skeletal muscle development, thereby enhancing RMR. It has been estimated that RMR increases approximately 50 kcal per day for each kilogram of increase in FFM. As already mentioned, Broeder and colleagues (1992) found no change in RMR following 12 weeks of resistance training. Van Etten and colleagues (1995) also observed no change in sleeping metabolic rate following a weight training program of similar duration. In both of these studies, the increase in FFM was rather small, perhaps because the duration of training was not sufficient to produce meaningful muscle hypertrophy. It should be pointed out also that the subjects in the study by Broeder and colleagues (1992) were under a negative caloric balance during the study period, which in theory should reduce RMR. Hence, the finding that RMR did not change following resistance training may be viewed as positive in the sense that the training program was successful in preventing attenuation in RMR.

The studies just discussed only used male subjects. Whether women would be more responsive to a resistance training program remains a question. As mentioned earlier, some studies have reported an increase in RMR in women following aerobic training that generated fat losses (Lawson et al. 1987; Lennon et al. 1984; Tremblay et al. 1986). However, much less information is available with regard to the effect of resistance training. After administering a combined aerobic and resistance training program in a group of adult women, Kraemer and colleagues (1997) found no differences in changes in FFM and RMR as compared to values with aerobic training alone. In fact, in this latter study, the authors failed to observe any change in RMR following either the aerobic program or the program that combined aerobic and resistance training.

Results appear to be more promising when a resistance training program is applied to older individuals, although the underlying mechanism remains to be determined. Most studies using older individuals have shown an increase in RMR concomitant with an increase in FFM following resistance training (Pratley et al. 1994; Campbell et al. 1994; Treuth et al. 1995a, 1995b). Resistance training in elderly people is particularly attractive because it can help to attenuate age-related declines in FFM. One may reason that the augmented RMR due to resistance training can be attributed to an increase in respiring tissues such as muscle mass. Nevertheless, this does not appear always to be the case, because some studies that demonstrated a training effect on RMR showed no or minimal changes in FFM. It has been proposed that a training-induced increase in protein turnover may contribute to an elevation in resting energy expenditure. However, Campbell and colleagues (1994) noted that the energy cost associated with increased protein metabolism was very small relative to the increase in RMR. Pratley and colleagues (1994) observed, in addition to an increase in RMR, an accompanying increase in plasma norepinephrine concentrations as a result of resistance training. In this context, one may speculate that a training-induced increase in RMR may also be due to increased activity of the sympathetic nervous system. This idea should be regarded with caution, however, as an increase in plasma norepinephrine following resistance training has not been a universal observation in older individuals (Ryan et al. 1995).

▶ **K E Y P O I N T** ◀

Whether physical training will augment RMR is an issue that remains debatable. It appears that a greater RMR due to aerobic training tends to occur in those with a significantly higher level of fitness or those who experience more profound training adaptations. On the other hand, an increase in RMR following resistance training appears to be more uniformly observed in older individuals, although this improvement often occurs without a concurrent change in FFM.

ROLE OF RESTING METABOLIC RATE IN THE PATHOGENESIS OF OBESITY

Obesity is a problem of imbalance between energy intake and energy expenditure. The relationship between these two forms of energy can be expressed as follows:

$$\text{energy intake} = \text{energy expenditure} + \text{energy storage}.$$

We store energy as fat because it is much denser than carbohydrate and because storage of fat does not require a large amount of water. Given the energy balance equation, it is clear that obesity results from either too high a food intake, too low an energy expenditure, or a combination of these. There has been sustained interest in whether or how RMR contributes to weight gain and thus obesity. This is so because RMR represents between 60% and 70% of the total energy expenditure each day, and an increase in RMR of only 1% to 2% could have a major effect on weight regulation in the long run. As discussed earlier, several cross-sectional studies have demonstrated higher RMR values expressed per kilogram of fat-free weight in both aerobically trained and resistance-trained individuals. However, other studies, especially those using a longitudinal approach, have been unable to confirm this relationship. On the other hand, evidence is clear that when caloric intake is restricted in either lean or obese people, their RMR as well as the total energy expenditure will fall. While this reduction in energy expenditure reflects a protective mechanism that allows people losing weight to conserve their energy stores and

avoid excessive depletion, it is a source of frustration for obese individuals trying to lose weight over the long term.

In the context of efforts to determine the kinds of interventions that may help in modulating RMR, perhaps the most fundamental question that needs to be answered is whether RMR plays a role in causing weight gain and thus obesity. According to the existing literature, answers to this question differ. Although it is a popular belief that a reduced level of energy expenditure leads to the development of obesity, this hypothesis remains controversial and has been difficult to prove. In a recent meta-analysis of studies on obese people who underwent weight loss, Astrup and colleagues (1999) concluded that in formerly obese subjects, RMR adjusted for FFM was 3% to 5% lower than that of control subjects. Using a metabolic chamber, Leibel and colleagues (1995) found that weight loss resulted in a compensatory decrease in RMR that served to return individuals to their previous body weight. In addition, using a similar assessment approach, Ravussin and colleagues (1988) noted that Pima Indians with a lower RMR showed greater weight gain than others with normal RMR. In the same study, Ravussin and colleagues also observed similarities in both RMR and the total energy expenditure within the same family, suggesting that there may be genetic determinants of energy expenditure. Indeed, using **dizygotic** and **monozygotic** twins, Bouchard and colleagues (1989) found that the heritability of RMR accounted for approximately 40% of the variance remaining after adjustment for age, gender, and FFM. Collectively, these results all point toward the notion that variations in RMR contribute to weight gain and that a lower RMR may be used to predict the future occurrence of obesity.

In contrast, a number of studies using a **postobese model** have failed to show a relationship between low RMR and weight gain (Amatruda et al. 1993; Weinsier et al. 1995, 2003). The purpose of the postobese model is to compare obesity-prone weight-reduced persons with obesity-resistant never-obese individuals. Such comparisons minimize the potential confounding effect associated with directly comparing obese and lean individuals who differ sharply in body composition (Weinsier et al. 2003). Since the body weight of postobese individuals is reduced to begin with, this model also enables researchers to study the pattern of weight regain in a prospective fashion. If formerly obese individuals have a lower metabolic rate than never-obese control subjects, this could explain why they were obese and why they may have a greater propensity to regain weight. With this model, accuracy of the data depends on subjects having attained energy balance after their weight loss before undergoing the metabolic assessment, as well as having their reduced weight well stabilized at an ideal level. Amatruda and colleagues (1993) studied 18 obese females before and after reduction to normal body weight and then followed 10 of them for another 18 to 34 months. Not only were their RMRs similar to those of 14 never-obese controls, but there was no evidence that the reduced energy expenditure in the 10 subjects predicted their weight regain during the one- to three-year period of follow-up. Weinsier and colleagues (1995, 2003) also failed to find a difference in RMR between postobese and never-obese women. In addition, they found no relationship between baseline RMR and amounts of weight regain at either the first year or subsequent years of follow-up. In all these investigations, subjects maintained a balance between energy intake and energy expenditure during the entire course of the study.

Of particular pertinence to RMR and obesity is the recent quest to develop a better understanding of uncoupling proteins (UCPs). Uncoupling proteins are a unique group of closely related mitochondrial carrier proteins involved in energy metabolism. According to the chemiosmotic theory discussed in chapter 1, cellular energy production takes place across the inner mitochondrial membrane. In this process, adenosine diphosphate (ADP) is phosphorylated to ATP using energy associated with a gradient of protons that is generated during electron transport. It is now known that via the action of UCPs, a significant proportion of mitochondrial respiration is dissipated as heat instead of being coupled to ATP synthesis. At least five uncoupling proteins have been identified, and they differ mainly in their location within body tissues. Of those identified, UCP1, UCP2, and UCP3 have been extensively investigated. UCP1, the first membrane protein discovered (in 1976), was found exclusively in mammalian brown fat tissue and is mainly involved in nonshivering thermogenesis. Brown fat, which differs from the white fat that composes most of the fat tissue in the body, is found in small amounts located near large blood vessels around the neck, back, and chest areas (Cannon and Nedergaard 2004). UCP2 and UCP3 are two homologs of UCP1 identified in 1997. Tissue distributions of UCP2 and UCP3 are markedly different. UCP2 has a much wider tissue distribution; it has been found in adipose tissue and in the heart, skeletal muscle, kidney, digestive tract, brain, and spleen. On the other hand, UCP3 is found primarily in skeletal muscle (Lowell and Spiegelman 2000).

Do obese individuals possess fewer UCPs and thus "burn off" less food into heat? This question remains unanswered at present. Using obese Zucker rats, Boss and colleagues (1998) found a 41% decrease in UCP1 expression in brown adipose tissue and a 42% decrease in UCP3 expression in skeletal muscle relative to normal-weight rats. This finding makes sense in that a decrease in UCPs would suggest a reduced RMR. However, using humans, Millet and colleagues (1997) found no differences in UCP2 and UCP3 expression in adipose tissue and skeletal muscle between obese and lean individuals. Interestingly, Simoneau and colleagues (1998) observed an overexpression of UCP2 in the skeletal muscle of obese as compared to lean individuals. Moreover, they also noted a positive correlation between UCP2 protein content and percent body fat. The direction of this relationship, however, appears to be opposite to that reported by Schrauwen and colleagues (1999), who found a negative correlation between UCP3 mRNA (messenger ribonucleic acid) and **body mass index** calculated as weight divided by height[2]. In light of these findings, it is tempting to suggest that obesity is associated with overexpression of UCP2 and underexpression of UCP3 in skeletal muscle. The more precise mechanism whereby each UCP mediates the development of obesity remains to be demonstrated.

In light of the controversies in the literature, it is premature to conclude that a change in RMR predicts weight gain and obesity. A reduced RMR is seen in some obese individuals, and a lower RMR has been shown to precede weight gain in infants and children as well as adult Pima Indians and Caucasians (Snitker et al. 2000). However, evidence is lacking to suggest that such a defect in RMR necessarily plays a causal role in the development of obesity. As discussed earlier, RMR shows substantial variation, some of which is due to genetic factors. It appears that even with a genetic predisposition to reduced RMR, obesity can occur not because of a reduced RMR, but rather because of variations associated with other energy components. There has been renewed interest in the question whether energy expenditure during planned or spontaneous physical activities may be a more important factor related to the etiology of obesity. There is no doubt that physical activity can tip the energy balance equation by augmenting the total energy expenditure, partly via stimulation of RMR.

Regular physical activity may also produce cellular adaptations that favor fat oxidation. Although we lack a definitive **prospective study** to show that a low level of physical activity is a risk factor for the development of obesity and that a high level of physical activity is protective against obesity, an overwhelming amount of indirect evidence suggests that this is the case. According to the World Health Organization (1998), physical activity plays a crucial role in protecting against obesity. In fact, the World Health Organization suggests that a physical activity level of 1.75, or a daily energy expenditure 1.75 times the RMR, is needed in order to provide this protective effect.

> ### ▶ KEY POINT ◀
>
> Although RMR composes the largest fraction of daily energy expenditure, concrete evidence is lacking for a link between a reduction in RMR and weight gain and obesity. Thus, future studies should also focus on other components of energy expenditure such as that associated with physical activity. Physical activity not only stimulates energy expenditure, including RMR, but also produces cellular adaptations that favor fat oxidation.

SUMMARY

Resting metabolic rate constitutes 60% to 75% of the total energy expenditure each day. This largest component of energy expenditure has drawn a great deal of attention particularly with regard to its role in the pathogenesis of obesity. The RMR can be viewed as the minimal energy expenditure necessary to sustain life. The RMR is directly proportional to FFM, and this relationship may explain why RMR is higher in men than in women and decreases with advancing age. This relationship also enables us to estimate RMR without having to use multiple gender- and age-specific equations. Resting metabolic rate is also influenced by body size, growth, climate conditions, and hormonal concentrations. In this context, it is important that RMR always be measured under standardized conditions and be expressed relative to body size in order to facilitate comparisons among individuals.

Whether physical training augments RMR and thereby prevents weight gain has been a topic of interest, but remains in debate at present. It is generally agreed that in order to increase RMR, people should train at more vigorous intensities for an extended period of time and that an increase in RMR is not always accompanied by an increase in FFM. Clearly an increase in RMR can help in creating the favorable energy balance required for weight loss. However, concrete evidence supporting a direct linkage between RMR and obesity is lacking. Efforts to unravel the etiology of obesity will continue, but they must also focus on the impact of other components of energy expenditure, such as that associated with physical activity, on the regulation of energy balance.

KEY TERMS

body mass index
dizygotic
ectothermic
endothermic

hyperthyroidism
logarithmic plot
monozygotic
postobese model

prospective study
thermogenesis
vasodilation

REVIEW QUESTIONS

1. Define the term *resting metabolic rate*. How does this term differ from basal metabolic rate?

2. How is RMR related to total daily energy expenditure? What is the RMR in average men and women?

3. List three factors that can influence RMR and explain how each one works.

4. From the current research perspective, what is the general consensus on the role that physical training plays in modulating resting energy expenditure?

5. What is the postobese model? What are the advantages of using this model for studying the association between RMR and obesity?

6. Describe how uncoupling proteins work in regulating energy metabolism.

11

THERMIC EFFECT OF FOOD

This chapter provides an in-depth discussion on the thermic effect of food (TEF) and its impact on total daily energy expenditure. In particular, the topics include what the TEF is and how it is determined, what factors can affect its magnitude, how it may respond to interventions such as physical activity, and the role it plays in the pathogenesis of obesity.

THERMIC EFFECT OF FOOD AND ITS MEASUREMENT

Diet-induced thermogenesis, or the thermic effect of food, is another important component of the total daily energy expenditure. This energy fraction is defined as the significant elevation of the metabolic rate that occurs after ingestion of a meal. Typically, this elevation reaches its peak within an hour and can last for a duration of 4 h after a meal. The TEF is proportional to the amount of energy being consumed and is estimated to be about 10% of energy intake. For example, an individual consuming 2000 kcal probably expends about 200 kcal on the TEF.

The TEF can be divided into two subcomponents: **obligatory** and **facultative** thermogenesis. The obligatory component of TEF is the energy cost associated with digestion, absorption, transport, and **assimilation** of nutrients, as well as the synthesis of protein, fat, and carbohydrate to be stored in the body. Several investigations have shown that the measured stimulation of energy expenditure is higher than the theoretical values of obligatory nutrient disposal and storage (Acheson et al. 1984a; Thiebaud et al. 1983b). The energy expended in excess of obligatory thermogenesis has been termed facultative thermogenesis.

This second component is thought to be mediated primarily by activation of the sympathetic nervous system, which functions to stimulate metabolic rate. This classification in essence provides underlying mechanisms that explain the increment in thermogenesis following a meal.

> **KEY POINT**
>
> Thermic effect of food can be divided into two subcomponents: obligatory and facultative thermogenesis. The obligatory component is the energy used for digestion, absorption, assimilation, and storage of absorbed nutrients. The facultative component is the additional increase in energy expenditure caused by activation of the sympathetic nervous system.

The value of TEF following each meal is rather small, ranging from about 50 to 100 kcal over a period of 4 to 8 h. Therefore, it is crucial that measurement of TEF be carried out with a great deal of care and precision and with use of a standard testing protocol and environment. To measure TEF, most studies have used indirect calorimetry in conjunction with a ventilated hood, a tight-fitting face mask, or a mouthpiece and nose clip system. The ventilated hood system differs from the face mask and mouthpiece systems in that a hood or canopy loosely covers the face of subjects while they are in a supine position, providing them a great deal of ease as metabolism is measured. It is often used in studies of resting metabolic rate or TEF. The results from measurement of resting metabolic rate suggest that these three methods seem to be equally accurate in determining energy cost, although the ventilated hood appears to be least stressful for subjects who are being studied for an extended period of time.

The thermic effect of a meal usually lasts about 4 h but can persist for up to 8 h, depending primarily on factors such as meal size and composition of the nutrients ingested. Houde-Nadeau and colleagues (1993) showed that a single 6 h measurement reliably represented the TEF of an 800 kcal meal. To measure the complete thermic effect of the meal, the measurement should be continued until oxygen consumption returns to the postabsorptive value. The postabsorptive state is a period when the gastrointestinal tract is empty and energy comes from the breakdown of the body's reserves. Although the measurement of TEF requires prolonged immobility of the person being measured, it has been suggested that data collection can occur in an intermittent fashion, in which the periods of testing are alternated with periods when the subject remains at rest but is able to talk and move. In fact, results obtained with this intermittent approach have been shown to be more reliable than those obtained with continuous measurement (Piers et al. 1992). Of course, the two more recently developed methods, the whole-room respiratory chamber and doubly labeled water, obviate the limitations associated with indirect calorimetry and allow investigators to assess the TEF under free-living conditions. Chapter 4 provides more details on methods used to measured energy expenditure.

INFLUENCE OF PROTEIN, CARBOHYDRATE, AND FAT CONSUMPTION

The TEF is affected by the type of food ingested. In other words, the level of TEF can be different depending upon whether protein, carbohydrate, or fat is consumed. This relationship may be attributed to differences in the chemical structure of these nutrients, which dictate the amount of the energy needed in order for them to be digested, absorbed, assimilated, and stored. For example, consumption of protein is considered the most thermogenic, partly because protein contains nitrogen that needs to be removed, which is energy costly. The following sections provide more details on how TEF is affected by consumption of protein, carbohydrate, and fat.

Thermic Effect of Food and Protein

The TEF was previously known as "specific dynamic action." This term was coined in 1902 by German

physiologist Max Rubner, who thought that the increased oxygen uptake after a meal was a specific response to the protein ingested. Rubner also discovered that the increased oxygen uptake was caused by oxidation of those amino acids not used for glucose synthesis. These conclusions were derived from an earlier study in which Rubner kept trained dogs in calorimeters at 33 °C under fasting conditions for more than 24 h. He then fed each dog 2 kg (4.4 lb) of fat-free meat containing 1926 kcal and found a 30.9% increase in metabolism. However, when the dogs were fed an equal number of calories of fat or carbohydrate, the metabolism increased by only 12.7% and 5.8%, respectively (James 1992). In subsequent decades, the relatively greater thermogenesis associated with protein consumption was generally confirmed, with values reported to be between 20% and 30% of energy intake. The term *specific dynamic action* was later replaced by *diet-induced thermogenesis* and more recently *thermic effect of food* in recognition of the fact that the augmented metabolism after eating does not depend exclusively on ingested protein.

Thermogenesis resulting from protein consumption has been ascribed mainly to the extra energy needed for absorption by the gastrointestinal tract, as well as deamination of amino acids and synthesis of protein in the liver. Ingested proteins are degraded in the gut into amino acids, a process that does not require energy (James 1992). However, amino acids are absorbed via an energy-requiring process in which 0.5 mol adenosine triphosphate (ATP) is consumed for each mole of amino acids absorbed (Tappy and Jequier 1993). After absorption, amino acids have two major metabolic fates. They can be deaminated so that their carbon skeleton can be further converted to glucose in the process of gluconeogenesis while the amino group that has been removed becomes urea in the liver. Once the newly synthesized glucose is oxidized, calculations show that the amount of energy required during the indirect process of amino acid oxidation is 25% of the energy content of amino acids. Absorbed amino acids can also be used for protein synthesis. In this process, energy is expended mainly in the synthesis of the peptide bonds and has been calculated to be also about 25% of the energy content of amino acids.

It is now widely accepted that the TEF produced by protein is about 20% to 30% of the energy intake, whereas the TEF for carbohydrate and fat approximates 5% to 10% and 0% to 5%, respectively. A meal rich in protein tends to elicit a greater and more prolonged TEF as compared to an isocaloric meal containing relatively more carbohydrate or fat. In a study in which healthy females were fed a meal containing about 70% energy from protein, carbohydrate, or fat on three separate occasions, Crovetti and colleagues (1998) found that the increase in energy expenditure following the high-protein meal was threefold greater than that following the high-carbohydrate or the high-fat meal, and that this increase persisted for over 7 h. The relatively large calorigenic effect of ingested protein has been used as evidence to promote high-protein diets for weight loss. This is based on the belief that since a greater amount of energy is needed during the process in which protein is digested, absorbed, and assimilated, fewer calories will become available to the body for storage with a high-protein meal than with a comparable meal consisting mainly of carbohydrate and fat. However, this notion should be viewed with caution, as it has been suggested that a high protein intake can lead to hypoglycemia and protein degradation as well as harmful strain on the kidney and liver over the long run.

Thermic Effect of Food and Carbohydrate

The impact of ingested carbohydrate and fat on TEF was not recognized until about 30 years after the 1902 publication by Max Rubner when the term "diet-induced thermogenesis" was introduced. Those who proposed the term diet-induced thermogenesis believed that the extra energy used in the process of absorption, assimilation, and storage was associated not only with protein but also with carbohydrate and fat. It has been estimated that the TEF for carbohydrate corresponds to approximately 5% to 10% of energy intake. The increase in energy expenditure due to carbohydrate consumption can be further attributed to the process of absorption and subsequent storage. Like amino acids, ingested glucose is absorbed by an energy-requiring process in which 0.5 mol ATP is consumed for each mole of glucose absorbed (Tappy and Jequier 1993). In the postabsorptive state, much of the absorbed glucose is stored as glycogen. The amount of energy that glucose possesses is equivalent to about 38 ATPs. This is the amount of ATP that is yielded once 1 mol of glucose is completely oxidized. On the other hand, the process in which glucose is stored as glycogen requires an energy consumption of 2 ATPs (Stryer 1988). Consequently, the energy cost associated with glucose absorption and storage corresponds to about 6% to 7% (i.e., 2.5/38), a value that is close to the value estimated for glucose-induced thermogenesis. Although some glucose can also be converted into fatty acids in the process of lipogenesis, this process is quantitatively

much less important except in the situation of marked carbohydrate and energy overfeeding (Acheson et al. 1982, 1984b).

It is interesting to note that ingestion of an oral load of fructose elicits a greater increase in energy expenditure than does a similar load of glucose (Tappy et al. 1986). The increase in energy expenditure is also greater after ingestion of equivalent loads of either sucrose or an equimolar mixture of fructose and glucose as compared to glucose alone (MacDonald 1984). Fructose, a sugar naturally present in fruit and honey, contributes as much as 10% to 15% of the total dietary carbohydrate intake in Westernized countries. Several mechanisms explain the large stimulation of energy expenditure after fructose consumption. Intestinal absorption of 1 mol of fructose requires consumption of 0.5 mol of ATP, just as with glucose. Most of the absorbed fructose is then stored as glycogen in the liver and muscle (Bergstrom and Hultman 1967). The process in which fructose is converted to glycogen or completely oxidized requires an energy consumption of 3 mol rather than the 2 mol of ATP seen for glucose (figure 11.1). An extra ATP is needed during the process of assimilating fructose into glucose before further metabolization (Tappy and Jequier 1993). Fructose yields the same amount of energy as glucose once it is oxidized. This suggests that the energy cost associated with fructose consumption is 9% (i.e., 3.5/38) of energy intake, a higher value than for glucose. Storage of absorbed fructose can also be metabolized in the liver or converted into fatty acids. However, these metabolic pathways have been found to be quantitatively much less important as compared to fructose being converted to glycogen (Tappy and Jequier 1993).

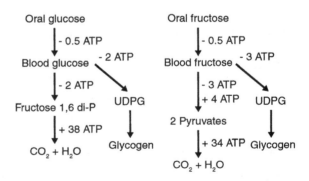

Figure 11.1 Utilization of adenosine triphosphate (ATP) in the initial steps of metabolism following oral fructose consumption. UDPG = Uridine diphosphoglucose, an intermediate in glycogen biosynthesis.

Reprinted, by permission, from L. Tappy and E. Jequier, 1993, "Fructose and dietary thermogenesis," *American Journal of Clinical Nutrition* 58(Suppl.): S766-S770.

Thermic Effect of Food and Fat

Of the three macronutrients, fat is the least thermogenic substrate; that is, the energy cost of depositing dietary lipids in the body is lowest compared to the costs for protein and carbohydrate. The increase in energy expenditure following fat consumption was reported to be around 2% of energy intake when subjects were infused with triglyceride (Thiebaud et al. 1983a). This small elevation in energy expenditure has been ascribed to ATP consumption in the process whereby free fatty acids are re-esterified into triglyceride. Ingesting excess carbohydrate, especially simple sugar, can lead to increased lipogenesis. This unwanted process, however, is thought to cost more energy than storing fat using fatty acids obtained directly from the diet (Rothwell and Stock 1983). The consumption of a high-fat diet has long been discouraged partly because this type of diet produces the lowest thermogenesis. In fact, a high-fat diet is considered most efficient in terms of energy storage and thus may be viewed as a viable option for those who are energy deprived or need to gain weight.

The conclusion that thermogenesis following fat consumption is minimal derives primarily from animal studies involving rats, in which thermogenic hormones such as thyroid hormone and sympathetic activity have been reported to be different from what is seen humans. Early studies using obese women reported no weight gain following a high-fat diet composed mainly of unsaturated fatty acids (Kasper at al. 1973). This finding suggests that even though a high-fat diet produces the smallest TEF, this type of diet may benefit weight loss via other mechanisms. For example, a high-fat diet has been associated with a reduction in the total caloric intake via the diet per se or via diet-related production of ketone bodies. The concept that a high-fat diet helps manage body weight is reflected in certain dietary programs such as the Atkins diet, and this type of diet may not necessarily be associated with a high level of blood cholesterol (Hays et al. 2003; Westman et al. 2006).

> **▶ KEY POINT ◀**
>
> Of the three macronutrients, protein is the most thermogenic, with an associated energy cost of 20% to 30% of the total energy intake, followed by carbohydrate (5-10%) and fat (0-5%). The greatest thermic effect associated with protein can be ascribed mainly to the extra energy needed for absorption by the gastrointestinal tract, as well as deamination of amino acids and synthesis of protein in the liver.

OTHER FACTORS INFLUENCING THERMIC EFFECT OF FOOD

A number of other factors also appear to affect TEF. As shown in table 11.1, these factors include macronutrient composition, dietary fiber content, age, environmental temperature, and alcohol consumption.

Macronutrient Composition

In an early study, Forbes and Swift (1944) compared TEF in rats that were fed protein, carbohydrate, and fat in single amounts and in combination as supplements to a standardized ration. They found that

Table 11.1   Factors That Influence the Thermic Effect of Food (TEF)

Factors	Effect
Type of macronutrient	Of the three macronutrients, fat is the least thermogenic, whereas protein is the most thermogenic.
Meal composition	Ingesting all three nutrients together produces a smaller increase in TEF as compared to the additive effect on TEF induced by protein, carbohydrate, and fat consumed separately.
Fiber content	A high-fiber meal is associated with a lower TEF.
Age	TEF declines with advancing age.
Environmental temperature	Consuming a meal in a cold environment may potentiate TEF.
Alcohol consumption	Alcohol consumption leads to an increase in TEF but has been found to reduce TEF in a cold environment.
Acute exercise	Exercise during the postprandial period potentiates TEF.
Training status	Individuals who train are associated with a decreased TEF.
Obesity	TEF is found to be lower in obese than in normal-weight individuals.

feeding all three nutrients at the same time produced a smaller increase in TEF as compared to the additive effect on TEF of protein, carbohydrate, and fat given separately. This finding implies that the body is more energetically efficient in handling a mixture of nutrients ingested simultaneously than when they are ingested separately. Although further studies are necessary to elucidate the underlying mechanism, it appears that absorbing and assimilating some nutrients can facilitate the disposition of other nutrients at the same time, thereby requiring less overall energy than would be projected based on the TEF for a single nutrient.

Dietary Fiber Content

Scalfi and colleagues (1987) compared **postprandial** thermogenesis over a 6 h period following the consumption of a **glucomannan**-supplemented meal, a high-fiber meal, and a lower-fiber meal of equivalent energy content. In contrast to the term postabsorptive, postprandial refers to the time period right after a meal during which digestion and absorption are taking place. Glucomannan is a polysaccharide; some parts of its chemical structure are similar to that of fiber, but its fiber content is lower than that of a typical high-fiber food. The authors observed a systemically lower rate of thermogenesis with both the glucomannan-supplemented and the high-fiber meal. The TEF was 6.9% after the lower-fiber meal, 5.1% after the glucomannan-supplemented meal, and 4.5% after the high-fiber meal. These differences were particularly apparent from 2 to 6 h after ingestion. It is thought that the lower thermogenesis associated with fiber is partly due to its slow absorption of glucose and thus to blunted insulin responses, which can reduce the energy required for insulin-mediated glucose disposal. These findings demonstrate that the fiber content of a meal, besides its well-known effect on energy intake, also affects energy expenditure.

Age

It is generally believed that TEF declines with advancing age. Visser and colleagues (1995) examined TEF for 3 h following a mixed meal composed of 300 kcal in older (~73 years of age) and younger (~25 years of age) men and women. No difference in TEF was observed between the younger and older women. However, TEF was lower in the older men than in the younger men. After having subjects ingest 75 g of glucose, Bloesch and colleagues (1988) also observed a lower TEF in older (~73 years) as compared to younger men (~25 years). The age-related reduction in TEF may be caused by a reduced ability to maintain glucose homeostasis (Felig 1984). A greater increase in blood glucose following a meal has been observed in elderly persons (Thorn and Wahren 1990). This can be explained by an age-related decrease in insulin-mediated glucose disposal, which leads to a corresponding decrease in the energy required to convert absorbed glucose into glycogen and thus a reduction in TEF. Given that muscle is a major depot for glucose storage, the impaired glucose disposal seen in those who are elderly can be further attributed to the loss of muscle mass associated with aging. This contention is underscored by the fact that in the studies of Visser and colleagues and Bloesch and colleagues, the age-related difference in TEF disappeared when energy expenditure was expressed relative to lean body mass. Among other explanations for the low TEF in elderly people is an age-related reduction in sympathetic nervous system activity. Thorne and Wahren (1990) observed a higher postprandial blood concentration of norepinephrine despite a reduced TEF in elderly subjects. This finding suggests that advancing age is associated with decreased tissue sensitivity to β-adrenergic stimulation, which is necessary in order for oxidative disposal of ingested energy substrates to proceed.

Environmental Temperature

It has long been known that metabolic rate increases as environmental temperature decreases. It has also been shown in animal studies that consuming a meal in a cold environment augments the diet-induced energy expenditure. For example, Rothwell and Stock (1980) observed that energy expenditure following a meal was greater in rats maintained in a cold environment (~4 °C) as compared to those kept in a warm environment (~28 °C). More recently, using humans, Kashiwazaki and colleagues (1990) also noted that energy expenditure after a meal was greater in subjects who stayed in a 20 °C room as compared to a 25 °C room. Interestingly, these cold-related elevations in thermal response following a meal appear to occur only when individuals have become acclimated to a cold environment. In the same study, Kashiwazaki and colleagues observed no difference in TEF between subjects in the two temperatures when measurements were taken in the summer. It seems that some degree of cold acclimatization is necessary in order for the protective increase in cold-related TEF to occur. The body's increased responsiveness to norepinephrine as a result of cold exposure is thought by some to play a major role in mediating cold-related increases in TEF.

Alcohol Consumption

Alcohol forms a significant portion of the total energy consumption in Western society. Although there is comparatively less research on the thermic effect of alcohol as compared to other nutrients, this nutrient has long been known to affect thermogenesis and substrate utilization. Alcohol increases short-term resting metabolism and suppresses fat oxidation (Yeomans et al. 2003). It also produces a greater TEF than most other carbohydrate foods (Stock and Stuart 1974). The increased energy expenditure due to alcohol consumption, however, can be easily offset by its stimulatory effect on appetite, which sometimes leads to overeating. Alcohol-induced thermogenesis is not always observed, especially when the quantity of alcohol consumed is quite small. Those who drink only occasionally often experience weight gain due to a small intake of alcohol concomitant with a resulting increase in appetite. On the other hand, chronic excess alcohol intake, frequently seen in those who are habituated to alcohol, normally leads to progressive weight loss.

INTERACTION BETWEEN PHYSICAL ACTIVITY AND THERMIC EFFECT OF FOOD

The interaction between TEF and physical activity has been examined in a number of different ways. It is clear that both physical activity and meal ingestion are thermogenic events. The question arises whether these two events would have a synergistic effect on energy expenditure if they occurred during same time period. In addition, reports indicate that obese individuals have blunted thermogenic responses following a meal, as discussed later in the chapter. Hence it is of interest to find out whether performing physical activity during the postprandial period potentiates diet-induced thermogenesis. A number of studies have examined the interactive effect of diet and physical activity on energy expen-

diture both in healthy individuals and in those with obesity. These studies have typically used a design in which subjects perform exercise with and without consumption of a test meal consisting of carbohydrate, fat, and protein so that investigators can determine the gross energy cost associated with exercise only and with exercise plus diet. In these studies, energy expenditure is also measured when subjects are not exercising but are under either the fasting or the postprandial condition. This permits quantification of the net energy cost associated with diet, exercise, or the combination of diet and exercise.

With this study design, Segal and Gutin (1983b) examined the interaction between exercise and diet in a group of lean and obese individuals. They demonstrated that in lean individuals, the energy expenditure during exercise under the fed condition was greater than the sum of the energy cost associated with exercise and the energy cost associated with the meal. This finding, which agreed with the results of an earlier study by Bray and colleagues (1974), suggests that combining exercise and a meal could be a strategy for enhancing energy expenditure. However, in obese individuals, for whom such a strategy might be desirable, it was found that exercise did not exert the same effect on TEF as in lean individuals. From a weight control standpoint, it would seem reasonable to encourage people to perform a bout of exercise not too long after eating in order to maximize energy expenditure. However, the fact that a synergistic effect of exercise and diet was not observed in obese subjects indicates that this reasoning requires further validation.

It seemed possible that the increase in energy cost during exercise following a meal that was seen in lean individuals might be due to a reduction in mechanical efficiency. To pursue this question, the same research group conducted additional studies not only to examine the synergistic effect of exercise and diet, but also to compare metabolic efficiency between lean and obese individuals (Segal and Gutin 1983a; Segal et al. 1984). The investigators found that although energy expenditure resulting from exercise and diet was greater in normal-weight than in obese individuals, mechanical efficiency remained the same between the two groups. This finding led to the conclusion that the thermogenic effect of food during exercise is related more to metabolic processes in tissues other than active skeletal muscle and that those metabolic processes may be impaired in obese individuals.

Both physical activity and meal ingestion are thermogenic events. When these two events occur during the same time period, they have a synergistic effect; that is, the amount of energy expended is greater than the sum of the thermic effects of the events considered separately. However, this is seen in normal-weight but not in obese individuals. For people of normal weight, performing a bout of exercise not too long after eating could be a strategy to maximize energy expenditure.

It appears that an acute bout exercise performed not too long after a meal can augment diet-induced thermogenesis in normal-weight adults. Can this finding be extrapolated to suggest that a synergistic effect of exercise and diet would be greater in those who exercise regularly? According to studies that compared trained and untrained individuals, this logic does not seem to hold. Tremblay and colleagues (1983) and LeBlanc and colleagues (1984b) reported that TEF was higher in untrained than in trained individuals. This difference was seen even when trained and untrained men were matched for body fat and fat-free mass (Poehlman et al. 1988). Tremblay and colleagues (1983) also showed a higher respiratory quotient in untrained than in trained individuals. It appears that those who are physically trained have a tendency to spare their postprandial energy expenditure and at the same time derive proportionally more of the postprandial energy expenditure from fat oxidation. In the study by LeBlanc and colleagues (1984a), plasma norepinephrine concentration increased significantly after meal consumption in untrained individuals, but remained unchanged in trained individuals. Given that norepinephrine is a thermogenic hormone, this finding explains at least part of the difference in TEF between trained and untrained groups.

THERMIC EFFECT OF FOOD AND OBESITY

Whether obesity is associated with impaired TEF is a question that has received sustained attention over the last several decades. Research on this topic can be traced back to 1924, when Wang and colleagues reported that the thermic response to protein was higher in thin than in normal-weight people, and that the response in both groups was higher than in obese individuals. To date, despite some controversy, a majority of studies have demonstrated smaller thermogenic responses not only to the ingestion of a meal, but also to stimuli such as cold exposure, norepinephrine infusion, and postprandial physical exercise in obese as compared to normal-weight adults. In addition, as discussed earlier, the synergistic effect of combined diet and exercise on energy output seems to be absent with obesity. Though TEF is a small portion of the total daily energy expenditure, some researchers argue that a defective TEF response could result in significant weight gain over time (Laville et al. 1993).

A study design involving the postobese model has been used to evaluate an inherent linkage between TEF and obesity. As mentioned in chapter 10, the postobese model is used to compare obesity-prone weight-reduced persons with obesity-resistant never-obese individuals. Such comparisons minimize the potential confounding effect associated with directly comparing obese and lean individuals who differ greatly in body composition. With this experimental approach, a number of studies have demonstrated an increase in TEF following weight loss in obese subjects such that their TEF became comparable with that of normal-weight individuals (Ravussin et al. 1983; Schwartz et al. 1983). Den Besten and colleagues (1988), however, found that weight reduction increased TEF only in women with gluteal-femoral obesity, not in women with abdominal obesity.

As discussed in chapter 10, resting metabolic rate decreases following weight loss due to dietary restriction. Several studies have shown that prolonged dietary restriction that produces weight loss also reduces TEF (Golay et al. 1989; Nelson et al. 1992; Schutz et al. 1987). It is unclear why TEF has responded to a weight loss regimen differently in different studies. As discussed in chapter 10, data errors can occur in studies using a postobese model because subjects may not have achieved stable energy equilibrium even though they have reached their target body weight. Indeed, in studies that yielded divergent findings on TEF, the weight stabilization period varied considerably, from seven days to three to four weeks. It appears that stabilization of the new weight for at least six months is the most stringent criterion that has been employed in studies using the postobese model. This effort toward experimental control should be enforced in all future studies of the impact of weight loss on various components of thermogenesis, including TEF and resting metabolic rate.

The purported defect in TEF associated with obesity may be further explained by insulin resistance, which has been shown to affect TEF even in the absence of obesity (Segal et al. 1992). Insulin resistance, defined as decreased ability of insulin to stimulate cellular glucose uptake and storage and to suppress hepatic glucose production, is commonly seen in persons with obesity. This condition can be assessed with use of the euglycemic, hyperinsulinemic glucose clamp technique; insulin resistance is said to be present if more insulin is needed in order to maintain a given level of blood glucose. Studies demonstrating an association between impaired TEF and insulin resistance have employed the clamp technique in conjunction with indirect calorimetry to measure energy expenditure. Given that a major part of the thermic response to a glucose infusion can be accounted for by glucose storage, this combined approach allows investigators to determine simultaneously the level of insulin-mediated glucose uptake and its impact on thermogenic responses. With this approach, Ravussin and colleagues (1983, 1985) found that a group of obese individuals showed a reduced glucose uptake in response to a given dose of insulin infusion, which suggests insulin resistance. They also found that this same obese group exhibited no increase in energy expenditure in response to glucose infusion, whereas glucose infusion resulted in a significant increase in energy expenditure in normal-weight individuals. These findings support the notion that insulin resistance mediates impairment in TEF.

The reduced TEF seen in people who are obese may also be attributable to decreased activity of the sympathetic nervous system. This notion has arisen from two lines of observation. First, it has been found that acute infusion of propranolol, a β-blocker, produces a decrease in TEF measured after the intravenous infusion of a glucose solution (Acheson et al. 1983; DeFronzo et al. 1984). Secondly, a reduced sympathetic response has been consistently demonstrated in studies using obese animal models (Bray 1986; Acheson et al. 1988).

However, there are questions about the validity of this idea. For example, it appears that in most studies, the administration of a β-blocker did not affect TEF measured after the oral ingestion of a mixed or carbohydrate-only meal (Thorne and Wahren 1989; Zed and James 1986; Zwillich et al. 1981). Also, in studies using obese humans, both increases and decreases in sympathetic activity have been reported (Peterson et al. 1986; Spraul et al. 1993; Troisi et al. 1991). Therefore, further studies are necessary to elucidate the role the sympathetic nervous system plays in mediating the obesity-related decrease in TEF. Again, these studies should be carefully designed in order to minimize the variation associated with measurement of TEF as discussed in the preceding sections of this chapter.

The theory that TEF is reduced in obesity appears plausible. However, the validity of the theory is still hampered by discrepant findings in the literature. This lack of consistency across studies can be largely ascribed to differences in the research methodology involved in measuring TEF. Despite the linkage between obesity and reduced TEF, it is still unclear whether the reduction in TEF occurs as a result of obesity or is a contributing factor to the development of obesity. Given that this energy component is quantitatively small with respect to the energy balance equation, it is unlikely that a defect in TEF serves as a cause of obesity. The prevalence of obesity has increased steadily over the last several decades. However, no evidence suggests that there has been a decline in TEF during the same time period (Hill and Melanson 1999).

> ### ▶ K E Y P O I N T ◀
>
> The TEF appears to be reduced in obese individuals, although it seems unlikely that this defect would cause obesity directly. The decrease in TEF has been ascribed to insulin resistance and the blunted sympathetic responses frequently associated with obesity. Proper responses to insulin and sympathetic stimulation are associated with an increase in energy expenditure.

SUMMARY

The TEF is a component that accounts for ~10% of the total daily energy expenditure. Although quantitatively small, its impact on energy balance can be significant over time. The TEF can be further divided into two subcomponents: obligatory and facultative thermogenesis. While the former represents the energy used for digestion, absorption, assimilation, and storage of absorbed nutrients, the latter is associated with energy expenditure caused by activation of the sympathetic nervous system. Consumption of protein produces the greatest TEF, followed by consumption of carbohydrate and of fat. The TEF appears to decrease with advancing age and with an increase in dietary fiber content, but to increase with consumption of alcohol and a decrease in environmental temperature. Of particular interest is that TEF can be augmented synergistically by the performance of exercise during the postprandial period, although this potentiation in energy expenditure may not occur in obese individuals. Persons with obesity demonstrate a blunted TEF. However, on the basis of the current literature, it seems unlikely that this defect plays a significant role in the pathogenesis of obesity.

KEY TERMS

assimilation

facultative thermogenesis

glucomannan

obligatory thermogenesis

postprandial

REVIEW QUESTIONS

1. What is the thermic effect of feeding? How does it differ with regard to consumption of protein, carbohydrate, and fat?

2. Distinguish the terms obligatory thermogenesis and facultative thermogenesis.

3. Distinguish the terms postabsporptive state and postprandial state.

4. How does performing an exercise bout during the postprandial period interact with TEF?

5. How would you explain the observation that obese individuals have a blunted TEF?

12

SELECTED PHARMACOLOGIC AND NUTRITIONAL SUBSTANCES

This chapter reviews selected nutritional substances and ergogenic products that have a direct impact on energy expenditure and fat utilization. Currently, a plethora of compounds or substances either are approved for use by prescription or are under pharmacological development. In addition, a broad spectrum of over-the-counter substances are claimed to be effective as slimming or ergogenic aids. In keeping with the theme of the book, this chapter reviews selected pharmacologic and nutritional substances whose use is supported by a plausible rationale and that have been extensively evaluated, including sibutramine, leptin, ephedrine, and caffeine. Each is discussed with particular reference to its impact on the management of body weight, the enhancement of physical performance, or both.

SIBUTRAMINE

Sibutramine, also known by the trade names Meridia and Reductil, is approved for the treatment of obesity in the United States as well as in Europe. It is used together with a reduced-calorie diet to help obese individuals to lose weight and to maintain weight loss. This drug is especially recommended for people who, in addition to being overweight, have other health problems such as hypertension, diabetes, or high cholesterol. Thus this medication helps people with comorbidities to reduce their weight and therefore their health risks.

Working Mechanism

The drug works by boosting levels of certain chemical messengers in the nervous system, including **serotonin, dopamine,** and norepinephrine, all of which have been shown to reduce energy intake. Controlled clinical trials in patients with obesity have consistently shown dose-related reductions in body weight with sibutramine, as seen in figure 12.1. Typically, weight loss was 3 to 5 kg (6.6 to 11 lb) greater than with placebo at 24 weeks, and this weight reduction was sustained for two years with continued treatment. The proportion of patients losing at least 5% of body weight over 12 months was 29% with placebo, 56% with 10 mg/day of sibutramine, and 65% with 15 mg/day of sibutramine (Bray et al. 1996). When weight loss was induced with diet, patients randomized to the sibutramine treatment continued to lose weight over a one-year period to a level 15% below baseline, whereas placebo-treated patients regained some weight (Apfelbaum et al. 1999). This drug can remain effective when used for a relatively long period of time. One of the biggest challenges in combating obesity with dietary intervention is maintaining the hard-earned weight loss.

Although sibutramine is primarily thought of as an appetite-suppressing agent, it seems that the mechanism for long-term weight loss with this medication in combination with diet may not be due entirely to an anorectic effect. It has been inferred from animal studies that the drug may

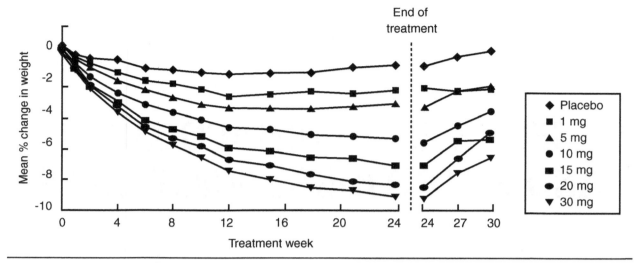

Figure 12.1   Dose-related weight loss with sibutramine.

Reprinted from *Nutrition*, Vol. 16, G.A. Bray, A concise review on the therapeutics of obesity, pp.953-960, Copyright 2000, with permission from Elsevier Limited.

also function to elevate energy expenditure. For example, Day and Bailey (1998) noted a decrease in body weight in obese mice treated with sibutramine unaccompanied by a corresponding decrease in food intake. This conclusion was subsequently confirmed by Connoley and colleagues (1999), who observed a 30% increase in oxygen consumption and an elevation in core temperature of ~1 °C following administration of sibutramine to rats. It is believed that the thermogenic effect of sibutramine is transmitted via the sympathetic nervous system and involves brown adipose tissue, which differs from most fat tissue in the body in that it is metabolically active and releases energy in the form of heat without adenosine triphosphate (ATP) production.

Demonstrating a thermogenic effect of sibutramine in humans appears to be more difficult. Seagle and colleagues (1998) failed to observe a change in energy expenditure during an eight-week study in spite of a clear reduction in body weight in overweight women. However, using healthy men, Hansen and colleagues (1999) found that administration of sibutramine increased energy expenditure and plasma norepinephrine levels during a 5 to 6 h postdose period in both the fed and the fasted state. It has been argued that even small effects on thermogenesis, which can be masked by the natural variation in energy expenditure in humans and below limits of detection using current methods, may still exert a clinically significant effect in the long-term treatment of obesity (Danforth 1999).

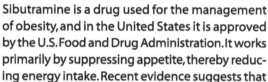

KEY POINT

Sibutramine is a drug used for the management of obesity, and in the United States it is approved by the U.S. Food and Drug Administration. It works primarily by suppressing appetite, thereby reducing energy intake. Recent evidence suggests that this anorectic agent can also raise body temperature and boost metabolism via sympathetic stimulation, although this thermogenic effect remains to be further substantiated in humans.

Potential Side Effects

With respect to its impact on the cardiovascular system, sibutramine is considered a **sympathomimetic** drug. This means that it has the potential to increase blood pressure and heart rate, which can in turn trigger undesirable cardiac events especially in those who are at risk of developing cardiovascular diseases. Thus people with uncontrolled high blood pressure should not take sibutramine or should take it in a carefully selected dosage. The potential side effects of the drug require careful evaluation. Indeed, some sympathomimetic drugs, such as phentermine and fenfluramine, have been withdrawn by the manufacturers due to their association with an increased risk of valvular disease and, more rarely, pulmonary hypertension. Among other side effects of sibutramine are **insomnia,** dry mouth, headache,

asthenia, and constipation. It is also recommended that sibutramine not be taken with other medicines for weight loss, including herbal and nonprescription weight loss drugs like ma huang or ephedra that in themselves can augment blood pressure and heart rate.

LEPTIN

Leptin (the term is derived from Greek *leptos,* meaning "thin") was discovered in 1994 following the isolation of the obese (or simply *ob*) gene (Zhang et al. 1994). It is thought that the *ob* gene, found primarily in adipose tissue, encodes and stimulates production of a body fat–signaling, hormone-like protein called *ob* protein or leptin, which then enters the bloodstream and travels to the hypothalamic area that controls appetite and metabolism.

Working Mechanism

As shown in figure 12.2, leptin functions as an afferent signal and binds with special receptors in the hypothalamus in order to execute its action, that is, to suppress eating drive. In theory, leptin blunts the urge to eat when caloric intake maintains ideal fat stores. However, its production decreases in accordance with a reduction in body fat, and this then helps augment or restore appetite and eating drive. In this context, leptin has been considered responsible for the great difficulty obese individuals have in sustaining significant fat loss. Leptin may also stimulate energy expenditure, and this thermogenic role appears to be mediated through the activation of the sympathetic nervous system (Hukshorn and Saris 2004). The production of leptin is regulated by the status of fat storage. More leptin is produced with increased body fat. Conversely, when body fat stores are low, the production of leptin decreases.

Leptin and Weight Management

Given the inhibitory effect of leptin on energy intake, numerous researchers during the last decade or so have examined whether this satiety-signaling molecule can be used to regulate energy balance and thus prevent obesity. Studies using an animal model have demonstrated a weight loss effect due to decreased food intake when animals were injected with **recombinant** leptin (Pelleymounter et al. 1995; Halaas et al. 1995; Campfield et al. 1995). These findings with animal models raised expectations that

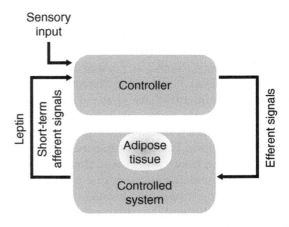

Figure 12.2 Body weight regulation can be viewed as a feedback system. In this system, the brain is the controller that serves as receiver and transducer of input signals and as an integrator of output signals to the neural and hormonal systems. The efferent signals modulate food consumption and regulate metabolism. The control system processes, stores, and metabolizes the nutrients obtained from food. The afferent signals comprise neural (vagal and sensory) stimulation and hormonal (leptin) stimulation that tell the controller about the state of the body and its surroundings.

Adapted by permission from Macmillan Publishers Ltd.: *Nature,* G.A. Bray and L.A. Tartaglia, "Medical strategies in the treatment of obesity," 404(6778): 672-677, Copyright 2000.

obesity in humans may be a state of leptin deficiency that is treatable with supplementation of exogenous leptin. However, most studies on humans have not only failed to demonstrate **mutations** in the gene encoding for leptin, but also shown that those who are obese actually have elevated serum leptin levels (Maffei et al. 1996; Considine et al. 1995, 1996). Some studies did demonstrate weight loss with the injection of recombinant leptin in humans (Farooqi et al. 1999; Heymsfield 1999; Mackintosh and Hirsch. 2001). However, in these studies, either the subjects were leptin deficient or the amount of weight loss was rather modest. Leptin levels and body fat are highly correlated, and it has been suggested that body fat accounts for approximately 50% to 60% of the variations in serum leptin concentrations. The high levels of leptin in obese individuals have led to the hypothesis that rather than being leptin deficient, these people may be leptin resistant or have reduced tissue sensitivity to leptin. Administration of more leptin may seem like a reasonable solution. However, Heymsfield (1999) and Mackintosh and colleagues (2001) showed that even supraphysiological leptin concentrations failed to significantly affect body weight in humans during weight maintenance or mild hypocaloric conditions.

Reduction of leptin resistance may theoretically be achieved through improving leptin transport across the **blood–brain barrier.** However, this idea remains exploratory and at present has not been definitively shown. It is possible that although obese individuals frequently have elevated blood leptin levels, they may not necessarily have elevated leptin levels in the cerebral spinal fluid compartment due to decreased leptin transport capacity. Currently, it remains unclear how leptin crosses the blood–brain barrier. Some evidence suggests that this occurs via specialized leptin transporters located along the brain capillary **endothelium,** but this contention remains to be further substantiated. It appears that increasing leptin transport across the blood–brain barrier may be a more important antiobesity strategy than increasing leptin per se.

▷ K E Y   P O I N T ◁

Leptin works by inhibiting energy intake. This tends to suggest that those who are obese may be leptin deficient. However, recent evidence suggests that obese individuals actually have elevated serum leptin levels. This leads to the hypothesis that obesity may be a condition associated with reduced tissue sensitivity to leptin, or impaired transport of leptin across the blood–brain barrier, or both.

Effect of Exercise on Leptin

The impact of exercise on leptin is another major area of interest within recent years. Exercise has been shown to be able to reduce body fat. Thus if leptin levels are affected as well, this may help provide some insight into how exercise exerts such a metabolic benefit. Exercise can also unbalance energy homeostasis by increasing energy expenditure. The question remains whether the blood leptin concentration can be used as a sensitive marker to reflect the derailment of energy homeostasis. In general, acute exercise of both short and long duration has been shown to have no or minimal impact on blood leptin levels. However, this appears to be the consequence of the relatively small amount of energy typically expended in a single bout of exercise. Exercise that produces a sufficient energy imbalance, however, such as a marathon or ultramarathon, seems to be able to reduce the 24 h mean and amplitude of the **diurnal**

rhythm of leptin, although this suppressive effect was found to be counterbalanced by feeding (Kremer et al. 2002). As with many other endocrine hormones, leptin concentrations in humans show a diurnal pattern, with the highest plasma concentration between midnight and early morning and the lowest around noon to midafternoon (Sinha et al. 1996; Schoeller et al. 1997).

Though the effect is questionable with short-term exercise, long-term training (>16 months) has been associated with a reduction in leptin; and it appears that the greater the volume of training, the lower the leptin level (Pasman et al. 1998). The training-induced reduction in leptin is in part due to a decrease in body fat. However, this adaptive change was also observed even after fat loss was accounted for, which suggests that factors other than improved body adiposity contribute to this training effect as well. It is thought that the training-induced reduction in leptin may reflect enhanced tissue sensitivity to this satiety hormone (Kremer et al. 2002). Reductions in leptin have also been attributed to alterations in energy balance, as well as improved insulin sensitivity and fat metabolism, although the exact mechanism whereby these metabolic improvements would affect the expression of leptin remains to be elucidated.

EPHEDRINE

Ephedrine is a sympathomimetic agent that is structurally related to catecholamine. It is found in several species of the plant ephedra and has been used for thousands of years as an herbal medicine. Ephedra, which is also called ma huang, is the dry stem of a plant indigenous to China, Pakistan, and northwestern India. Ephedra has two major active components, ephedrine and pseudoephedrine, although ephedrine is found in substantially higher concentrations in the plant and on an equal-weight basis is about twice as potent as pseudoephedrine. While ephedrine occurs naturally in some botanicals, it can also be synthetically derived. Ephedrine is widely available in over-the-counter remedies for nasal congestion and hay fever. It has also been used as a central stimulant to treat depression and sleep disorders.

Working Mechanism

Ephedrine is a potent chemical stimulant with a variety of peripheral and central effects. It acts by enhancing the release of norepinephrine from sympathetic neurons and is also a potent β-**adrenergic agonist.**

In this regard, ephedrine has been known for its effect of stimulating bronchodilation, increasing heart rate and cardiac contraction, and augmenting energy expenditure and fat oxidation. Ephedrine also functions to suppress appetite and food intake through adrenergic pathways in the hypothalamus.

Effect on Weight Loss

Ephedrine is recognized as an antiobesity drug mainly due to its stimulating effect on the sympathetic nervous system. Ephedrine's potential for weight loss was first reported in 1972. It was noted that asthmatic patients who were being treated with a compound containing ephedrine, caffeine, and **phenobarbital** experienced unintentional weight loss. This unexpected discovery led to a series of investigations aimed at substantiating whether ephedrine is indeed effective in facilitating weight loss and whether it warrants safety concerns despite its efficacy. In the early studies in which ephedrine was the only substance tested, results on energy expenditure and weight loss were controversial. Astrup and colleagues (1985) noticed a sizable increase in resting oxygen consumption and a significant reduction in body weight in obese women who were treated with ephedrine at 60 mg/day. However, Pasquali and colleagues (1987a) found that administration of ephedrine at a dose of 75 or 150 mg/day produced essentially no effect compared to the placebo condition in obese individuals. In this study, side effects such as agitation, insomnia, headache, palpitation, giddiness, tremor, and constipation were reported in the group treated with 150 mg/day. This same research group, however, observed that when ephedrine was given at 150 mg/day in conjunction with a more stringent hypocaloric diet (~1000 kcal/day), obese women did experience a significant weight loss (Pasquali et al. (1987b). It appears that ephedrine's efficacy is questionable especially when it is given alone.

Function as an Ergogenic Aid

Ephedrine has also been promoted as a performance-enhancing or ergogenic aid. However, most studies have not demonstrated any kind of improvement in athletic performance following ingestion of ephedrine alone at a dose generally considered safe, that is, <120 mg. DeMeersman and colleagues (1987) found no significant effects of ephedrine administered in a dose of 40 mg on sustained aerobic exercise. Swain and colleagues (1997) also failed to observe increases in $\dot{V}O_2$max and endurance time to exhaustion following consumption of pseudoephedrine at doses of

1 mg · kg^{-1} or 2 mg · kg^{-1}. Ephedrine also seems to be ineffective with respect to its impact on anaerobic performance. Chu and colleagues (2002) reported that ingestion of 120 mg pseudoephedrine 2 h before testing did not improve muscular strength as measured by intermittent isometric contraction or anaerobic performance as measured by the Wingate cycling test (30 s maximal cycling against a predetermined resistance). It may be argued that a threshold dosage level may exist at which the ergogenic effects of ephedrine become manifest, as peak weightlifting performance was improved after subjects took a high dose (180 mg) of pseudoephedrine (Gill et al. 2000). Nevertheless, in a more recent study in which subjects took 240 mg of pseudoephedrine, Chester and colleagues (2003) failed to demonstrate an ergogenic effect, although the testing protocol in this study involved aerobic rather than anaerobic performance.

Interactive Effect of Ephedrine and Caffeine

Ephedrine taken along with caffeine may be more effective in facilitating energy expenditure and weight loss. In theory, this hypothesis is plausible. It has been suggested that while ephedrine stimulates the release of norepinephrine, caffeine acts to delay the degradation of norepinephrine in the neural cleft, thereby prolonging its effect (Atkinson 1997). The effectiveness of combined use of ephedrine and caffeine appears to be supported by a majority of studies (Astrup et al. 1991, 1992; Dulloo and Miller 1986; Malchow-Moller et al. 1981; Toubro et al. 1993). The supporting evidence may be best illustrated by the study of Astrup and colleagues (1992), who compared placebo, caffeine at 200 mg three times a day, ephedrine at 20 mg three times a day, and the combination of caffeine (200 mg three times a day) and ephedrine (20 mg three times a day) in 180 obese subjects. As shown in figure 12.3, weight loss was significantly greater in the group receiving both caffeine and ephedrine, whereas weight losses in the caffeine-only and the ephedrine-only group were no different from those in the placebo group.

More recently, a great deal of research has been undertaken to investigate the ergogenic effect of ephedrine in combination with caffeine. There is a general consensus that the combined use of ephedrine and caffeine is of greater ergogenic benefit than use of each compound alone. For instance, Bell and colleagues (2001) demonstrated a significant increase in power output during a 30 s Wingate test following

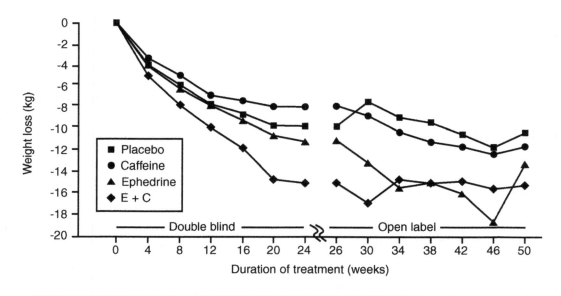

Figure 12.3 Effect of the combination of 20 mg ephedrine and 200 mg caffeine three times per day on weight loss and weight maintenance for one year. The first 24 months involved a randomized, double-blind, placebo-controlled study, whereas all groups received E + C from week 26 to 50.

Reprinted from *Nutrition*, Vol. 16, G.A. Bray, A concise review on the therapeutics of obesity, pp.953-960, Copyright 2000, with permission from Elsevier Limited.

the ingestion of ephedrine (1 mg · kg⁻¹) alone and in combination with caffeine (5 mg · kg⁻¹) as compared with caffeine alone and placebo. Using the same treatment paradigm, this research group also found a significant reduction in times for a 10K race following ingestion of ephedrine (0.8 mg · kg⁻¹) alone and in combination with caffeine (4 mg · kg⁻¹) as compared with caffeine alone and placebo (Bell and colleagues 2002). The mechanism responsible for the performance advantages of this combined approach is unclear. It may relate to the hypothesis mentioned in connection with weight loss, that caffeine serves to prolong the ephedrine-induced adrenergic effect. It should be noted that the amount of ephedrine used by Bell and colleagues is about twice as potent as the amount of pseudoephedrine used in many early studies. This may have contributed to the positive findings shown in most of Bell's publications.

Potential Side Effects

The most common side effects for ephedrine are agitation, insomnia, headache, palpitations, dizziness, tremor, and constipation, many of which are quite similar to those associated with sympathomimetic drugs like phentermine and fenfluramine as mentioned earlier. Ephedrine has also been considered able to trigger cardiovascular events such as **tachycardia,**

cardiac arrhythmias, angina, and vasoconstriction with hypertension, although these cardiovascular side effects have been found to be rather infrequent and tend to diminish on repeated dosing (Waluga et al. 1998). Ephedrine itself has never been illegal in the United States or in many other parts of the world. However, societal concerns about the side effects associated with abuse of ephedrine have steadily increased. In 1997, the Food and Drug Administration (FDA) in the United States proposed a regulation that limited an ephedra dose to 8 mg, with no more than 24 mg/day for a maximum of seven days. In 2004, the FDA created a ban on ephedrine alkaloids that are marketed for reasons other than asthma, colds, allergies, other diseases, or traditional Asian use.

> **▷ K E Y   P O I N T ◁**
>
> Ephedrine when used in conjunction with caffeine has proven effective in stimulating weight loss and improving athletic performance. This herbal derivative, however, has many side effects, some of which can be dangerous especially with use at high doses. Thus, despite its efficacy, the use of this substance has been banned by the FDA for purposes other than medical treatment.

CAFFEINE

Caffeine occurs naturally in a variety of beverages and foods including coffee, tea, and chocolate. It is consumed by most adults worldwide. It has been estimated that the daily intake of caffeine for an average adult is approximately 3 mg · kg^{-1}, 80% of which is consumed in the form of coffee (Barone and Roberts 1996). Caffeine is recognized as a food and a drug in both the scientific and regulatory domains. It is, however, not a typical nutrient and is not essential for health. Several over-the-counter medications contain caffeine in amounts from 30 to 100 mg. These include cold remedies, diuretics, weight loss products, and preparations that help people stay awake. Caffeine can be rapidly absorbed in the digestive tract and distributed to all tissues. It can also easily cross the blood–brain barrier to reach the tissues of the central nervous system.

Working Mechanism

Caffeine serves as an antagonist of the adenosine receptor. Adenosine is an intermediate derivative in a wide range of metabolic pathways and is a constituent of ATP and nucleic acids. Administration of adenosine can produce sedation, **bradycardia,** hypotension, hypothermia, and attenuation of the responses of the heart, blood vessels, and adipose tissue to sympathetic stimulation. Caffeine is very similar in structure to adenosine and can bind to membrane receptors for adenosine, thereby blocking its actions. As noted earlier, caffeine facilitates the actions of hormones such as epinephrine and norepinephrine. This occurs via an effect on the intracellular response of cyclic adenosine monophosphate (cyclic AMP), whose production increases due to increased hormonal secretion. As a secondary messenger, cyclic AMP activates protein kinases, which then catalyze a cascade of enzyme phosphorylation necessary to execute hormonal actions. Caffeine also inhibits phosphodiesterase, an enzyme responsible for degrading cyclic AMP. This inhibition leads to maintenance of the cellular level of cyclic AMP, thereby potentiating the effects of thermogenic hormones.

Effect on Cognitive Function

Compared to the situation with ephedrine, the existing literature appears to be more definitive in support of the ergogenic effects of caffeine on cognitive and physical performance. On the other hand, caffeine alone is not as effective in modulating weight loss. Caffeine's cognitive and behavioral effects have been documented in a number of well-controlled studies using males, females, and young and elderly volunteers. Effects on particular aspects of cognitive function, as well as effects on mood state, are generally consistent with the lay perception of caffeine as a compound that increases mental energy and performance. For example, using a double-blind, placebo-controlled protocol, Fine and colleagues (1994) reported that a single dose of 200 mg of caffeine improved visual vigilance in rested volunteers. Similar effects have been documented with doses equivalent to a single serving of a cola beverage (~40 mg) up to multiple cups of coffee (Lieberman et al. 1987a, 1987b). However, with a very high dose of caffeine (i.e., ~500 mg), cognitive performance was reported to decrease (Kaplan et al. 1997). This finding indicates that a dosage achievable via the diet should be sufficient to produce the cognitive benefits of caffeine. Caffeine has also been shown to enhance aspects of cognitive performance such as reasoning and memory when an individual's mental ability is decreased due to sleep deprivation (Penetar et al. 1993).

Effect on Sport Performance

With respect to sport performance, the most consistent observation is that caffeine can increase the time to exhaustion during submaximal exercise bouts lasting approximately 30 to 60 min, though results are more inconsistent when activities of shorter duration are examined. Caffeine as an ergogenic substance has been evaluated over several decades. Despite overwhelming agreement on caffeine's ability to improve endurance performance, the precise mechanism whereby this compound exerts its ergogenic effect remains elusive. For years it has been postulated that caffeine causes glycogen sparing and therefore prolongs endurance performance. However, in terms of the exercise duration for which caffeine has been found to be effective, it is unlikely that muscle glycogen would be depleted and thus serve as a limiting factor. Graham and colleagues (2000), who employed a muscle biopsy technique, found that ingestion of caffeine at 5 mg · kg^{-1} did not alter muscle glycogen utilization. Most studies have used a rather small sample size (i.e., ~8 subjects), which may have reduced the statistical power needed to detect differences. Nevertheless, after pooling together studies conducted in the same laboratory but at different times and involving a total of 37 subjects, Graham (2001) failed to observe a significant difference in muscle glycogen utilization between the caffeine and placebo conditions.

Effect on Fat Utilization

Another metabolic claim associated with caffeine is that it increases fat utilization. This notion was derived from early studies in which caffeine resulted in a larger decrease in muscle triglycerides concomitant with less utilization of muscle glycogen as compared to a placebo condition (Essig et al. 1980). It has been speculated that caffeine increases fat oxidation by augmenting lipolysis, a process stimulated by the release of epinephrine and norepinephrine. However, according to recent studies by Raguso and colleagues (1996) and Graham and colleagues (2000), variables reflecting intracellular utilization of fatty acids such as fat oxidation and uptake of fatty acids by muscle remained unaffected by caffeine. These findings provide little support for the theory that caffeine increases fat oxidation. It appears that the physiological processes involved in the actions of caffeine may go well beyond the traditional explanation. The fact that caffeine blocks the action of adenosine is a well-established phenomenon, and this may serve as a more compelling direction for future research aimed at unraveling the mechanism whereby caffeine functions as an ergogenic aid.

> ### ▶ KEY POINT ◀
>
> Caffeine's ability to increase time to exhaustion during aerobic exercise has long been attributed to its stimulating effect on lipolysis and fat oxidation and consequent sparing of muscle glycogen. However, recent studies have failed to detect caffeine-related alterations in substrate utilization. It appears that the physiological processes involved in the ergogenic effect of caffeine may go well beyond the traditional explanation.

SUMMARY

Sibutramine, leptin, ephedrine, and caffeine are pertinent to bioenergetics in that they are able to alter energy intake, energy expenditure, or both. Both sibutramine and leptin function primarily by inhibiting appetite, thereby producing a negative caloric balance favorable for weight loss. However, they differ in that sibutramine is obtained by prescription whereas leptin is a naturally produced substance that can be acquired as a weight loss supplement without prescription. Ephedrine, on the other hand, is a potent stimulator of the sympathetic nervous system and, when used in conjunction with caffeine, can augment energy expenditure, thereby facilitating weight loss or enhancing athletic performance.

As with any substance or medication, potential side effects are always an issue of interest and are often evaluated against the intended actions. Among the substances discussed here, ephedrine appears to be of most concern because of its association with many serious side effects including cardiac arrhythmias, angina, and high blood pressure. In this regard, the Food and Drug Administration has placed tough restrictions on the use of ephedrine in the United States.

Although caffeine alone is not as effective in producing weight loss, its ergogenic effect on cognitive function and mental performance has been well evidenced. Caffeine can also prolong endurance, but the underlying mechanism is currently being debated. The traditional explanation that caffeine stimulates fat utilization and thus spares glycogen has not been shown in recent studies.

KEY TERMS

asthenia

β-adrenergic agonist

blood–brain barrier

bradycardia

diurnal rhythm

dopamine

endothelium

insomnia

mutations

phenobarbital

recombinant

serotonin

sympathomimetic

tachycardia

REVIEW QUESTIONS

1. Describe the working mechanism of sibutramine in treating obesity. What are the side effects associated with this drug?

2. What is leptin? How is this substance related to obesity?

3. Explain how caffeine functions in potentiating the effect of ephedrine.

4. List potential side effects associated with ephedrine. What are the governmental restrictions relating to the use of this substance?

Answers
to Review Questions

CHAPTER 1

1. The law states that energy is neither created nor destroyed; it is merely transformed from one state to another. Kinetic energy is defined as the energy possessed by a body because of its motion. Potential energy is defined as the energy possessed by a body as a result of its position or condition. In a high jump or pole vault, jumpers accelerate to maximize their kinetic energy before takeoff. The kinetic energy they possess allows them to ascend and leap over the bar, and at the height of the leap they gain potential energy.

2. Using Atwater factors, the total energy is computed as $15 \text{ g} \times 4 \text{ kcal/g} + 9 \text{ g} \times 9 \text{ kcal/g} + 4 \text{ g} \times 4 \text{ kcal/g} = 60 \text{ kcal} + 81 \text{ kcal} + 16 \text{ kcal} = 157$ kcal.

3. The three energy systems can be compared as shown in appendix table 1.

4. The terms are defined as follows:

   Glycogenolysis: the biochemical breakdown or hydrolysis of glycogen to glucose

   Lipolysis: the biochemical breakdown or hydrolysis of triglycerides or other compound fats into fatty acids and glycerol

   Deamination: the process of removing an amino group, that is, NH_2, from an organic compound such as amino acid

   Transamination: the process of transferring an amino group from one compound to another

5. The hypothesis states that electron transport and adenosine triphosphate (ATP) synthesis are coupled by a proton gradient across the inner mitochondrial membrane. In particular, the energy released as electrons are transferred along the respiratory chain leads to the pumping of H^+ protons from the matrix to the other side of the inner mitochondrial membrane. As a result, there is a higher concentration of H^+ within the intermembrane space compared to that in the matrix. It is this proton gradient that serves as a source of energy for ATP synthesis.

CHAPTER 2

1. Several advantages are associated with the use of carbohydrate as an energy source. First, stored carbohydrate such as glycogen can be degraded into glucose units quickly. Second, glucose can be metabolized for energy with and without oxygen. A process such as glycolysis that doesn't

Appendix Table 1   Comparison of ATP-PCr, Glycolytic, and Oxidative Energy Systems

Characteristic	ATP-PCr	Glycolysis	Oxidative
Complexity	Simple	Moderate	Complex
Cellular location	Cytoplasm	Cytoplasm	Mitochondria
End products	ADP, P_i, Creatine	Lactic acid	H_2O and CO_2
Oxygen requirement	No	No	Yes
Rate of ATP production	High	Moderate	Slow
Capacity of ATP production	Low	Moderate	High
Sample sporting event	100 m	800 m	10K

depend on oxygen provides a rather quick means of furnishing energy. Third, oxidation of glucose needs to occur in order to burn off stored fat. Finally, carbohydrate is the only fuel that can be used by the central nervous system for its survival and functioning.

2. As intensity increases, utilization of carbohydrate increases. This explains why carbohydrate is essential to many sporting events in which activities are often performed at high intensity. On the other hand, to maximize the utilization of fat, one may choose to exercise at a lower intensity for a longer duration, although performing exercise that is too mild can hamper utilization of the absolute quantity of fat.

3. As an exercise continues for a prolonged period, there is progressively less dependence on carbohydrate energy and a concomitant increase in fat oxidation. Together these changes help maintain an adequate energy supply. In extreme cases in which glycogen stores of both muscle and liver decrease significantly, there can be an increase in protein degradation that leads to the formation of amino acids. Amino acids can then provide additional energy via gluconeogenesis and the Krebs cycle.

4. Gluconeogenesis is the process of producing glucose from nonglucose molecules such as amino acids, lactate, and glycerol. The Cori cycle and the glucose-alanine cycle are pathways that have been identified as examples of gluconeogenesis. The Cori cycle is a pathway whereby lactate is recycled to form new glucose molecules, whereas the glucose-alanine cycle is a pathway in which muscle protein is converted into new glucose molecules.

5. β-oxidation is the oxidative pathway in which long-chain fatty acids are degraded into two-carbon units that can enter the Krebs cycle. It takes place in mitochondria. β-oxidation differs from lipolysis in that lipolysis is the process of degrading triglycerides into fatty acids and glycerol and occurs outside of the mitochondria. For a given triglyceride to be used for energy, lipolysis must take place first. This is then followed by β-oxidation.

6. In order for proteins to be used for energy, they must be degraded into amino acids. Some amino acids can then be converted into intermediates of the Krebs cycle, whereas others can be transformed into glucose via gluconeogenesis.

Because of their molecular structure, branched-chain amino acids are considered more capable of producing energy than other amino acids. They have also been found to benefit tissue repair. Branched-chain amino acids can also function as neurotransmitters, contributing to a reduction in the feeling of fatigue.

CHAPTER 3

1. Homeostasis is defined as the maintenance of a constant or unchanging internal environment. The term is often used in relation to an unstressful condition such as resting. Steady state is the term frequently used by exercise physiologists to denote a stable physiological environment maintained during exercise. In most cases, the term refers to a match between oxygen demand and oxygen supply.

2. A typical control system consists of (1) receptors, (2) afferent pathway, (3) integrating center, (4) efferent pathway, and (5) effectors. A control system operates in a negative feedback fashion. This can be illustrated in the regulation of blood pressure. An increase in blood pressure stimulates pressure-sensitive receptors within major arteries, generating nerve impulses to be transmitted to the medulla of the brain. As a result, this cardiovascular control center decreases the number of impulses transmitted to the heart, thereby lowering the amount of blood being ejected and causing blood pressure to return to normal.

3. The Randle cycle provides a theoretical explanation for how an increase in fat utilization reduces carbohydrate utilization. With an increase in plasma fatty acid concentration, there is an increase in fatty acid entry into the cell and a subsequent increase in β-oxidation in which fatty acids are broken down to acetyl-CoA. An increased concentration of acetyl-CoA then inhibits the pyruvate dehydrogenase complex that breaks down pyruvate to acetyl-CoA. In addition, the increased production of acetyl-CoA increases the concentration of citrate, an intermediate of the Krebs cycle. This increased citrate level then inhibits phosphofructokinase (PFK), a rate-limiting enzyme of glycolysis, as well as hexokinase, which regulates cellular glucose uptake. These reduced enzymatic activities ultimately decrease carbohydrate utilization.

4. The metabolic role and secreting gland for each hormone is shown in appendix table 2.

5. Blood glucose concentration can be maintained during most prolonged exercises despite the fact that muscle glycogen decreases significantly. This is accomplished due to a series of concerted adjustments made in an effort to maintain energy homeostasis. These adjustments include (1) mobilization of glucose from liver glycogen stores, (2) increased fat utilization to spare plasma glucose, (3) synthesis of new glucose from nonglucose molecules via gluconeogenesis, and (4) blocking the entry of glucose from plasma to prompt more use of fat as a fuel.

C H A P T E R   4

1. A calorie is defined as the amount of heat needed to raise the temperature of 1 gram of water by 1 °C. A kilocalorie or Cal is defined as the amount of heat needed to raise the temperature of 1 kilogram of water by 1 °C. The calorie equivalent is a measure of the amount of heat produced as a result of using 1 L of oxygen. For each liter of oxygen used, approximately 4.7 to 5.0 kcal are produced, and this measure differs depending on the relative uses of carbohydrate and fat.

2. Both direct calorimetry and indirect calorimetry are techniques for measuring energy expenditure. Direct calorimetry measures the amount of heat produced as a result of metabolism, whereas indirect calorimetry measures the amount of oxygen being used during energy transformation. Closed-circuit and open-circuit spirometry are two techniques used to determine pulmonary gas exchange including oxygen uptake, which is the difference between the oxygen inspired and the oxygen expired. The two techniques differ in that closed-circuit spirometry requires the subject to inhale oxygen gas, whereas open-circuit spirometry allows the subject to inhale ambient air.

3. The respiratory quotient is a ratio of carbon dioxide production to oxygen consumption. As this ratio is influenced by the kinds of substrates being oxidized, it has been used to reflect patterns of fuel utilization. For example, a ratio of 0.7 suggests that all energy is being derived from fat oxidation; a ratio of 1.0 suggests that all energy is being derived from carbohydrate oxidation; and a ratio of 0.85 suggests that energy is being derived equally from fat and carbohydrate.

4. First, determine the respiratory quotient (RQ) as follows:

$$RQ = CO_2 \text{ production} \div O_2 \text{ uptake}$$
$$= 1.2 \text{ L/min} \div 1.5 \text{ L/min} = 0.8.$$

According to the RQ table, a ratio of 0.8 suggests that the caloric equivalent is 4.8 kcal/L; thus the rate of energy expenditure is 1.5 L/min × 4.8 kcal/L = 7.2 kcal/min. This RQ of 0.8 also indicates that the energy derived from carbohydrate and fat is 33% and 67%, respectively. Thus carbohydrate energy is 2.4 kcal/min (7.2 kcal/min × 0.33) and fat energy is 4.8 kcal/min (7.2 kcal/min × 0.67).

Appendix Table 2   Sources and Metabolic Roles of Epinephrine, Insulin, Glucagon, Cortisol, and Growth Hormone

	Endocrine gland	Role in energy metabolism
Epinephrine	Medulla of adrenal gland	Stimulates glycogenolysis Stimulates lipolysis
Insulin	Pancreas	Stimulates glycogen synthesis Stimulates lipogenesis
Glucagon	Pancreas	Stimulates glycogenolysis and gluconeogenesis Stimulates lipolysis
Cortisol	Cortex of adrenal gland	Stimulates protein degradation Stimulates gluconeogenesis Stimulates lipolysis
Growth hormone	Anterior pituitary gland	Stimulates gluconeogenesis Stimulates lipolysis

5. Pedometers track bodily motion or, more precisely, the steps being taken, whereas accelerometers track changes in movement velocity. Information from these motion sensors is then converted into energy expenditure using appropriate mathematical equations. Heart rate monitors work based on the fact that there is a direct linear relationship between heart rate and oxygen consumption. In this case, heart rates recorded are first converted to oxygen consumption via a regression equation. Oxygen consumption is then transformed into caloric expenditure using an appropriate caloric equivalent of oxygen.

6. The subjective methods are relatively inexpensive compared to motion sensors or heart rate monitoring. Therefore they can be used for large population surveys or epidemiological studies. This method can also furnish more detailed information on intensity, duration, and frequency, which are important in quantifying energy expenditure. With this technique, recording errors can occur and activities can be over- or underestimated. This is especially the case when a survey is administered to children or when the activities listed in a questionnaire are not specific enough. Questionnaires and logs may also influence the pattern of activity of subjects who are recording their ongoing activities. For example, they may increase their activity level simply because of the survey.

C H A P T E R   5

1. Oxygen deficit is the lag in oxygen uptake at the onset of exercise. Oxygen debt and excess postexercise oxygen consumption are terms that can be used interchangeably and are defined as a persistent elevation in oxygen uptake above resting levels following exercise. The oxygen slow component refers to a small but progressive increase in oxygen uptake during constant-load exercise especially when exercise intensity exceeds one's lactate threshold.

2. Gross oxygen uptake refers to the total oxygen consumption during a given exercise. It is the sum of oxygen consumption due to rest and oxygen consumption due to exercise. The net oxygen uptake refers to the oxygen consumption brought about by the exercise per se; to obtain this value, the resting oxygen uptake is subtracted from the gross oxygen uptake. Oxygen uptake is commonly expressed in L/min. This measure can also be transformed to ml $\cdot$ kg^{-1} $\cdot$ min^{-1}. To convert from L/min to ml $\cdot$ kg^{-1} $\cdot$ min^{-1}, you multiply a $\dot{V}O_2$ value by 1000 first and then divide the product by body weight in kilograms. For example, if an individual who weighs 154 lb (70 kg) consumes oxygen at 2 L/min, his $\dot{V}O_2$ in ml $\cdot$ kg^{-1} $\cdot$ min^{-1} will be 2 $\times$ 1000 $\div$ 70 = 28.6 ml $\cdot$ kg^{-1} $\cdot$ min^{-1}. This conversion removes the influence of body mass on the measure of $\dot{V}O_2$. This effort is especially important if $\dot{V}O_2$ values are to be compared between two groups of subjects that differ in body mass.

3. Metabolic efficiency = mechanical work accomplished $\div$ energy expended. In the case of cycling, mechanical work is obtained through determination of the power output of cycling, which is the product of brake resistance and pedal frequency. Energy expended is derived from measuring oxygen consumption during the same exercise period. Compared to cycling, arm cranking is generally less efficient in that it produces less work and at the same time recruits disproportionally more muscle mass, especially from muscles in the upper body that serve as stabilizers.

4. According to the power equation of force times velocity, the facts in this question can be expressed mathematically as 600 kgm/min = linear velocity $\times$ 2 kg. Given that each pedal revolution on a Monark bicycle produces 6 m in distance, the linear velocity can be further expressed as pedal rate $\times$ 6 m. Thus, the power equation can be rewritten as 600 kgm/min = pedal rate $\times$ 6 m $\times$ 2 kg. From this equation, pedal rate is 50 rev/min.

5. To allow use of the metabolic equation, walking speed needs to be first converted to metric system units: 3.4 $\times$ 26.8 = 91.1 m $\cdot$ min^{-1}. According to the equation for walking (provided in chapter 5),

gross $\dot{V}O_2$ in ml $\cdot$ kg^{-1} $\cdot$ min^{-1}
= 0.1(speed) + 1.8(speed) $\times$ %grade + resting $\dot{V}O_2$
= 0.1(91.1) + 1.8(91.1) $\times$ 0.05 + 3.5
= 9.1 + 8.2 + 3.5 = 20.8 ml $\cdot$ kg^{-1} $\cdot$ min^{-1}.

6. According to the equation for jogging/running (provided in chapter 5),

gross $\dot{V}O_2$ in ml $\cdot$ kg^{-1} $\cdot$ min^{-1}
= 0.2(speed) + 0.9(speed $\times$ %grade + resting $\dot{V}O_2$
= 0.2(5.5) (26.8) + 0.9(5.5) (26.8) $\times$ 0.12 + 3.5
= 29.5 + 15.9 + 3.5
= 48.9 ml $\cdot$ kg^{-1} $\cdot$ min^{-1}.

Since 1MET = 3.5 ml $\cdot$ kg^{-1} $\cdot$ min^{-1}
$\dot{V}O_2$ in METs is 48.9 ÷ 3.5 = 14.

CHAPTER 6

1. Physical activity is defined as any bodily movement produced by skeletal muscle that results in energy expenditure. In this context, it includes not only physical exercises or conditioning, but also leisure and household activities such as gardening and chores. Exercise is defined as bodily movement that is planned, structured, and repetitive and is done to improve or maintain physical fitness. It includes activities such as running, swimming, cycling, and weightlifting. A typical exercise can be characterized by mode, intensity, and duration.

2. Daily energy expenditure can be divided into three components: (1) resting metabolic rate, (2) physical activity, and (3) thermic effect of food. Of these three components, resting metabolic rate usually represents the largest quantity (~60-75%). A negative caloric balance results when energy expenditure is greater than intake. In theory, if this negative energy balance persists, it will bring about weight loss. Thus, for those who are overweight or obese, it is important to strive for or maintain this type of energy balance. People can produce a negative energy balance by either increasing energy expenditure while maintaining the same energy intake, reducing energy intake while maintaining the same energy expenditure, or increasing energy expenditure and reducing energy intake at the same time.

3. Intensity refers to the state of exertion or degree of strenuousness and can be expressed using heart rate, oxygen consumption, blood lactate concentration, or ratings of perceived exertion. Duration refers to the length of time a physical activity continues and can also be expressed in terms of the total number of calories expended. Frequency refers to the rate of occurrence of an activity within a given period of time or the number of exercise sessions completed every week. Progression refers to a gradual increase in intensity and duration of exercise in an effort to provide sufficient stimulation to bring about desirable adaptations. All these parameters are essential to exercise prescription. A proper exercise prescription for achieving weight loss should aim at maximizing energy expenditure and may entail aerobic forms of exercise to be performed at low to moderate intensity for a comparatively longer period of time, no less than five times a week.

4. A typical circuit weight training program allows individuals to work on 8 to 10 major muscle groups (e.g., gluteals, quadriceps, hamstrings, pectorals, latissimus dorsi, deltoids, and abdominal muscles) in one session that lasts for as long as 60 min. This training format involves multiple sets of low intensity (i.e., 40% 1-RM) and high repetitions (e.g., 10-15 repetitions) on each muscle group, coupled with a relatively shorter rest interval between sets (e.g., ~15 s). In addition to its positive effect on strength gain, this type of weight training has been found to increase energy expenditure, reduce cardiovascular risk factors, and improve cardiorespiratory fitness.

CHAPTER 7

1. The concept of training specificity refers to the types of adaptations occurring in muscle as a result of training. If a muscle is engaged in endurance types of exercise, the primary adaptations are in the systemic circulation as well as in capillary and mitochondrial number, which together increase the capacity of muscle to generate energy aerobically. If a muscle is engaged in heavy resistance training, the primary adaptations are increases in the quantity of contractile proteins and overall muscle mass or hypertrophy; the mitochondrial and capillary densities actually decrease.

2. Aerobic training induces adaptations such as increases in mitochondrial and capillary density along with improved oxygen transport. These adaptations allow more energy to be derived aerobically. As a result, less acid is produced. This reduced blood H^+ concentration, coupled with an improved oxygen supply, enables proportionally more fat to be used as an energy source. Being able to utilize more fat helps spare bodily glycogen stores, and this can

translate into a delayed onset of fatigue during prolonged endurance exercise.

3. Oxygen deficit refers to a lag in the oxygen supply at the onset of exercise. Because of the oxygen deficit, energy comes from hydrolysis of stored adenosine triphosphate (ATP) followed by degradation of phosphocreatine during the rest-to-exercise transition. Consequently, adenosine diphosphate (ADP) concentration increases, and this triggers a subsequent increase in oxygen transport and utilization. It has been found that muscle cells with few mitochondria must have a high ADP concentration to stimulate limited mitochondria to consume oxygen. However, with a training-induced increase in mitochondrial content, less of an increase in ADP concentration is necessary to provide the same stimulation. This quicker response in the mitochondria, together with improved capillary density, leads to earlier activation of oxidative phosphorylation. This then translates into a fast rise in the oxygen uptake curve at the onset of exercise, or a smaller oxygen deficit.

4. As $\dot{V}O_2$max increases, any given workload represents relatively less metabolic strain on the body. This translates into the derivation of proportionally more energy from aerobic metabolism and less of a rise in blood H^+ concentration. Compared to energetic processes that occur without oxygen, energy transformation using oxygen is always more productive in yielding ATP. Increased $\dot{V}O_2$max also allows a better and more sustained oxygen supply during submaximal exercise. This increased oxygen availability, coupled with an attenuated blood H^+ concentration, provides a metabolic environment that is conducive to fat utilization. An increase in fat utilization helps in sparing glycogen.

5. Ultra-short-term activities generally last fewer than 10 s and are powered primarily by energy produced from the ATP-PCr (adenosine triphosphate–phosphocreatine) system. The category of ultra-short-term includes shot put, high jump, long jump, and 40 yd dash, as well as the majority of plays and movements in sports such as football and baseball. Short-term activities generally last longer than 10 s but no more than 3 min. Short-term activities include the 400 and 800 m run, 100 and 200 m swimming, and wrestling and boxing. These

activities are also anaerobic in nature but are powered primarily by energy derived from anaerobic glycolysis.

6. Training with ultra-short-term activities increases muscle mass and improves motor unit recruitment patterns. However, it has little or no effect on the ATP-PCr system. On the other hand, training with short-term activity increases the activity of several key enzymes involved in the glycolytic pathway including phosphorylase, phosphofructokinase (PFK), and lactate dehydrogenase. Interestingly, this type of training has also been found to increase oxidative enzymes such as succinate dehydrogenase and citrate synthase. A decrease in capillary and mitochondrial density, frequently reported following resistance training, can be largely attributed to hypertrophy or an increase in cell volume, which dilutes these two cellular measures.

C H A P T E R   8

1. The notion that women are better able to use fat during exercise can be supported by two lines of observation. One is through the use of indirect calorimetry, whereby investigators have repeatedly reported a lower respiratory exchange ratio (RER) during endurance exercise in women than in men of similar fitness. A lower RER represents a greater percentage of energy derived from fat oxidation. Other observations come from studies that used more invasive techniques such as muscle biopsy and isotopic tracer method to determine utilization of bodily glycogen stores. In these studies, the investigators reported reduced utilization of muscle glycogen in women compared to men during aerobic exercise.

2. Some studies have failed to show that women are better able than men to use fat during exercise. The discrepancy may be attributed to the form of exercise used. Investigators, who previously observed lower glycogen utilization in women during running, failed to confirm this finding when exercise was conducted on a stationary cycle ergometer. Another explanation is related to the control of experiments. The existing literature reflects large variations in terms of the phases of the menstrual cycle at the time measurements were taken. Fat utilization is influenced by estrogen level, which fluctuates throughout the cycle.

3. Impaired utilization of carbohydrate and fat has been considered a metabolic hallmark of the aging process. A reduction in carbohydrate utilization can be mainly ascribed to age-related reductions in insulin sensitivity. Insulin is the hormone that allows carbohydrate energy substrates to become available within the cells for use as energy or to be stored as glycogen. Impaired fat utilization can be largely attributed to a loss of size and oxidative capacity of lean tissue, such as muscle, as well as to reduced sympathetic stimulation, which is necessary in order for lipolysis to take place.

4. Generally speaking, aerobic capacity as measured by $\dot{V}O_2$max in L/min is smaller in children and adolescents than in adults. However, if this measure is expressed in ml $\cdot$ kg^{-1} $\cdot$ min^{-1}, which takes body mass into consideration, $\dot{V}O_2$max in adolescents is virtually similar to that of adults. Anaerobic capacity, on the other hand, has been found to be generally lower in children especially in events that depend primarily on glycolysis for energy transfer, and this is seen despite the adjustment for body mass. The reduction in glycolytic capacity can be further ascribed to a lower glycogen storage capacity as well as reduced catecholamine release.

5. Children are generally lower in areas such as muscular strength and anaerobic capacity. As would be commonly perceived, many differences may be attributable to the fact that children are smaller in size. However, some differences exist even when body size is corrected for. For example, glycogen storage capacity for a given body mass was found to be lower in children as compared to adults. Additionally, it has been found that some physiological parameters such as blood concentration of oxygen and glucose remain unchanged despite a gain in body size over time. As the age-related differences in physiological function are not always proportional to body size, children should not be viewed as miniature adults.

6. Given that $\dot{V}O_2$ in ml $\cdot$ kg^{-1} $\cdot$ min^{-1} is generally comparable between children and adults, a prescription of supervised aerobic conditioning such as running, swimming, or cycling is appropriate for children. A caution, however, relates to the fact that the general activity pattern of children is intermittent. Thus some aerobic routines may be slightly altered to include multiple bouts of comparatively shorter duration. In addition, despite the similar mass-specific $\dot{V}O_2$max between children and adults, children can be more prone to fatigue primarily because they are less metabolically efficient. Therefore instruction in proper techniques should be a part of any conditioning program because these techniques can help children become more efficient. A well-monitored resistance training program can also be considered, since it has been found that children can gain strength independent of muscle size and anabolic hormones. Such programs should emphasize high repetitions and low intensity.

CHAPTER 9

1. Both insulin-dependent (IDDM) and non-insulin-dependent (NIDDM) diabetes are manifested by the condition of hyperglycemia. However, the cause of this symptom differs in the two cases. In IDDM, hyperglycemia is brought about by a lack of insulin due to pancreatic dysfunction, whereas in NIDDM the hyperglycemia is due to impairment in the tissues' ability to respond to insulin. Non-insulin-dependent diabetes is often associated with hyperinsulinemia in addition to hyperglycemia. Among other differences are that IDDM usually emerges before age 30 and tends to come on suddenly; NIDDM is far more common than IDDM, develops more gradually, and usually starts after age 30. In addition, the majority of those who have NIDDM are obese.

2. Both insulin sensitivity and insulin responsiveness are established from clinical measurement using the euglycemic, hyperinsulinemic glucose clamp technique. In this technique, the blood glucose concentration is kept constant by glucose infusion regulated according to repeated, rapid blood glucose measurements. The blood insulin concentration is initially raised and then maintained constant via prime-continuous infusion of insulin. Under these steady-state conditions of euglycemia and hyperinsulinemia, the rate of glucose infusion, which equals the rate of glucose uptake by cells or glucose disposal, can then be used to determine insulin sensitivity and responsiveness. Insulin sensitivity is determined as the insulin concentration that produces half of the maximal response of insulin-mediated glucose disposal; and the smaller the insulin concentration, the greater the insulin sensitivity. Insulin responsiveness is determined as the peak

rate of glucose disposal achieved during the euglycemic, hyperinsulinemic clamp procedure; and the greater the peak glucose disposal, the greater the insulin responsiveness.

3. The glucose tolerance test is another commonly used clinical approach for assessing insulin resistance. It is less invasive than the glucose clamp technique. The patient consumes a given dose of a high-glucose solution. This is then followed by 2 to 3 h of repeated measurements of blood glucose concentrations (i.e., every 30 min). An overall greater than normal glucose response throughout the measurement period suggests reduced insulin sensitivity. It has been counterargued that this interpretation can be incorrect in that glucose concentrations can also be affected by factors other than insulin sensitivity, such as gastrointestinal glucose absorption and pancreatic insulin secretion.

4. It has been found that stored fat or triglycerides from the abdominal region, particularly within visceral compartments, have a greater tendency to be broken down into free fatty acids. As free fatty acids formed in this region are directly released into the portal circulation that drains to the liver, it is thought that the liver in individuals with central obesity may have been more frequently exposed to high concentration of free fatty acids, which can ultimately decrease hepatic insulin sensitivity. An excess of fatty acids in the systemic circulation derived from the abdominal region can also inhibit skeletal muscle glucose uptake and metabolism, according to the theory of the Randle cycle, and this has been considered a cause of insulin resistance manifested in the peripheral region such as in skeletal muscle.

5. Aerobic exercise can produce an acute reduction in blood glucose levels of people with diabetes. In IDDM, the reduction in blood glucose has been attributed to the mass action of hyperglycemia. In other words, during exercise, more glucose molecules rush into tissue from the blood due to a greater concentration gradient, despite the absence of insulin. This exercise-induced favorable glycemic response seen in IDDM is also observed in NIDDM. In fact, this glucose-lowering effect has been found to be significantly greater in patients with NIDDM than in healthy controls. The effect has been attributed to the combination of muscle contraction, hyperglycemia, and hyperinsulinemia.

6. Long-term exercise training has been found to reduce the production of insulin at a given blood glucose concentration, suggesting augmented insulin sensitivity. This training-induced improvement can be ascribed to an increase in glucose transporters in muscles, which helps in facilitating insulin-mediated glucose disposal. It can also be attributed to the loss of body fat especially within the region of the abdomen, as fat tissue in this region has been linked directly to insulin resistance. Cellular adaptations with aerobic training also allow adipose tissue to become more sensitive to hormones that stimulate lipolysis and muscle tissue to become more capable of oxidizing stored fat as an energy source.

CHAPTER 10

1. Resting metabolic rate (RMR) refers to the minimum number of calories your body needs to support its basic physiological functions, which include breathing, circulating blood, and biochemical reactions required to keep you alive. In concept, RMR is similar to basal metabolic rate (BMR). However, the two parameters differ with respect to how they are measured. The determination of BMR requires the subject to fast for at least 12 h and to remain in a reclined position. In addition, the measurement of BMR must take place in a light- and sound-controlled environment that imposes minimal or no stress or disturbances on the subject. If such measurement takes place under conditions that do not meet these criteria, the results are then referred to as RMR. Resting metabolic rate is usually 5% to 10% higher than BMR.

2. Resting metabolic rate generally accounts for 60% to 75% of the total daily energy expenditure. The value changes in accordance with factors such as age, gender, and body composition. On an average, RMR has been estimated to be about 1680 kcal · day^{-1} for men and 1340 kcal · day^{-1} for women. The gender difference can be largely ascribed to the proportionally greater lean body mass associated with men.

3. Factors such as body composition, age, and climate can influence resting metabolic rate (RMR). Body composition refers to the proportion of metabolically active lean tissue relative to

the total body mass. Given two people with the same body weight, the person with the greater lean body mass will have the greater RMR. When adults and children are compared, RMR adjusted for body size is higher in children. This age-related difference is attributed to the fact that the size of metabolically active organs such as the brain is proportionally greater in children. With advancing age from adulthood on, RMR progressively decreases due to a loss of lean body mass such as muscle. Exposure to a cold or hot environment raises RMR as well. The increase in RMR associated with a cold environment can be explained largely by involuntary muscle shivering, whereas the increase in hot weather can be attributed mainly to the cardiovascular adjustments necessary to dissipate excessive heat.

4. Whether physical training augments RMR is an issue that remains in debate. It appears that a greater RMR due to aerobic training tends to occur in individuals who have a significantly higher $\dot{V}O_2$max or experience more profound training adaptations. On the other hand, an increase in RMR following resistance training appears to be more uniformly observed in older individuals, although such an improvement can occur without a concurrent increase in fat-free mass.

5. The postobese model is used to compare obesity-prone weight-reduced persons (i.e., those who were previously obese) with obesity-resistant never-obese individuals. Such comparisons minimize the potential confounding effect associated with directly comparing obese and lean persons who differ extremely in body weight and body composition. The model also helps to reveal the role that RMR may play in causing obesity. If weight-reduced obese subjects have a resting metabolic rate similar to or lower than that of never obese-control subjects, it can be concluded that RMR is not a major contributor to the occurrence of obesity.

6. Uncoupling proteins (UCPs) are a unique group of closely related carrier proteins located in the inner membrane of mitochondria. They are involved in the process of the electron transport chain, which yields energy or adenosine triphosphate (ATP) from NADH and FADH (the reduced forms of nicotinamide adenine dinucleotide and flavin adenine dinucleotide). According to the chemiosmotic theory mentioned in chapter 1, energy production takes place across the inner mitochondrial membrane. In this process, adenosine diphosphate (ADP) is phosphorylated to adenosine triphosphate (ATP) using energy associated with a gradient of protons that is generated during electron transport. It is now known that via the action of UCPs, a significant proportion of mitochondrial respiration is dissipated as heat instead of being coupled to ATP synthesis. In this context, UCPs can be viewed as molecules that function to facilitate energy dissipation rather than ATP production.

CHAPTER 11

1. The thermic effect of food (TEF) represents another component of the total daily energy expenditure. It is defined as an elevation of the metabolic rate that is brought about by ingestion of a meal, and accounts for approximately 10% of the total daily energy expenditure. This increase in energy expenditure is associated with digestion, absorption, assimilation, and storage. The magnitude of TEF can be different depending upon whether protein, carbohydrate, or fat is being consumed. The TEF will generally be 20-30%, 5-10%, and 0-5% of the total energy consumption for protein, carbohydrate, and fat, respectively.

2. Obligatory and facultative thermogenesis are the two subcomponents of TEF. The obligatory component of TEF is the energy cost associated with digestion, absorption, transport, and assimilation of nutrients, as well as the synthesis of protein, fat, and carbohydrate to be stored in the body. Facultative thermogenesis is the energy cost associated with meal-induced activation of the sympathetic nervous system, which stimulates metabolic rate and thus energy expenditure.

3. The postabsorptive state is the time period when the gastrointestinal tract is empty and energy comes from the breakdown of the body's reserves such as glycogen and triglycerides. The postprandial state is the time period that begins immediately after a meal and lasts 4 to 6 h. During this period, digestion, absorption, and assimilation are taking place.

4. Both physical activity and meal ingestion are thermogenic events. When these two events are administered close to one another in time, they

produce a synergistic effect; that is, the amount of energy expended is greater than the sum of the thermic effect associated with each event separately. Though the mechanism underlying this synergistic effect has yet to be elucidated, it appears that performing a bout of exercise not too long after eating could be a strategy to maximize energy expenditure.

5. The TEF has been reported to decrease in obese individuals. The decrease can be explained by (1) insulin resistance and (2) blunted sympathetic responses. Insulin is the hormone necessary in order for cellular storage of energy substrates to take place. This storage process requires energy, which partially explains energy expenditure increases following a meal. Insulin resistance, which has been frequently associated with obesity, suggests reduced insulin-mediated cellular activities related to storing energy, and this can translate into a decrease in TEF. Proper sympathetic responses, which involve activation of the cardiovascular system and release of catecholamines, have been associated with an increase in metabolism and thus energy expenditure. The sympathetic response has been found to be reduced in obese individuals compared to normal-weight individuals following a meal.

CHAPTER 12

1. Sibutramine, also known by the trade names Meridia and Reductil, is a drug approved by the U.S. Food and Drug Administration (FDA) and used for the treatment of obesity. The drug works by boosting levels of certain chemical messengers in the nervous system including serotonin, dopamine, and norepinephrine, all of which have been shown to reduce energy intake. Although to a lesser extent, consumption of this drug is also believed to be able to raise energy expenditure due to its stimulation of the sympathetic nervous system. The drug can increase heart rate and blood pressure, which can lead to undesirable cardiac events especially in individuals with hypertension, hyperlipidemia, or other risks for developing cardiovascular diseases. It has been recommended that this drug not be taken with other medicines for weight loss such as ma huang or ephedra that in themselves can augment blood pressure and heart rate. Other side effects of this drug include insomnia, dry mouth, headache, asthenia, and constipation.

2. Leptin is a peptide hormone produced by fat cells. It functions as a neurotransmitter to affect the hypothalamus, thereby suppressing appetite. It also activates the sympathetic nervous system, thereby stimulating energy expenditure. Leptin levels have not been found to be deficient in obese individuals. In fact, leptin levels and body fat are highly correlated; that is, the greater the amount of the body fat, the higher the serum leptin concentrations. The high levels of leptin seen in obesity have led to the hypothesis that rather than being leptin deficient, individuals who are obese may be leptin resistant or have reduced tissue sensitivity to leptin.

3. Ephedrine taken along with caffeine has been found to be more effective in facilitating energy expenditure and weight loss than ephedrine taken alone. The additive effect of using both substances can be explained by the fact that while ephedrine stimulates the release of norepinephrine, caffeine acts to delay the degradation of norepinephrine in the neural cleft, thereby prolonging its effect. It has also been found that caffeine inhibits phosphodiesterase, an enzyme responsible for degrading cyclic adenosine monophosphate (AMP). Cyclic AMP is a second messenger needed to execute the actions of norepinephrine. This inhibition leads to the maintenance of cellular levels of cyclic AMP, thereby potentiating the thermogenic effect of norepinephrine.

4. The common side effects for ephedrine are agitation, insomnia, headache, palpitations, dizziness, tremor, and constipation; many of these are quite similar to those associated with sympathomimetic drugs such as phentermine and fenfluramine, which have been withdrawn from the market in the United States by the Food and Drug Administration (FDA). Ephedrine can also trigger cardiovascular events such as tachycardia, cardiac arrhythmias, angina, and vasoconstriction with hypertension, although these side effects have been found to be rather infrequent and tend to diminish on repeated dosing. In 1997, the FDA proposed a regulation on ephedra that limited an ephedra dose to 8 mg, with no more than 24 mg per day for a maximum of seven days. In 2004, the FDA created a ban on ephedrine alkaloids that are marketed for reasons other than asthma, colds, allergies, other disease, or traditional Asian use.

Glossary

acceleration—The rate of change of velocity, measured in meters/second^2.

acetylcholine—A neurotransmitter that is released at the ends of nerve fibers in the somatic and parasympathetic nervous systems.

acetyl-CoA—A compound that functions as a coenzyme in many biological reactions and is formed as an intermediate in the oxidation of carbohydrates, fats, and proteins.

adenosine diphosphate—Product of hydrolysis of ATP (adenosine triphosphate) that can be reversibly converted to ATP for energy storage by the addition of a high-energy phosphate group.

adenosine monophosphate—Product of hydrolysis of ATP or ADP (adenosine diphosphate) that can be reversibly converted to ADP or ATP for energy storage by the addition of a high-energy phosphate group.

adenosine triphosphatase—An enzyme that catalyzes the reaction when ATP joins with water, abbreviated as ATPase.

adenosine triphosphate—A high-energy compound or energy currency used for various cellular reactions.

adenylate cyclase—An enzyme that catalyzes the conversion of ATP to cyclic AMP (adenosine monophosphate).

adenylate kinase—An enzyme that catalyzes the phosphorylation of one molecule of ADP by another, yielding ATP and AMP.

adipocytes—Cells in loose connective tissue that are specialized for the synthesis and storage of fat.

adolescence—The period in which a child matures into an adult.

adrenal gland—One of a pair of ductless glands, located above the kidneys, consisting of a cortex, which produces steroidal hormones, and a medulla, which produces epinephrine and norepinephrine.

airway resistance—The opposition to air flow caused by the forces of friction; changes dramatically with changes in airway diameter.

aldosterone—A steroid hormone produced from the adrenal cortex that helps regulate sodium and potassium reabsorption by the cells of the tubular portion of the kidney.

algorithm—A procedure or formula for solving a problem; often includes steps that repeat or require decisions involving logic or comparison.

amino acids—Basic organic molecules that combine to form proteins and consist of hydrogen, carbon, oxygen, and nitrogen. Examples of amino acids are lysine, phenylalanine, and tryptophan.

assimilation—A process in which nutrients ingested are converted into substances usable in the body.

asthenia—Lack or loss of bodily strength.

autonomic—Relating to or controlled by the autonomic nervous system.

β-adrenergic agonist—A molecule that selectively binds to a β receptor, thereby facilitating the action of epinephrine and norepinephrine.

basal metabolic rate—The amount of energy expended during rest in a neutrally temperate environment and in the postabsorptive state, meaning that the digestive system is inactive, which requires about 12 h of fasting in humans.

biosynthesis—The formation of chemical compounds by a living organism. For example, in response to a meal, many glucose molecules are bound together to form polymers of glycogen to be stored in the muscle and liver.

blood–brain barrier—A layer of tightly packed cells that make up the walls of brain capillaries and prevent substances in the blood from diffusing freely into the brain.

blood lipid profile—A group of tests that determine various lipid-related variables in the blood including total cholesterol, HDL (high-density lipoprotein) cholesterol, LDL (low-density lipoprotein) cholesterol, and triglycerides.

body mass index—A measure of body weight in relation to height: Weight in kilograms is divided by height in meters squared to provide an index of overweight or obesity.

β-oxidation—The process in which a long-chain fatty acid is reduced to multiple units of acetyl-CoA in a sequence of steps.

bradycardia—A resting heart rate of under 60 beats per minute; in athletes not considered abnormal if the individual has no associated symptoms.

branched-chain amino acids—Three essential amino acids that have their own unique functions—leucine, isoleucine, and valine.

caloric equivalent of oxygen—Amount of energy yielded from 1 L of oxygen used.

calorie or kilocalorie—Unit of heat energy; 1 calorie is the quantity of heat required to raise the temperature of 1 gram of pure water 1 °C. 1 kilocalorie = 1000 calories.

calorimetry—Measurement of the amount of heat evolved or absorbed in a chemical reaction.

capillary density—Defined as the number of capillaries per unit cross-sectional area of muscle; related to bodily oxygen consumption.

cardiac output—Amount of blood ejected by the ventricle per minute; equal to stroke volume times heart rate.

carnitine—A compound present in the liver and in skeletal muscle that is synthesized from the amino acids lysine and methionine and is responsible for fatty acid transport across mitochondrial membranes.

carnitine acyltransferase—An enzyme complex that functions to facilitate the carnitine-mediated transport of fatty acids across mitochondrial membranes.

catabolism—Subdivision of metabolism involving all degradative chemical reactions in the living cell. For example, large molecules such as proteins are first split into their constituent monomeric units such as amino acids, after which the monomers themselves can be broken down into simple cellular metabolites like ammonia and urea.

catecholamines—A group of amines derived from catechol that have important physiological effects as neurotransmitters and hormones; includes epinephrine, norepinephrine, and dopamine.

childhood—A term usually applied to the phase of development in humans between infancy and adolescence.

circuit training—See *circuit weight training*.

circuit weight training—A weight training routine consisting of 10 to 12 resistance exercises for the upper and lower body, executed in a planned fashion with limited rest between exercises.

citrate synthase—An enzyme that catalyzes the first step in the Krebs cycle, in which oxaloacetate and acetyl-CoA are combined to form citrate.

comorbidity—The presence of one or more disorders in addition to a primary disease.

concentric—Referring to muscle contraction in which muscle shortens while producing force.

conformational change—A change in protein shape that is necessary in order for a protein to undertake a new function.

corpus luteum—A yellow, progesterone-secreting mass of cells that forms from an ovarian follicle after the release of an ovum.

cortex—Outer portion of a gland or of certain organs.

cortisol—A steroid hormone released from the adrenal cortex and active in carbohydrate, fat, and protein metabolism.

creatine kinase—An enzyme that catalyzes the reversible conversion of ADP and phosphocreatine into ATP and creatine.

cytochrome c oxidase—A large transmembrane protein complex found in the mitochondrion that functions to receive electrons and transfer them to oxygen molecules during electron transport.

deamination—The removal of the amino radical from an amino acid or other amino compound, often accompanied by a production of urea.

disaccharides—A class of sugars composed of two glycosidically linked monosaccharides; the most common examples are sucrose, lactose, and maltose.

diurnal rhythm—Biological circadian rhythm that is synchronized with the day–night cycle.

dizygotic—Referring to twins produced when two fertilized eggs are implanted in the uterine wall at the same time.

dopamine—A catecholamine neurotransmitter that is formed during the synthesis of norepinephrine and is essential to the normal functioning of the central nervous system.

doubly labeled water—Water in which both the hydrogen and the oxygen have been partly or completely labeled with an uncommon isotope of these elements.

duration—The length of time physical activity continues; can also be expressed in terms of the total number of calories expended.

eccentric—Referring to muscle contraction in which a muscle lengthens while producing force.

ectothermic—Referring to cold-blooded animals, which regulate their body temperature largely by exchanging heat with the surroundings.

efferent—Referring to the outward direction, as when nerve impulses are carried away from the central nervous system to effectors such as muscles or glands.

efficiency—A measure of how much of liberated energy is utilized to accomplish mechanical work, as in moving the body over a given distance; determined as a ratio of the actual work accomplished to the amount of energy required to perform the work.

electrocardiography—A method of measuring electrical potentials on the body surface that generates a record of the electrical currents associated with heart muscle activity.

endogenous—Arising from within an organism, tissue, or cell.

endothelium—A type of epithelium composed of a single layer of smooth, thin cells; lines the inside of the heart, blood vessels, lymphatics, and serous cavities.

endothermic—Referring to animals that are able to generate heat internally to moderate body temperature.

energy—Ability to produce change; measured by the amount of work performed during a given change.

epinephrine—A hormone secreted by the adrenal medulla upon stimulation by the central nervous system in response to stress, for example from anger or fear, and acting to increase heart rate, blood pressure, cardiac output, and carbohydrate metabolism.

ergogenic—Enhancing physical or mental performance.

estrogen—A major female sex hormone produced primarily by the ovarian follicles and responsible for developing and maintaining secondary female sex characteristics and preparing the uterus for the reception of a fertilized egg.

euglycemic—Normoglycemic, or referring to a condition in which blood glucose concentrations are within a normal range.

excess postexercise oxygen consumption—A term similar to oxygen debt, but also used to recognize the fact that the excess oxygen consumption can last longer than originally thought.

exogenous—Arising from outside an organism or derived externally.

expiration—Emission of air from the lungs due to a reduction of thoracic volume.

extracellular—Referring to the space outside the cell that is occupied by fluid.

facultative thermogenesis—Referring to a part of diet-induced thermogenesis mediated by the activation of the sympathetic nervous system.

fat-free mass—Body mass that includes all parts (muscle, bone, skin, and organs) except fat.

fatty acid—A carboxylic acid, often with a long unbranched chain of carbons, derived from the degradation of triglycerides.

flavin adenine dinucleotide—A coenzyme derived from the B vitamin riboflavin; serves as a reductant or reducing agent in various metabolic processes.

follicular—Pertaining to the phase when follicles are formed during the menstrual cycle, or the time period from onset of menstruation to ovulation.

frequency—Rate of occurrence within a given period of time or a number of exercise sessions completed every week.

gait—A means of locomotion achieved using human limbs.

gestational diabetes—A form of diabetes that occurs only during pregnancy.

glucagon—A polypeptide hormone produced in the alpha cells of the islets of Langerhans of the pancreas that acts in opposition to insulin in the regulation of blood glucose levels and cellular metabolism of glucose and other nutrients.

glucokinase—A hexokinase isozyme that also facilitates phosphorylation of glucose to glucose-6-phosphate, although its action is much less potent than that of hexokinase.

glucomannan—A water-soluble fiber derived from the roots or corm of the konjac plant and that has been marketed as a treatment for constipation, obesity, and high cholesterol.

gluconeogenesis—The generation of glucose from nonsugar carbon substrates like pyruvate, lactate, glycerol, and amino acids.

glucose—A monosaccharide sugar with the formula $C_6H_{12}O_6$; occurs widely in most plant and animal tissue. It is the principal circulating sugar in the blood and the major energy source of the body.

glucose tolerance—A parameter measured by a test that determines how quickly ingested glucose is cleared from the blood.

glucose transporters—A family of membrane proteins found in most mammalian cells that function to transport glucose molecules across the cell membrane.

glycerol—An important component of triglycerides and phospholipids that is usually produced from the degradation of triglyceride when the body uses stored fat as a source of energy.

glycogen—A form of stored carbohydrate that is present primarily in the liver and muscle tissue and is readily converted to glucose as needed by the body to satisfy its energy needs.

glycogenolysis—The biochemical breakdown or hydrolysis of glycogen to glucose.

glycogen phosphorylase—An enzyme that breaks up glycogen into glucose subunits.

glycolysis—A metabolic process that breaks down glucose through a series of reactions to either pyruvic acid or lactic acid and releases energy for the body in the form of ATP.

gross energy expenditure—A measure of energy expenditure that includes energy used during physical activity and energy expended at rest.

growth hormone—A polypeptide hormone, synthesized, stored, and secreted by the anterior pituitary gland, that stimulates growth and cell reproduction.

heart rate reserve—A term used to describe the difference between a person's maximum heart rate and resting heart rate, reflective of fitness; for example, as fitness increases, resting heart rate drops and thus the heart rate reserve increases.

hepatic glucose output—Glucose produced by the liver and released into blood.

hepatic-portal vein—A branch of the venous system in the body that drains blood from the digestive system and its associated glands into the liver.

hexokinase—An enzyme that phosphorylates a six-carbon sugar, a hexose, to a hexose phosphate, which can then be used in either glycolysis or glycogen synthesis.

homeostasis—Maintenance of a constant or unchanging internal environment, especially during unstressful conditions such as rest.

hydrolysis—A process in which a compound is split into other compounds by reacting with water; can be viewed as an example of catabolism.

hyperglycemia—Abnormally high blood sugar concentrations, usually associated with diabetes.

hyperinsulinemic—Referring to the presence of excess insulin in the blood.

hyperlipidemia—A condition in which blood lipids, including cholesterol and triglycerides, are abnormally elevated.

hyperthyroidism—The clinical syndrome caused by an excess of circulating free thyroxine (T4) or free triiodothyronine (T3) or both.

hypoglycemia—A pathologic state produced by a lower than normal amount of sugar (glucose) in the blood.

infancy—The stage of human development lasting from birth to approximately one year.

insomnia—Chronic inability to fall asleep or remain asleep for an adequate length of time.

inspiration—The drawing of air into the lungs as a result of an increase of thoracic volume.

insulin—A polypeptide hormone, produced in the β-cells of the islets of Langerhans of the pancreas, that helps regulate blood levels of glucose and cellular metabolism of glucose and other nutrients.

insulin-like growth factors—Referred to as somatomedins; see *somatomedins* for their actions.

insulin resistance—A condition in which normal amounts of insulin are inadequate to produce a normal insulin response from fat, muscle, and liver cells.

insulin responsiveness—A measure of the ability of tissues to respond to insulin; determined as the peak rate of insulin-mediated glucose disposal using the euglycemic, hyperinsulinemic clamp technique.

insulin sensitivity—A measure of tissue's ability to respond to insulin; can be determined as the insulin concentration that produces half of the maximal response in insulin-mediated glucose disposal using the euglycemic, hyperinsulinemic clamp technique.

intensity—The strenuousness of exertion; can be expressed using heart rate, oxygen consumption, or blood lactate concentration.

intermittent—Referring to exercise that alternately stops and resumes.

intramuscular triglycerides—Triglycerides stored within muscle tissues and considered a more immediate source of energy than fat stored in regions outside of muscle such as the abdomen.

islets of Langerhans—Clusters of endocrine cells scattered throughout the tissue of the pancreas that secrete insulin and glucagon.

isocitrate dehydrogenase—An enzyme that catalyzes the oxidation of isocitric acid to alpha-ketoglutaric acid during the Krebs cycle.

isoform—A protein that has the same function as another and a similar sequence, but is derived from a different gene.

isotopes—Any of the several different forms of an element, each having different atomic mass.

joule or kilojoule—Unit of energy; 1 joule is the energy required to exert a force of 1 Newton for a distance of 1 meter. 1 kilojoule = 1000 joules.

ketones—A by-product of fat oxidation that usually occurs when carbohydrate oxidation is impaired.

kilocalorie—See *calorie*.

kilojoule—See *joule*.

kinetic energy—The energy of a body or a system with respect to the motion of the body or of the particles in the system.

lactate—An ester or salt of lactic acid.

lactate dehydrogenase—An enzyme that catalyzes the interconversion of pyruvate and lactate, with concomitant interconversion of NAD^+ (nicotinamide adenine dinucleotide) and NADH (the reduced form of NAD).

lactate threshold—A level of exercise intensity above which the blood lactate concentration increases sharply.

law of the conservation of energy—Also known as the first law of thermodynamics, which states that energy cannot be created or destroyed and that energy can be changed from one form to another.

lipase—An enzyme that catalyzes the hydrolysis of fats into glycerol and fatty acids.

lipolysis—The breakdown of fat stored in fat cells, resulting in the production of free fatty acids and glycerol.

lipoprotein—A biochemical complex that contains both proteins and lipids and a method of transporting lipids in the blood.

lipoprotein lipase—An enzyme that cleaves one fatty acid from a triglyceride.

logarithmic plot—A graph in which both the horizontal and vertical axis are plotted on a logarithmic scale in order to help one visualize data that change with a power law.

luteal—Pertaining to the phase during the which the corpus luteum secretes progesterone, which prepares the uterus for the implantation of an embryo, or the second half of the menstrual cycle (from ovulation to the beginning of the next menstrual flow).

malonyl-CoA—A compound that is formed by carboxylation of acetyl-CoA and is an intermediate in fatty acid synthesis.

mass spectrometer—An instrument used to measure the mass-to-charge ratio of ions and often used to determine the isotopic composition of elements in a compound.

maximal aerobic capacity—The amount of oxygen consumed during exercise of maximal intensity; often used to represent one's cardiorespiratory fitness.

mechanical energy—Energy in mechanical form, or capacity for mechanical work.

medulla—The inner portion of a gland or of certain organs.

membrane permeability—The ability of a cell membrane to allow substances to pass through it.

metabolic equivalent—A measure of the amount of oxygen consumed by an organism per minute in an activity relative to the resting metabolic rate. A single unit of metabolic equivalent or 1 MET equals resting oxygen consumption, which is approximately 3.5 ml per kilogram of body weight per minute.

metabolic inflexibility—The inability to switch the utilization of lipids and carbohydrates in peripheral tissue (e.g., muscle) based on substrate availability.

micro-electromechanical system—A technology that combines computers with tiny mechanical devices such as sensors, used to detect mechanical changes such as acceleration.

mitochondria—Organelles in the cytoplasm of nearly all eukaryotic cells, containing many enzymes important for cell metabolism, including those responsible for the conversion of food to biologically usable energy.

monosaccharide—The simplest unit of carbohydrate; the most common examples are glucose, fructose, and galactose.

monozygotic—Referring to twins produced when a single egg is fertilized to form one zygote that then divides into two separate embryos.

motor unit—A motor neuron and the group of muscle fibers innervated by its axon.

muscle biopsy—A medical procedure in which a small piece of muscle tissue is removed for examination.

muscle hypertrophy—Scientific term for the growth and increase of muscle size, which can be caused by an increase in the size or number of individual muscle fibers.

mutation—A change of the DNA sequence within a gene or chromosome of an organism, resulting in the creation of a character or trait not found in the parental type.

myocardial—Referring to the muscular tissue of the heart.

myoglobin—An iron-containing protein found in muscle fibers that functions to carry oxygen within muscle fibers.

negative caloric balance—A condition in which the caloric intake is less than calorie expenditure, desirable for weight loss.

negative feedback—A type of feedback in which the system responds in a direction opposite to that of the perturbation.

net energy expenditure—Energy attributed solely to physical activity, calculated as the difference between gross energy expenditure and energy expended at rest.

neurotransmitters—Chemical substances that are produced and secreted by a neuron and that then diffuse across a synapse to cause excitation or inhibition of another neuron. Acetylcholine, norepinephrine, dopamine, and serotonin are examples of neurotransmitters.

nicotinamide adenine dinucleotide—A coenzyme that is derived from the B vitamin nicotinic acid and serves as a reductant or reducing agent in various metabolic processes.

nitrogen balance—A simple index of whether protein requirements are being met; a negative nitrogen balance is said to occur when urinary nitrogen losses are greater than oral nitrogen intake.

norepinephrine—A neurotransmitter released by adrenergic nerve terminals in the autonomic nervous system that has such effects as constricting blood vessels, raising blood pressure, and dilating bronchi.

obligatory thermogenesis—The part of diet-induced thermogenesis that is due to digestion and absorption as well as to the synthesis of protein, fat, and carbohydrate to be stored in the body.

oxidative phosphorylation—The synthesis of ATP by phosphorylation of ADP, for which energy is obtained by electron transport within mitochondria during aerobic respiration.

oxygen debt—The amount of extra oxygen required by the body during recovery from prior exercise.

oxygen deficit—A lag in oxygen consumption at the onset of exercise, computed as the difference between oxygen uptake during early stages of exercise and oxygen uptake for a similar duration in a steady state of exercise.

parasympathetic—Referring to a branch of the autonomic nervous system consisting of nerves that arise from the thoracic and lumbar regions of the spinal cord and functioning in opposition to the sympathetic system, as in inhibiting heartbeat or contracting the pupil of the eye.

parathyroid hormone—A polypeptide hormone produced in the parathyroid glands that helps regulate blood levels of calcium and phosphate.

peptides—Chemical compounds that are composed of a chain of two or more amino acids and are usually smaller than a protein.

phenobarbital—A sedating drug or barbiturate that acts as a central nervous system depressant, often used to control seizures.

phosphocreatine—A compound found chiefly in muscle and formed by the enzymatic interaction of an organic phosphate and creatine; its breakdown provides energy for muscle contraction.

phosphofructokinase—An enzyme that functions in glycolysis by catalyzing the transfer of a phosphate (e.g., from ATP) to fructose.

phosphorylase—An enzyme that catalyzes the production of glucose-1-phosphate from a polyglucose such as glycogen or starch, which is the beginning step of degrading carbohydrate for energy.

phosphorylation—A process of adding a phosphate group to an organic molecule, for example the formation of ATP from ADP and P_i.

piezoceramic material—Special material used in making sensors for devices such as accelerometers.

pituitary gland—A small oval endocrine gland attached to the base of the brain and consisting of an anterior and a posterior lobe; its secretions control the other endocrine glands and influence growth, metabolism, and maturation.

placenta—A membranous vascular organ in the uterus developed during pregnancy; provides oxygen and nutrients to and transfers wastes from the developing fetus.

placental lactogen—A polypeptide hormone that has structure and function similar to those of human growth hormone; important in facilitating the energy supply of the fetus.

portal circulation—Circulation of blood to the liver from the gastrointestinal tract via the portal vein.

positive energy balance—A condition in which caloric intake is greater than calorie expenditure, accompanied by an increase in body fat and body weight.

postabsorptive state—The period when the gastrointestinal tract is empty and energy comes from the breakdown of the body's reserves such as glycogen and triglycerides.

postobese model—A research design that compares obesity-prone weight-reduced persons with obesity-resistant never-obese individuals in order to minimize the potential confounding effect associated with directly comparing obese and lean individuals with extremely different body composition.

postprandial—Referring to the time period after any meal.

potential energy—The energy of a body or a system with respect to the position of the body or the arrangement of the particles of the system.

progesterone—A steroid hormone that prepares the uterus for the fertilized ovum and maintains pregnancy.

proinsulin—A prohormone precursor to insulin, made in the β-cells of the islets of Langerhans, that is converted into mature insulin by enzymatic action.

prolactin—A peptide hormone secreted by the pituitary gland and primarily associated with lactation.

prospective study—A study in which the subjects are identified and then followed forward in time.

puberty—The period of time when a child begins to mature biologically and become an adult capable of reproduction.

pulmonary ventilation—A measure of the total exchange of air between the lungs and the ambient air each minute.

pyruvate dehydrogenase complex—A group of enzymes that function to catalyze the oxidation of pyruvate to form acetyl-CoA.

recombinant—Resulting from new combinations of genetic material.

respiratory exchange ratio—A measure similar to the respiratory quotient except that both oxygen consumption and carbon dioxide production are measured at the lungs.

respiratory quotient—A ratio of oxygen consumption to carbon dioxide production, which is influenced by the utilization of carbohydrate and fat as energy fuels.

resting metabolic rate—The rate at which heat is produced by an individual at rest, often expressed as the calories released per kilogram of body weight or per square meter of body surface per hour.

R-R wave—Measurement of the time elapsed between two consecutive heartbeats and often used to analyze cardiac rhythm.

running economy—A measure of metabolic efficiency during running, typically defined as oxygen consumption for a given velocity of submaximal running.

second messengers—Molecules that relay signals received at receptors on the cell surface to target molecules in the cytosol, nucleus, or both.

serotonin—A neurotransmitter derived from tryptophan that is involved in sleep, depression, memory, and other neurological processes.

slow component of oxygen uptake—A small and gradual increase in oxygen consumption despite the maintenance of exercise intensity; usually occurs at a high intensity or an intensity that is above one's lactate threshold.

somatomedins—Polypeptides synthesized in the liver and other tissues in response to growth hormone; capable of stimulating growth processes, especially in bone, cartilage, and muscle.

specificity—The idea that physiological and biochemical adaptations are specific to the type of training being used; for example, endurance training produces an increase in mitochondrial content, whereas resistance training does not.

spirometry—Instrumentation for measuring the volume of air entering and leaving the lungs.

splanchnic—Pertaining to internal organs within the region of the abdomen and pelvis.

steady state—A condition in which oxygen demand is met by oxygen supply.

stroke volume—The amount of blood ejected by the ventricle during each heartbeat.

subcutaneous—Beneath or under the skin.

succinate dehydrogenase—An enzyme that oxidizes succinate to fumarate using FAD (flavin adenine dinucleotide) as a hydrogen carrier.

sympathetic—Referring to a branch of the autonomic nervous system consisting of nerves that arise from the thoracic and lumbar regions of the spinal cord and functioning in opposition to the parasympathetic system, as in stimulating heartbeat, dilating the pupil of the eye, and so on.

sympathomimetic—Relating to stimulation of the sympathetic nervous system or the action of epinephrine.

synthase—An enzyme that catalyzes a process of synthesis. For example, glycogen synthase converts glucose residues one by one into a polymeric chain for storage as glycogen.

tachycardia—An abnormally fast heart rate commonly caused by a normal physiological process, such as during exercise or after the ingestion of drugs like caffeine and amphetamine, or by disorders such as serious infections or hyperthyroidism.

testosterone—A steroid hormone secreted by the testes that stimulates the development of male sex organs, secondary sexual traits, and sperm.

thermic effect of food—An increase in energy expenditure associated with digestion and absorption of food ingested.

thermogenesis—The process by which the body generates heat or energy via physiological processes.

transamination—The transfer of an amine group from an amino acid to another acid molecule so that a new amino acid is formed.

transducer—An electrical device that converts one type of energy or signal into another.

triglycerides—The chief lipids constituting fats and oils; function to store chemical energy in plants and animals.

type I muscle fibers—Type I muscle fibers or slow-oxidative fibers use primarily cellular respiration and, as a result, have relatively high endurance. To support their high oxidative metabolism, these muscle fibers typically have many mitochondria, much myoglobin, and many capillaries and thus appear red.

type II muscle fibers—Type II muscle fibers use primarily anaerobic metabolism, have relatively low endurance, and cannot sustain contractions for significant lengths of time. These muscle fibers are typically used during tasks requiring short bursts of strength, such as sprints or weightlifting.

uncoupling proteins—A group of unique proteins found in mitochondria that function to reduce the coupling of energy stored in food to ATP synthesis.

urea—A water-soluble product resulting from protein metabolism; occurs in urine and other body fluids.

vasodilation—Widening of blood vessels following relaxation of the smooth muscle in the vessel wall.

vastus lateralis—The largest part of the quadriceps femoris.

visceral—Referring to internal organs of the body, specifically those within the chest (such as the heart or lungs) or within the abdomen (such as the liver, pancreas, or intestines).

References

Acheson, K.J., J.P. Flatt, and E. Jequier. 1982. Glycogen synthesis versus lipogenesis after a 500 gram carbohydrate meal in man. *Metabolism,* 31:1234-1240.

Acheson, K., E. Jequier, and J. Wahren. 1983. Influence of β-adrenergic blockade on glucose-induced thermogenesis in man. *Journal of Clinical Investigation,* 72:981-986.

Acheson, K.J., E. Ravussin, D.A. Schoeller, L. Christin, L. Bourquin, P. Baertschi, E. Danforth, Jr., and E. Jequier. 1988. Two-week stimulation of blockade of the sympathetic nervous system in man: influence on body weight, body composition, and 24-H energy expenditure. *Metabolism,* 37:91-98.

Acheson, K.J., E. Ravussin, J. Wahren, and E. Jequier. 1984a. Thermic effect of glucose in man: Obligatory and facultative thermogenesis. *Journal of Clinical Investigation,* 74:1572-1580.

Acheson, K., Y. Schutz, T. Bessard, E. Ravussin, E. Jequier, and J.P. Flatt. 1984b. Nutritional influences on lipogenesis and thermogenesis after a carbohydrate meal. *American Journal of Physiology,* 246:E62-E70.

Achten, J., M. Gleeson, and A.E. Jeukendrup. 2002. Determination of the exercise intensity that elicits maximal fat oxidation. *Medicine and Science in Sports and Exercise,* 34:72-97.

Achten, J., and A.E. Jeukendrup. 2004. Relation between plasma lactate concentration and fat oxidation rates over a wide range of exercise intensities. *International Journal of Sports Medicine,* 25:32-37.

Ahlborg, G., P. Felig, L. Hagenfeldt, R. Hendler, and J. Wahren. 1974. Substrate turnover during prolonged exercise in man: Splanchnic and leg metabolism of glucose, free fatty acids, and amino acids. *Journal of Clinical Investigation,* 53:1080-1090.

Ahlquist, L.E., D.R. Bassett, R. Sufit, F.J. Nagle, and D.P. Thomas. 1992. The effects of pedaling frequency on glycogen depletion rates in type I and II quadriceps muscle fibers during submaximal cycling exercise. *European Journal of Applied Physiology,* 65:360-364.

Ahmaidi, S., P. Granier, Z. Taoutaou, J. Mercier, H. Dubouchaud, and C. Prefaut. 1996. Effects of active recovery on plasma lactate and anaerobic power following repeated intensive exercise. *Medicine and Science in Sports and Exercise,* 28:450-456.

Ainslie, P.N., T. Reilly, and K.R. Westerterp. 2003. Estimating human energy expenditure: A review of techniques with particular reference to doubly labeled water. *Sports Medicine,* 33:683-698.

Ainsworth, B.E., W.L. Haskell, A.S. Leon, D.R. Jacobs, Jr., H.J. Montoye, J.F. Sallis, and R.S. Paffenbarger, Jr. 1993. Compendium of physical activities: classification of energy costs of human physical activities. *Medicine and Science in Sports and Exercise,* 25:71-80.

Ainsworth, B.E., W.L. Haskell, M.C. Whitt, M.L. Irwin, A.M. Swartz, S.J. Strath, W.L. O'Brien, D.R. Bassett, Jr., K.H. Schmitz, P.O. Emplaincourt, D.R. Jacobs, Jr., and A.S. Leon. 2000. Compendium of physical activities: an update of activity codes and MET intensities. *Medicine and Science in Sports and Exercise,* 32 (Suppl):S498-S516.

Ainsworth, B.E., S.R. Leggett, C.A. Mathien, J.A. Main, D.C. Hunter, and G.E. Duncan. 1996. Accuracy of five electronic pedometers for measuring distance walked. *Medicine and Science in Sports and Exercise,* 28:1071-1077.

Ainsworth, B.E., and A.S. Leon. 1991. Gender differences in self-reported physical activity. *Medicine and Science in Sports and Exercise,* 23:S105.

Almuzaini, K.S., J.A. Potteiger, and S.B. Green. 1998. Effects of split exercise sessions on excess post-exercise oxygen consumption and resting metabolic rate. *Canadian Journal of Applied Physiology,* 23:433-443.

Amatruda, J.M., M.C. Statt, and S.L. Welle. 1993. Total and resting energy expenditure in obese women reduced to ideal body weight. *Journal of Clinical Investigation,* 92:1236-1242.

American College of Obstetricians and Gynecologists. 1994. Exercise during pregnancy and post-partum period. *American College of Obstetricians and Gynecologists Technical Bulletin,* 189:2-7.

American College of Sports Medicine. 1998. The recommended quantity and quality of exercise for developing and maintaining cardiorespiratory and muscular fitness, and flexibility in healthy adults. *Medicine and Science in Sports and Exercise,* 30:975-991.

American College of Sports Medicine. 2001. Appropriate intervention strategies for weight loss and prevention of weight regain for adults. *Medicine and Science in Sports and Exercise,* 33:2145-2156.

American College of Sports Medicine. 2006. *ACSM's guidelines for exercise testing and prescription.* 7th ed. Baltimore: Lippincott Williams & Wilkins.

Andersen, R.E. 1999. Exercise, as active lifestyle, and obesity. *Physician and Sportsmedicine,* 27:41-50.

Andersen, R.E., C.J. Crespo, S.J. Bartlett, L.J. Cheskin, and M. Pratt. 1998. Relationship of physical activity and television watching with body weight and level of fatness among children. *Journal of the American Medical Association,* 279:938-942.

Anderson, T. 1996. Biomechanics and running economy. *Sports Medicine*, 22:76-89.

Apfelbaum, M., P. Vague, O. Ziegler, C. Hanotin, F. Thomas, and E. Leutenegger. 1999. Long-term maintenance of weight loss after a very low caloric diet: efficacy and tolerability of sibutramine. *American Journal of Medicine*, 106:179-184.

Armon, Y., D.M. Cooper, R. Flores, S. Zanconato, and T.J. Barstow. 1991. Oxygen uptake dynamics during high-intensity exercise in children and adults. *Journal of Applied Physiology*, 70:841-848.

Arner, P., E. Kriegholm, P. Englfeldt, and J. Bolinder. 1990. Adrenergic regulation of lipolysis in situ at rest and during exercise. *Journal of Clinical Investigation*, 85:893-898.

Åstrand, I. 1960. Aerobic work capacity in men and women with a special reference to age. *Acta Physiologica Scandinavica*, 49 (Suppl 169):1-92.

Åstrand, P.-O., and K. Rodahl. 1986. *Textbook of work physiology: physiological basis of exercise.* 3rd ed. New York: McGraw-Hill.

Astrup, A., L. Breum, S. Toubro, P. Hein, and F. Quaade. 1992. The effect and satiety of an ephedrine and caffeine compound compared to ephedrine, caffeine, and placebo in obese subjects on an energy restricted diet: A double blind trial. *International Journal of Obesity Related Metabolic Disorders*, 16:269-277.

Astrup, A., P.C. Gotzsche, K. van de Werken, C. Ranneries, S. Toubro, A. Raben, and B. Buemann. 1999. Meta-analysis of resting metabolic rate in formerly obese subjects. *American Journal of Clinical Nutrition*, 69:1117-1122.

Astrup, A., C. Lundsgaard, J. Madsen, and N.J. Christensen. 1985. Enhanced thermogenic responsiveness during chronic ephedrine treatment in man. *American Journal of Clinical Nutrition*, 42:83-94.

Astrup, A., S. Toubro, S. Cannon, P. Hein, and J. Madsen. 1991. Thermogenic synergism between ephedrine and caffeine in healthy volunteers: A double blind placebo controlled study. *Metabolism*, 40:323-329.

Atkinson, R.L. 1997. Use of drugs in the treatment of obesity. *Annual Review of Nutrition*, 17:383-403.

Atwater, W.O., and F.G. Benedict. 1903. *Experiments on the metabolism of matter and energy in the human body, 1900-1902.* U.S. Department of Agriculture Office of Experiment Stations, Bulletin 136. Washington, DC: U.S. Government Printing Office.

Babij, P., S.M. Matthews, and M.J. Rennie. 1983. Changes in blood ammonia, lactate, and amino acids in relation to workload during bicycle ergometer exercise in man. *European Journal of Applied Physiology*, 50:405-411.

Baecke, J.A.H., W.A. van Staveren, and J. Burema. 1983. Food consumption, habitual physical activity, and body fatness in Dutch adults. *American Journal of Clinical Nutrition*, 37:278-286.

Bahr, R., P.K. Opstad, J.I. Medbo, and O.M. Sejersted. 1991. Strenuous prolonged exercise elevates resting metabolic rate and causes reduced mechanical efficiency. *Acta Physiologica Scandinavica*, 141:555-563.

Bailey, S.P., and R.R. Pate. 1991. Feasibility of improving running economy. *Sports Medicine*, 12:228-236.

Baldwin, K.M., G.H. Klinkerfuss, R.L. Terjung, P.A. Mole, and J.O. Holloszy. 1972. Respiratory capacity of white, red, and intermediate muscle: adaptive response to exercise. *American Journal of Physiology*, 222:373-378.

Banerji, M., R. Chaiken, D. Gordon, and H. Lebowitz. 1995. Does intra-abdominal adipose tissue in black men determine whether NIDDM is insulin-resistant or insulin sensitive? *Diabetes*, 44:141-146.

Barone, J.J., and H.R. Roberts. 1996. Caffeine consumption. *Food and Chemical Toxicology*, 34:119-129.

Bar-Or, O., and T.W. Rowland. 2004. Physiologic and perceptual responses to exercise in the healthy child. In: *Pediatric exercise medicine*, 3-59. Champaign, IL: Human Kinetics.

Barringer, T.B., Jr. 1916. The effect of exercise upon the carbohydrate tolerance in diabetes. *American Journal of Medical Science*, 151:181-184.

Barstow, T.J., A.M. Jones, P.H. Nguyen, and R. Casaburi. 1996. Influence of muscle fiber type and pedal frequency on oxygen uptake kinetics of heavy exercise. *Journal of Applied Physiology*, 81:1642-1650.

Bassett, D., Jr. 2000. Validity and reliability issues in objective monitoring of physical activity. *Research Quarterly for Exercise and Sport*, 71:30-36.

Bassett, D.R. Jr., B.E. Ainsworth, S.R. Leggett, C.A. Mathien, J.A. Main, D.C. Hunter, and G.E. Duncan. 1996. Accuracy of five electronic pedometers for measuring distance walked. *Medicine and Science in Sports and Exercise*, 28:1071-1077.

Bell, D.G., I. Jacobs, and K. Ellerington. 2001. Effect of caffeine and ephedrine ingestion on anaerobic exercise performance. *Medicine and Science in Sports and Exercise*, 33:1399-1403.

Bell, D.G., T.M. McLellan, and C.M. Sabiston. 2002. Effect of ingesting caffeine and ephedrine on 10-km run performance. *Medicine and Science in Sports and Exercise*, 34:344-349.

Bell, G.J., G.D. Snydmiller, D.S. Davis, and H.A. Quinney. 1997. Relationship between aerobic fitness and metabolic recovery from intermittent exercise in endurance trained athletes. *Canadian Journal of Applied Physiology*, 22:78-85.

Bell, R.D., J.D. MacDougall, R. Billeter, and H. Howald. 1980. Muscle fiber types and morphometric analysis of skeletal muscle in six year old children. *Medicine and Science in Sports and Exercise*, 12:28-31.

Bergstrom, J., and E. Hultman. 1967. Synthesis of muscle glycogen in men after glucose and fructose infusion. *Acta Medica Scandinavia*, 182:93-107.

Billat, V.L., B. Flechet, B. Petit, G. Muriaux, and J.P. Koralsztein. 1999. Interval training at $\dot{V}O_2$max effects on aerobic performance and overtraining markers. *Medicine and Science in Sports and Exercise*, 31:156-163.

Binzen, C.A., P.D. Swan, and M.M. Manore. 2001. Postexercise oxygen consumption and substrate use after resistance exercise in women. *Medicine and Science in Sports and Exercise*, 33:932-938.

Blaak, E.E., D.P.C. van Aggel-Leijssen, A.J.M. Wagenmakers, W.H.M. Saris, and M.A. Baak. 2000. Impaired oxidation of plasma-derived fatty acids in type 2 diabetic subjects during moderate-intensity exercise. *Diabetes*, 49:2102-2107.

Blachford, F.K., R.G. Knowlton, and D. Schneider. 1985. Plasma FFA responses to prolonged walking in untrained men and women. *European Journal of Applied Physiology,* 53:343-347.

Black, A.E., A.M. Prentice, and W.A. Coward. 1986. Use of food quotients to predict respiratory quotients for the doubly labeled water method of measuring energy expenditure. *Human Nutrition and Clinical Nutrition,* 40:381-391.

Blaxter, K. 1989. *Energy metabolism in animal and man.* Cambridge: Cambridge University Press.

Bloesch, D., Y. Schutz, E. Breitenstein, E. Jequier, and J.P. Felber. 1988. Thermogenesis response to an oral glucose load in men: comparison between young and elderly subjects. *Journal of the American College of Nutrition,* 7:471-483.

Bogardus, C., P. Thuillez, E. Ravussin, B. Vasquez, M. Narimiga, and S. Azhar. 1983. Effect of muscle glycogen depletion on in vivo insulin action in men. *Journal of Clinical Investigation,* 72:1605-1610.

Boobis, L., C. William, and S.A. Wooton. 1982. Human muscle metabolism during brief maximal exercise. *Journal of Physiology,* 338:21-22.

Borsheim, E., and R. Bahr. 2003. Effect of exercise intensity, duration, and mode on post-exercise oxygen consumption. *Sports Medicine,* 33:1037-1060.

Boss, O., S. Samec, F. Kuhne, P. Bijlenga, F. Assimacopoulos-Jeannet, J. Seydoux, J.P. Giacobino, and P. Muzzin. 1998. Uncoupling protein-3 expression in rodent skeletal muscle is modulated by food intake but not by changes in environmental temperature. *Journal of Biological Chemistry,* 273:5-8.

Bouchard, C., F.T. Dionne, J.-A. Simoneau, and M.R. Boulay. 1992. Genetics of aerobic and anaerobic performance. *Exercise and Sport Sciences Reviews,* 20:27-58.

Bouchard, C., A. Tremblay, J.P. Despres, G. Theriault, M.R. Boulay, G. Lortie, C. LeBlanc, and G. Fournier. 1989. Genetic effect in resting and exercise metabolic rates. *Metabolism,* 38:364-370.

Bouten, C., W. Verboeket-van de Venne, K. Westerterp, M. Verduin, and J. Janssen. 1996. Daily physical activity assessment: comparison between movement registration and doubly labeled water. *Journal of Applied Physiology,* 81:1019-1026.

Bouten, C.V., K.R. Westerterp, M. Verduin, and J.D. Janssen. 1994. Assessment of energy expenditure for physical activity using a triaxial accelerometer. *Medicine and Science in Sports and Exercise,* 26:1516-1523.

Boyd, A.E., S.R. Giamber, M. Mager, and H.E. Lebovitz. 1974. Lactate inhibition of lipolysis in exercising man. *Metabolism,* 23:531-542.

Brage, S., N. Brage, P.W. Franks, U. Ekelund, M. Wong, L.B. Anderson, K. Froberg, and N.J. Wareham. 2003. Branched equation modeling of simultaneous accelerometry and heart rate monitoring improves estimate of directly measured physical activity energy expenditure. *Journal of Applied Physiology,* 96:343-351.

Braun, B., and T. Horton. 2001. Endocrine regulation of exercise substrate utilization in women compared to men. *Exercise and Sport Sciences Reviews,* 29:149-154.

Bray, G.A. 1983. The energetics of obesity. *Medicine and Science in Sports and Exercise,* 15:32-40.

Bray, G.A. 1986. Autonomic and endocrine factors in the regulation of energy balance. *Federation Proceedings,* 45:1404-1410.

Bray, G.A., D.H. Ryan, D. Gordon, S. Heidingsfelder, F. Cerise, and K. Wilson. 1996. A double-blind randomized placebo-controlled trial of Sibutramine. *Obesity Research,* 4:263-270.

Bray, G.A., and L.A. Tartaglia. 2000. Medicinal strategies in the treatment of obesity. *Nature,* 404:672-677.

Bray, G.A., B.J. Whipp, and S.N. Koyal. 1974. The acute effects of food on energy expenditure during cycle ergometry. *American Journal of Clinical Nutrition,* 27:254-259.

Bray, G.A., B. Zachary, W.T. Dahms, R.L. Atkinson, and T.H. Oddie. 1978. Eating patterns of the massively obese individual. *Journal of the American Dietetic Association,* 72:24-27.

Bray, M.S., W.W. Wong, J.R. Morrow Jr., N.F. Butte, and J.M. Pivarnik. 1994. Caltrac versus calorimeter determination of 24-h energy expenditure in female children and adolescents. *Medicine and Science in Sports and Exercise,* 26:1524-1530.

Broeder, C.E., K.A. Burrhus, L.S. Svanevik, and J.H. Wilmore. 1992. The effects of either high intensity resistance or endurance training on resting metabolic rate. *American Journal of Clinical Nutrition,* 55:892-810.

Brooks, G.A., and C.M. Donovan. 1983. Effect of endurance training on glucose kinetics during exercise. *American Journal of Physiology,* 244:E505-E512.

Brooks, G.A., T.D. Fahey, and K.M. Baldwin. 2005. *Exercise physiology: human bioenergetics and its applications.* 4th ed. New York: McGraw-Hill.

Brooks, G.A., K.J. Hittelman, J.A. Faulkner, and R.E. Beyer. 1971a. Temperature, liver mitochondrial respiratory functions, and oxygen debt. *Medicine and Science in Sports,* 2:72-74.

Brooks, G.A., K.J. Hittelman, J.A. Faulkner, and R.E. Beyer. 1971b. Temperature, skeletal muscle mitochondrial functions, and oxygen debt. *American Journal of Physiology,* 220:1053-1059.

Buch, I., P.J. Hornnes, and C. Kuhl. 1986. Glucose tolerance in early pregnancy. *Acta Endocrinologica,* 112:263-266.

Buemann, B., and A. Tremblay. 1996. Effect of exercise training on abdominal obesity and related metabolic complications. *Sports Medicine,* 21:191-212.

Burgomaster, K.A., S.C. Hughes, G.J.F. Heigenhauser, S.N. Bradwell, and M.J. Gibala. 2005. Six sessions of sprint interval training increases muscle oxidative potential and cycle endurance capacity in humans. *Journal of Applied Physiology,* 98:1985-1990.

Burleson, Jr., M.A., H.S. O'Bryant, M.H. Stone, M.A. Collins, and T. Triplett-McBride. 1998. Effect of weight training exercise and treadmill exercise on post-exercise oxygen consumption. *Medicine and Science in Sports and Exercise,* 30:518-522.

Burstein, R., Y. Epstein, Y. Shapiro, I. Charuzi, and E. Karnieli. 1990. Effect of an acute bout of exercise on glucose disposal in human obesity. *Journal of Applied Physiology,* 69:299-304.

Butte, N.F. 2000. Carbohydrate and lipid metabolism in pregnancy: normal compared with gestational diabetes mellitus. *American Journal of Clinical Nutrition,* 71 (Suppl):1256S-1261S.

Campbell, S.E., and M.A. Febbraio. 2001. Effect of ovarian hormones on mitochondrial enzyme activity in fat oxidation pathway of skeletal muscle. *American Journal of Physiology,* 281:E803-E808.

Campbell, W.W., M.C. Crim, V.R. Young, and W.J. Evans. 1994. Increased energy requirements and changes in body composition with resistance training in older adults. *American Journal of Clinical Nutrition,* 60:167-175.

Campfield, L.A., F.J. Smith, Y. Guisez, R. Devos, and P. Burn. 1995. Recombinant mouse OB protein: evidence for a peripheral signaling linking adiposity and central neural networks. *Science,* 269:546-549.

Cannon, B., and J. Nedergaard. 2004. Brown adipose tissue: function and physiology significance. *Physiology Review,* 84:277-359.

Capelli, C., G. Rosa, F. Butti, G. Ferretti, A. Veicsteinas, and P.E. di Prampero. 1993. Energy cost of and efficiency of riding aerodynamic bicycles. *European Journal of Applied Physiology,* 67:144-149.

Carlson, M.G., W.L. Snead, J.O. Hill, N. Nurjahan, and P.J. Campbell. 1991. Glucose regulation of lipid metabolism in humans. *American Journal of Physiology,* 261:E815-E820.

Carter, H., A.M. Jones, and J.H. Doust. 1999. Effect of six weeks of endurance training on the lactate minimum speed. *Journal of Sports Science,* 17:957-967.

Carter, S., S. McKenzie, M. Mourtzakis, D.J. Mahoney, and M.A. Tarnopolsky. 2001a. Short-term 17β-estradiol decreases glucose Ra but not whole body metabolism during endurance exercise. *Journal of Applied Physiology,* 90:139-146.

Carter, S.L., C. Rennie, and M.A. Tarnopolsky. 2001b. Substrate utilization during endurance exercise in men and women after endurance training. *American Journal of Physiology,* 280:E898-E907.

Casaburi, R., T.W. Storer, I. Ben-Dov, and K. Wasserman. 1987. Effect of endurance training on possible determinants of VO_2 during heavy exercise. *Journal of Applied Physiology,* 62:199-207.

Caspersen, C.J., K.E. Powell, and G.M. Christensen. 1985. Physical activity, exercise, and physical fitness: definitions and distinctions for health-related research. *Public Health Reports,* 100:126-131.

Cavanagh, P.R., and R. Kram. 1985. The efficiency of human movement: a statement of the problem. *Medicine and Science in Sports and Exercise,* 17:304-308.

Cerretelli, P., D. Shindell, D.P. Pendergast, P.E. Di Prampero, and D.W. Rennie. 1977. Oxygen uptake transients at the onset and offset of arm and leg work. *Respiration Physiology,* 30:81-97.

Cheetham, M.E., L.H. Boobis, S. Brooks, and C. Williams. 1986. Human muscle metabolism during sprint running. *Journal of Applied Physiology,* 61:54-60.

Chester, N., T. Reilly, and D.R. Mottram. 2003. Physiological, subjective and performance effects of pseudoephedrine and phenylpropanolamine during endurance running exercise. *International Journal of Sports Medicine,* 24:3-8.

Chi, M.M.-Y., C.S. Hintz, E.F. Coyle, W.H. Martin III, J.L. Ivy, P.M. Nemeth, J.O. Holloszy, and O.H. Lowry. 1983. Effects of detraining on enzymes of energy metabolism in individual human muscle fibers. *American Journal of Physiology,* 244: C276-C287.

Choi, D., K.J. Cole, B.H. Goodpaster, W.J. Fink, and D.L. Costill. 1994. Effect of passive and active recovery on resynthesis of muscle glycogen. *Medicine and Science in Sports and Exercise,* 26:992-996.

Christensen, E.H., and O. Hansen. 1939. Arbeitsfähigkeit and Ehrnährung. *Scandinavia Archive Physiologic,* 13:160-175.

Chu, K.S., T.J. Doherty, G. Parise, J.S. Milheiro, and M.A. Tarnopolsky. 2002. A moderate dose of pseudoephedrine does not alter muscle contraction strength or anaerobic power. *Clinical Journal of Sports Medicine,* 12:387-390.

Clapp, J.F., M. Wesley, and R.H. Sleamaker. 1987. Thermoregulatory and metabolic responses to jogging prior to and during pregnancy. *Medicine and Science in Sports and Exercise,* 19:124-130.

Cloherty, E.K., L.A. Sultzman, R.J. Zottola, and A. Carruthers. 1995. Net sugar transport is a multistep process: Evidence for cytosolic sugar binding sites in erythrocytes. *Biochemistry,* 34:15395-15406.

Coggan, A.R., and E.F. Coyle. 1987. Reversal of fatigue during prolonged exercise by carbohydrate infusion or ingestion. *Journal of Applied Physiology,* 63:2388-2395.

Coggan, A.R., D.L. Habash, L.A. Mendenhall, S.C. Swanson, and C.L. Kien. 1993. Isotopic estimation of CO_2 production during exercise before and after endurance training. *Journal of Applied Physiology,* 75:70-75.

Coggan, A.R., W.M. Kohrt, R.J. Spina, D.M. Bier, and J.O. Holloszy. 1990. Endurance training decreases plasma glucose turnover and oxidation during moderate intensity exercise in men. *Journal of Applied Physiology,* 68:990-996.

Coggan, A.R., C.A. Raguso, B.D. Williams, L.S. Sidossis, and A. Gastaldelli. 1995. Glucose kinetics during high-intensity exercise in endurance-trained and untrained humans. *Journal of Applied Physiology,* 78:1203-1207.

Colberg, S.R., J.M. Hagberg, S.D. McCole, J.M. Zumda, P.D. Thompson, and D.E. Kelley. 1996. Utilization of glycogen but not plasma glucose is reduced in individuals with NIDDM during mild-intensity exercise. *Journal of Applied Physiology,* 81:2027-2033.

Collins, M.H., D.J. Pearsall, G.S. Zavorsky, H. Bateni, R.A. Turcotte, and D.L. Montgomery. 2000. Acute effects of intense interval training on running mechanics. *Journal of Sports Science,* 18:83-90.

Conley, D.L., and G.S. Krahenbuhl. 1980. Running economy and distance running performance of highly trained athletes. *Medicine and Science in Sports and Exercise,* 12:357-360.

Conley, K.E., S.A. Jubrias, and P.C. Esselman. 2000. Oxidative capacity and ageing in human muscle. *Journal of Physiology,* 526:203-210.

Connoley, I.P., Y.L. Liu, I. Frost, I.P. Reckless, D.J. Heal, and M.J. Stock. 1999. Thermogenic effects of sibutramine and its metabolites. *British Journal of Pharmacology,* 126:1487-1495.

Considine, R.V., E.L. Considine, C.J. William, M.R. Nyce, S.A. Magosin, T.L. Bauer, E.L. Rosato, J. Colberg, and J.F. Caro. 1995. Evidence against either a premature stop codon or the absence of obese gene mRNA in human obesity. *Journal of Clinical Investigation,* 95:2986-2988.

Considine, R.V., M.K. Sinha, M.L. Heiman, A. Kriauciunas, T.W. Stephens, M.R. Nyce, J.P. Ohannesian, C.C. Marco, L.J. McKee, T.L. Bauer, and J.F. Caro. 1996. Serum immunoreactive-leptin concentrations in normal weight and obese humans. *New England Journal of Medicine*, 334:292-295.

Costill, D.L. 1967. The relationship between selected physiological variables and distance running performance. *Journal of Sports Medicine and Physical Fitness*, 7:61-66.

Costill, D.L., E. Coyle, G. Dalsky, W. Evens, W. Fink, and D. Hoopes. 1977. Effects of elevated plasma FFA and insulin on muscle glycogen usage during exercise. *Journal of Applied Physiology*, 43:695-699.

Costill, D.L., E.F. Coyle, W.F. Fink, G.R. Lesmes, and F.A. Witzmann. 1979. Adaptations in skeletal muscle following strength training. *Journal of Applied Physiology*, 46:96-99.

Costill, D.L., W.J. Fink, and M.L. Pollock. 1976. Muscle fiber composition and enzyme activities of elite distance runners. *Medicine and Science in Sports and Exercise*, 8:96-100.

Costill, D.L., H. Thomason, and E. Roberts. 1973. Fractional utilization of the aerobic capacity during distance running. *Medicine and Science in Sports*, 5:248-252.

Coyle, E.F. 2000. Physical activity as a metabolic stressor. *American Journal of Clinical Nutrition*, 72 (Suppl):S512-S520.

Coyle, E.F., A.R. Coggan, M.K. Hemmert, and J.L. Ivy. 1986. Muscle glycogen utilization during prolonged strenuous exercise when fed carbohydrate. *Journal of Applied Physiology*, 61:165-172.

Coyle, E.F., M.T. Hamilton, J. Gonzalez-Alonso, S.J. Montain, and J.L. Ivy. 1991. Carbohydrate metabolism during intense exercise when hyperglycemic. *Journal of Applied Physiology*, 70:834-840.

Coyle, E.F., A.E. Jeukendrup, A.J. Wagenmakers, and W.H. Saris. 1997. Fatty acids oxidation is directly regulated by carbohydrate metabolism during exercise. *American Journal of Physiology*, 273:E268-E275.

Crawford, D., R.W. Jeffery, and S.A. French. 2000. Can anyone successfully control their weight? Findings of a three year community-based study of men and women. *International Journal of Obesity Related Metabolic Disorders*, 9:1107-1110.

Crovetti, R., M. Porrini, A. Santangelo, and G. Testolin. 1998. The influence of thermic effect of food on satiety. *European Journal of Clinical Nutrition*, 52:482-488.

Cunningham, J.J. 1982. Body composition and resting metabolic rate: the myth of feminine metabolism. *American Journal of Clinical Nutrition*, 36:721-726.

Danforth, E. 1999. Sibutramine and thermogenesis in humans. *International Journal of Obesity*, 23:1007-1008.

Daniels, J.T. 1985. A physiologist's view of running economy. *Medicine and Science in Sports and Exercise*, 17:332-338.

Dauncey, M.J. 1990. Thyroid hormones and thermogenesis. *Proceedings of the Nutrition Society*, 49:203-215.

Davidson, M. 1979. The effect of aging on carbohydrate metabolism: a review of diabetes mellitus in the elderly. *Metabolism*, 28:688-705.

Davies, C.T. 1980. Effect of wind assistance and resistance on the forward motion of a runner. *Journal of Applied Physiology*, 48:702-709.

Davies, C.T.M., C. Barnes, and S. Godfrey. 1972. Body composition and maximal exercise performance in children. *Human Biology*, 44:195-214.

Day, C., and C.J. Bailey. 1998. Effect of the antiobesity agent sibutramine in obese-diabetic ob/ob mice. *International Journal of Obesity and Related Metabolic Disorders*, 22:619-623.

DeFronzo, R.A., E. Ferrannini, Y. Sato, P. Felig, and J. Wahren. 1981. Synergistic interaction between exercise and insulin on peripheral glucose uptake. *Journal of Clinical Investigation*, 68:1468-1474.

DeFronzo, R., R. Gunnarsson, D. Bjorkman, M. Olsson, and J. Warren. 1985. Effects of insulin on peripheral and splanchnic glucose metabolism in noninsulin-dependent (type II) diabetes mellitus. *Journal of Clinical Investigation*, 76:149-155.

DeFronzo, R.A., D. Thorin, J.P. Felber, D.C. Simonson, D. Thiebaud, E. Jequier, and A. Golay. 1984. Effect of beta- and alpha-adrenergic blockade in glucose-induced thermogenesis in man. *Journal of Clinical Investigation*, 73:633-639.

DeFronzo, R.A., J.D. Tobin, and R. Andres. 1979. Glucose clamp technique: a method for quantifying insulin secretion and resistance. *American Journal of Physiology*, 237:E214-E223.

De Glisezinski, I., I. Harant, F. Crampes, F. Trudeau, A. Felez, J.M. Cottet-Emard, M. Garrigues, and D. Riviere. 1998. Effects of carbohydrate ingestion on adipose tissue lipolysis during long-lasting exercise in trained men. *Journal of Applied Physiology*, 84:1627-1632.

DeMeersman, R., D. Gatty, and D. Schaffer. 1987. Sympathomimetics and exercise enhancement: all in the mind? *Pharmacology, Biochemistry, and Behavior*, 28:361-365.

den Besten, C., G. Vansant, J. Westrate, and P. Deurenberg. 1988. Resting metabolic rate and diet-induced thermogenesis in abnormal and gluteal-femoral obese women before and after weight reduction. *American Journal of Clinical Nutrition*, 47:840-847.

D'Eon, T.M., and B. Braun. 2002. The roles of estrogen and progesterone in regulating carbohydrate and fat utilization at rest and during exercise. *Journal of Women's Health and Gender-Based Medicine*, 11:225-237.

Devlin, J.T., M. Hirshman, E.D. Horton, and E.S. Horton. 1987. Enhanced peripheral and splanchnic insulin sensitivity in NIDDM men after single bout of exercise. *Diabetes*, 36:434-439.

Devlin, J.T., and E.S. Horton. 1985. Effect of prior high intensity exercise on glucose metabolism in normal and insulin-resistant men. *Diabetes*, 34:973-979.

Dill, D.B. 1965. Oxygen cost of horizontal and grade walking and running on the treadmill. *Journal of Applied Physiology*, 20:19-22.

Di Pietro, L. 1999. Physical activity in the prevention of obesity: current evidence and research issues. *Medicine and Science in Sports and Exercise*, 31:S542-S546.

Di Prampero, P.E., C. Capelli, P. Pagliaro, G. Antonutto, M. Girardis, P. Zamparo, and R.G. Soule. 1993. Energetics of best performance in middle-distance running. *Journal of Applied Physiology*, 74:2318-2324.

Dohm, G.L. 1986. Protein as a fuel for endurance exercise. *Exercise and Sport Sciences Reviews,* 14:143-173.

Dohm, G.L., G.J. Kasperek, E.B. Tapscott, and H.A. Barakat. 1987. Protein degradation during endurance exercise and recovery. *Medicine and Science in Sports and Exercise,* 19:S166-S171.

Dolny, D., and P. Lemon. 1988. Effect of ambient temperature on protein breakdown during prolonged exercise. *Journal of Applied Physiology,* 64:550-555.

Donnelly, J.E., D.J. Jacobsen, K. Snyder Heelan, R. Seip, and S. Smith. 2000. The effects of 18 months of intermittent vs. continuous exercise on aerobic capacity, body weight and composition, and metabolic fitness in previously sedentary, moderately obese females. *International Journal of Obesity,* 24:566-572.

Donovan, C.M., and M.J. Pagliassotti. 1990a. Endurance training enhances lactate clearance during hyperlactatemia. *American Journal of Physiology,* 257:E169-177.

Donovan, C.M., and M.J. Pagliassotti. 1990b. Enhanced efficiency of lactate removal after endurance training. *Journal of Applied Physiology,* 68:1053-1058.

Donovan, C.M., and K.D. Sumida. 1997. Training enhanced hepatic gluconeogenesis: the importance for glucose homeostasis during exercise. *Medicine and Science in Sports and Exercise,* 29:628-634.

Dudley, G.A. 1988. Metabolic consequences of resistive-type exercise. *Medicine and Science in Sports and Exercise,* 20 (Suppl):S158-S161.

Dulloo, A.G., and D.S. Miller. 1986. The thermogenic properties of ephedrine/methylxanthine mixtures: human studies. *International Journal of Obesity,* 10:467-481.

Dyck, D.J., S.A. Peters, P.S. Wendling, A. Chesley, E. Hultman, and L.L. Spriet. 1996. Regulation of muscle glycogen phosphorylase activity during intense aerobic cycling with elevated FFA. *American Journal of Physiology,* 265:E116-E125.

Dyck, D.J., C.T. Putman, G.J.F. Heigenhauser, E. Hultman, and L.L. Spriet. 1993. Regulation of fat-carbohydrate interaction in skeletal muscle during intense aerobic cycling. *American Journal of Physiology,* 265:E852-859.

Ekblom, B. 1969. Effect of physical training on oxygen transport system in man. *Acta Physiologica Scandinavica* (Suppl), 328:1-45.

Elia, M. 1992. Organ and tissue contribution to metabolic rate. In: *Energy metabolism: tissue determinants and cellular corollaries,* ed. J.M. Kinney and H.N. Tucker, 61-79. New York: Raven Press.

Elia, M., P. Ritz, and R.J. Stubbs. 2000. Total energy expenditure in the elderly. *European Journal of Clinical Nutrition,* 54 (Suppl 3):S92-S103.

Ellis, G.S., S. Lanza-Jacoby, A. Gow, and Z.V. Kendrick. 1994. Effect of estradiol on lipoprotein lipase activity and lipid availability in exercised male rats. *Journal of Applied Physiology,* 77:209-215.

Eriksson, B.O., P.D. Gollnick, and B. Saltin. 1973. Muscle metabolism and enzyme activities after training in boys 11-13 years old. *Acta Physiologica Scandinavica,* 87:485-497.

Eriksson, B.O., J. Karlsson, and B. Saltin. 1971. Muscle metabolites during exercise in pubertal boys. *Acta Paediatrica Scandinavica,* 217 (Suppl):154-157.

Essig, D., D.L. Costill, and P.J. Van Handel. 1980. Effect of caffeine ingestion on utilization of muscle glycogen and lipid during leg ergometry cycling. *International Journal of Sports Medicine,* 1:86-90.

Eston, R.G., A.V. Rowlands, and D.K. Ingledew. 1998. Validity of heart rate, pedometry, and accelerometry for predicting the energy cost of children's activities. *Journal of Applied Physiology,* 84:362-371.

Fain, J.N., and J.A. Sainz-Garcia. 1983. Adrenergic regulation of adipocyte metabolism. *Journal of Lipid Research,* 24:945-966.

Faria, E.W., D.L. Parker, and I.E. Faria. 2005. The science of cycling: physiology and training - part 2. *Sports Medicine,* 35:313-337.

Farooqi, I.S., S.A. Jebb, G. Langmack, E. Lawrence, C.H. Cheetham, A. Prentice, I.A. Hughes, M.A. McCarmish, and S. O'Rahilly. 1999. Effects of recombinant leptin therapy in a child with congenital leptin deficiency. *New England Journal of Medicine,* 341:879-884.

Fawkner, S.G., and N. Armstrong. 2003. Oxygen uptake kinetic response to exercise in children. *Sports Medicine,* 33:651-669.

Felig, P. 1984. Insulin is the mediator of feeding-related thermogenesis: insulin resistance and/or deficiency results in thermogenic defect which contributes to the pathogenesis of obesity. *Clinical Physiology,* 4:267-273.

Felig, P., A. Cherif, A. Minagawa, and J. Wahren. 1982. Hypoglycemia during prolonged exercise in normal men. *New England Journal of Medicine,* 306:895-900.

Felig, P., and J. Wahren. 1971. Amino acids metabolism in exercising man. *Journal of Clinical Investigation,* 50:2703-2714.

Fellingham, G.W., E.S. Roundy, A.G. Fisher, and G.R. Bryce. 1978. Calorie cost of walking and running. *Medicine and Science in Sports,* 10:132-136.

Ferrannini, E. 1988. The theoretical basis of indirect calorimetry: a review. *Metabolism,* 37:287-301.

Ferrannini, E., E.J. Barrett, S. Bevilacqua, and R. DeFronzo. 1983. Effects of fatty acids on glucose production and utilization in man. *Journal of Clinical Investigation,* 72:1737-1747.

Fielding, R.A., and J. Parkington. 2002. What are the dietary protein requirements of physically active individuals? New evidence on the effects of exercise on protein utilization during post-exercise recovery. *Nutrition in Clinical Care,* 5:191-196.

Fine, B.J., J.L. Kobrick, H.R. Lieberman, B. Marlowe, R.H. Riley, and W.J. Tharion. 1994. Effects of caffeine or diphenhydramine on visual vigilance. *Psychopharmacology* (Berl), 114:233-238.

Fitts, R.H., F.W. Booth, W.W. Winder, and J.O. Holloszy. 1975. Skeletal muscle respiratory capacity, endurance, and glycogen utilization. *American Journal of Physiology,* 228:1029-1033.

Flechtner-Mors, M., H. Ditschuneit, I. Yip, and G. Adler. 1999. Sympathetic modulation of lipolysis in subcutaneous adipose tissue: effects of gender and energy restriction. *Journal of Laboratory and Clinical Medicine,* 134:33-41.

Flegal, K.M., M.D. Carroll, R.L. Kuczarki, and C.L. Johnson. 1998. Overweight and obesity in the United States: prevalence and trends, 1960-1994. *International Journal of Obesity,* 22:39-47.

Forbes, E.B., and R.W. Swift. 1944. Associative dynamic effects of protein, carbohydrate and fat. *Journal of Nutrition,* 27:453-468.

Forbes, G.B. 1987. *Growth, aging, nutrition, and activity.* New York: Springer.

Franch, J., K. Madsen, M.S. Djurhuus, and P.K. Pedersen. 1998. Improved running economy following intensified training correlates with reduced ventilatory demands. *Medicine and Science in Sports and Exercise,* 30:1250-1256.

Francis, P.R., A.S. Witucki, and M.J. Buono. 1999. Physiological response to a typical studio cycling session. *ACSM Health Fitness Journal,* 3:30-36.

Frayn, K.N. 1983. Calculation of substrate oxidation rates in vivo from gaseous exchange. *Journal of Applied Physiology,* 55:628-634.

Freedson, P.S., and K. Miller. 2000. Objective monitoring of physical activity using motion sensors and heart rate. *Research Quarterly for Exercise and Sport,* 71:21-29.

Frexes-Steed, M., D.B. Lacy, J. Collins, and N.N. Abumard. 1992. Role of leucine and other amino acids in regulating protein metabolism in vivo. *American Journal of Physiology,* 262: E925-E935.

Frey, G.C., W.C. Byrnes, and R.S. Mazzeo. 1993. Factors influencing post-exercise oxygen consumption in trained and untrained women. *Metabolism,* 42:822-828.

Friedlander, A.L., G.A. Casazza, M.A. Horning, M.J. Huie, M.F. Piacentini, J.K. Trimmer, and G.A. Brooks. 1998. Training-induced alterations of carbohydrate metabolism in women: women respond differently from men. *Journal of Applied Physiology,* 85:1175-1186.

Froberg, K., and P.K. Pedersen. 1984. Sex differences in endurance capacity and metabolic response to prolonged heavy exercise. *European Journal of Applied Physiology,* 52:446-450.

Fruin, M.L., and J. Walberg Rankin. 2004. Validity of multi-sensor armband in estimating rest and exercise energy expenditure. *Medicine and Science in Sports and Exercise,* 36:1063-1069.

Gabbay, R.A., and H.A. Lardy. 1984. Site of insulin inhibition of cAMP-stimulated glycogenolysis: cAMP-dependent protein kinase is affected independent of cAMP changes. *Journal of Biological Chemistry,* 259:6052-6055.

Gaesser, G.A., and G.A. Brooks. 1980. Glycogen repletion following continuous and intermittent exercise to exhaustion. *Journal of Applied Physiology,* 49:722-728.

Gaesser, G.A., and D.C. Poole. 1996. The slow component of oxygen uptake kinetics in humans. *Exercise and Sport Sciences Reviews,* 24:35-70.

Garrow, J.S. 1995. Exercise in the treatment of obesity: a marginal contribution. *International Journal of Obesity,* 19 (Suppl 4): S126-S129.

Gatalano, P.M., E.D. Tyzbir, and N.M. Roman. 1991. Longitudinal changes in insulin release and insulin resistance in non-obese pregnant women. *American Journal of Obstetrics and Gynecology,* 165:1667-1672.

Gatalano, P.M., E.D. Tyzbir, R.R. Wolfe, N.M. Roman, S.B. Amini, and E.A.H. Sims. 1992. Longitudinal changes in basal hepatic glucose production and suppression during insulin infusion in normal pregnant women. *American Journal of Obstetrics and Gynecology,* 167:913-919.

Gill, N.D., A. Shield, A.J. Blazevich, S. Zhou, and R.P. Weatherby. 2000. Muscular and cardiorespiratory effects of pseudoephedrine in human athletes. *British Journal of Clinical Pharmacology,* 50:205-213.

Gillette, C.A., R.C. Bullough, and C.L. Melby. 1994. Post-exercise energy expenditure in response to acute aerobic or resistive exercise. *International Journal of Sports Nutrition,* 4:347-360.

Gladden, B., W. Stainsby, and B. Macintosh. 1982. Norepinephrine increases canine muscle O2 during recovery. *Medicine and Science in Sports and Exercise,* 14:471-476.

Gnehm, P., S. Reichenbach, E. Altpeter, H. Widmer, and H. Hoppeler. 1997. Influence of different racing position on metabolic costs in elite cyclists. *Medicine and Science in Sports and Exercise,* 29:818-823.

Going, S., D. Williams, and T. Lohman. 1995. Aging and body composition: biological changes and methodological issues. In: *Exercise and Sport Sciences Reviews,* ed. J.O. Holloszy, 411-458. Baltimore: Williams & Wilkins.

Golay, A., Y. Schutz, J.P. Felber, D. Jallut, and E. Jequier. 1989. Blunted glucose-induced thermogenesis in "overweight" patients: a factor contributing to relapse of obesity. *International Journal of Obesity,* 13:767-775.

Goldberg, A.L., and T.W. Chang. 1978. Regulation and significance of amino acid metabolism in skeletal metabolism. *Federation Proceedings,* 37:2301-2307.

Goldsmith, R., and T. Hale. 1971. Relationship between habitual physical activity and physical fitness. *American Journal of Clinical Nutrition,* 24:1489-1493.

Gollnick, P. 1985. Metabolism of substrates: energy substrate metabolism during exercise and as modified by training. *Federation Proceedings,* 44:353-356.

Gollnick, P.D., R.B. Armstrong, C.W. Saubert, K. Piehl, and B. Saltin. 1972. Enzyme activity and fiber composition in skeletal muscle of untrained and trained men. *Journal of Applied Physiology,* 33:312-319.

Gollnick, P.D., K. Piehl, and B. Saltin. 1974. Selective glycogen depletion pattern in human muscle fibers after exercise of varying intensity and at varying pedal rates. *Journal of Physiology,* 241:45-57.

Gollnick, P.D., and B. Saltin. 1982. Significance of skeletal muscle oxidative enzyme enhancement with endurance training. *Clinical Physiology,* 2:1-12.

Goodpaster, B.H., D.E. Kelley, R.R. Wing, A. Meier, and F.L. Thaete. 1999. Effects of weight loss on regional fat distribution and insulin sensitivity in obesity. *Diabetes,* 48:839-847.

Goodpaster, B.H., F.L. Thaete, and D.E. Kelley. 2000. Thigh adipose tissue distribution is associated with insulin resistance in obesity and in type 2 diabetes mellitus. *American Journal of Clinical Nutrition,* 71:885-892.

Goodpaster, B.H., R.R. Wolfe, and D.E. Kelley. 2002. Effect of obesity on substrate utilization during exercise. *Obesity Research,* 10:575-584.

Goran, M.I., and E.T. Poehlman. 1992. Total energy expenditure and energy requirements in healthy elderly persons. *Metabolism,* 41:744-753.

Gore, C., and R. Witters. 1990. Effect of exercise intensity and duration on post-exercise metabolism. *Journal of Applied Physiology,* 68:2362-2368.

Gortmaker, S.L., A. Must, A.M. Sobol, K. Peterson, G.A. Colditz, and W.H. Dietz. 1996. Television viewing as a cause of increasing obesity among children in the United States, 1986-1990. *Archives of Pediatrics and Adolescent Medicine,* 150:356-362.

Graham, T.E. 2001. Caffeine, coffee and ephedrine: impact on exercise performance and metabolism. *Canadian Journal of Applied Physiology,* 26 (Suppl):S103-119.

Graham, T.E., J.W. Helge, D.A. MacLean, B. Kiens, and E.A. Richter. 2000. Caffeine ingestion does not alter carbohydrate or fat metabolism in human skeletal muscle during exercise. *Journal of Physiology,* 529:837-847.

Graham, T.E., and D.A. MacLean. 1992. Ammonia and amino acid metabolism in human skeletal muscle during exercise. *Canadian Journal of Physiology and Pharmacology,* 70:132-141.

Graham, T.E., J.W.E. Rush, and D.A. MacLean. 1995. Skeletal muscle amino acid metabolism and ammonia production during exercise. In: *Exercise metabolism,* ed. M. Hargreaves, 131-175. Champaign, IL: Human Kinetics.

Green, H.J., S. Jones, M. Ball-Burnett, B. Farrance, and D. Ranney. 1995. Adaptations in muscle metabolism to prolonged voluntary exercise and training. *Journal of Applied Physiology,* 78:138-145.

Griggs, R.C., W. Kingston, R.F. Jozefowicz, B.E. Herr, G. Forbes, and D. Halliday. 1989. Effect of testosterone on muscle mass and muscle protein synthesis. *Journal of Applied Physiology,* 66:498-503.

Grundy, S.M., G. Blackburn, M. Higgins, R. Lauer, M.G. Perri, and D. Ryan. 1999. Roundtable consensus statement: physical activity in the prevention and treatment of obesity and it comorbidities. *Medicine and Science in Sports and Exercise,* 31:S502-S508.

Gueli, D., and R.J. Shephard. 1976. Pedal frequency in bicycle ergometry. *Canadian Journal of Applied Sports Science,* 1:137-141.

Hackney, A.C. 1990. Effects of menstrual cycle on resting muscle glycogen content. *Hormone and Metabolic Research,* 22:647.

Hackney, A.C., M.A. McCracken-Compton, and B. Ainsworth. 1994. Substrate responses to submaximal exercise in the midfollicular and midluteal phase of the menstrual cycle. *International Journal of Sports Nutrition,* 4:299-308.

Hagberg, J.M., R.C. Hickson, A.A. Ehsani, and J.O. Holloszy. 1980. Faster adjustment to and recovery from submaximal exercise in the trained state. *Journal of Applied Physiology,* 48:218-224.

Hagberg, J., J. Mullin, and F. Nagle. 1978. Oxygen consumption during constant load exercise. *Journal of Applied Physiology,* 45:381-384.

Hakim, A.A., J.D. Curb, H. Petrovich, et al. 1999. Effects of walking on coronary heart disease in elderly men: the Honolulu Heart Study. *Circulation,* 100:9-13.

Hakkinen, K., P.V. Komi, and P.A. Tesch. 1981. Effect of combined concentric and eccentric strength training and detraining on force-time, muscle fiber, and metabolic characteristics of leg extensor muscles. *Scandinavian Journal of Sports Science,* 3:50-58.

Halaas, J.L., K.S. Gajiwala, M. Maffei, S.L. Cohen, B.T. Chait, D. Rabinowitz, R.L. Lallone, S.K. Burley, and J.M. Friedman. 1995. Weight-reducing effects of the plasma protein encoded by the obese gene. *Science,* 269:543-546.

Halton, R.W., R.R. Kraemer, R.A. Sloan, E.P. Hebert, K. Frank, and J.L. Tryniecki. 1999. Circuit weight training and its effect on excess postexercise oxygen consumption. *Medicine and Science in Sports and Exercise,* 31:1613-1618.

Hamilton, A.L., M.E. Nevill, S. Brooks, and G. Williams. 1991. Physiological responses to maximal intermittent exercise: differences between endurance trained and runners and game players. *Journal of Sports Science,* 9:371-382.

Hamilton, K.S., F.K. Gibbons, D.P. Lacy, A.D. Cherrington, and D.H. Wasserman. 1996. Effect of prior exercise on the partitioning of an intestinal glucose load between splanchnic bed and skeletal muscle. *Journal of Clinical Investigation,* 98:125-135.

Hammer, R.L., C.A. Barrier, E.S. Roundy, J.M. Bradford, and A.G. Fisher. 1988. Calorie-restricted low fat diet and exercise in obese women. *American Journal of Clinical Nutrition,* 1:77-85.

Hansen, D.L., S. Toubro, M.J. Stock, I.A. Macdonald, and A. Astrup. 1999. The effect of sibutramine on energy expenditure and appetite during chronic treatment without dietary restriction. *International of Obesity and Related Metabolic Disorders,* 23:1016-1024.

Hansen, F.M., N. Fahmy, and J.H. Nielsen. 1980. The influence of sexual hormones on lipogenesis and lipolysis in rat cells. *Acta Endocrinologica,* 95:566-570.

Hargreaves, M. 2006. Skeletal muscle carbohydrate metabolism during exercise. In: *Exercise metabolism,* ed. M. Hargreaves and L. Spriet, 29-54. Champaign, IL: Human Kinetics.

Hargreaves, M., and J. Proietto. 1994. Glucose kinetics during exercise in trained men. *Acta Physiologica Scandinavica,* 150:221-225.

Haskell, W.L., M.C. Yee, A. Evans, and P.J. Irby. 1993. Simultaneous measurement of heart rate and body motion to quantitate physical activity. *Medicine and Science in Sports,* 25:109-115.

Hatano, Y. 1993. Use of the pedometer for promoting daily walking exercise. *International Council for Health, Physical Education, and Recreation,* 29:4-8.

Haymes, E.M., and W.C. Byrnes. 1993. Walking and running energy expenditure estimated by Caltrac and indirect calorimetry. *Medicine and Science in Sports and Exercise,* 25:1365-1369.

Hays, J.H., A. DiSabatino, R.T. Gorman, S. Vincent, and M.E. Stillabower. 2003. Effect of a high saturated fat and no-starch diet on serum lipid subfractions in patients with documented atherosclerotic cardiovascular disease. *Mayo Clinics Proceedings,* 78:1331-1336.

Heath, G.W., J.R. Gavin, III, J.M. Hinderliter, J.M. Hagberg, S.A. Bloomfield, and J.O. Holloszy. 1983. Effects of exercise and lack of exercise on glucose tolerance and insulin sensitivity. *Journal of Applied Physiology,* 55:512-517.

Hebert, D.N., and A. Carruthers. 1992. Glucose transporter oligomeric structure determines transporter function: reversible redox-dependent interconversions of tetrameric and dimeric GLUT1. *Journal of Biological Chemistry,* 267:23829-23838.

Hebestreit, H., S. Kriemler, R.L. Hughson, and O. Bar-Or. 1998. Kinetics of oxygen uptake at the onset of exercise in boys and men. *Journal of Applied Physiology*, 85:1833-1841.

Hendelman, D., K. Miller, C. Baggett, E. Debold, and P. Freedson. 2000. Validity of accelerometry for the assessment of moderate intensity physical activity in the field. *Medicine and Science in Sports and Exercise*, 32 (Suppl):S442-S449.

Henriksson, J. 1977. Training induced adaptation of skeletal muscle and metabolism during submaximal exercise. *Journal of Physiology*, 270:661-667.

Henriksson, J. 1995. Muscle fuel selection: effect of exercise and training. *Proceedings of the Nutrition Society*, 54:125-138.

Hermansen, L., and M. Wachtlova. 1971. Capillary density of skeletal muscle in well-trained and untrained men. *Journal of Applied Physiology*, 30:860-863.

Heyman, M.B., P. Fuss, V.R. Young, W.J. Evans, and S.B. Roberts. 1991. Prediction of total energy expenditure using the Caltrac activity monitor. *International Journal of Obesity*, 15 (Suppl):1-23.

Heymsfield, S.B., A.S. Greenberg, K. Fujioka, R.M. Dixon, R. Kushner, T. Hunt, J.A. Lubina, J. Patane, B. Self, P. Hunt, and M. McCamish. 1999. Recombinant leptin for weight loss in obese and lean adults: a randomized, controlled, dose-escalation trial. *Journal of the American Medical Association*, 282:1568-1575.

Hickson, R.C. 1980. Interference of strength development by simultaneously training for strength and endurance. *European Journal of Applied Physiology*, 45:255-263.

Hickson, R.C., H.A. Bomze, and J.O. Holloszy. 1978. Faster adjustment of oxygen uptake to the energy requirement of exercise in the trained state. *Journal of Applied Physiology*, 44:877-881.

Hill, A.V., and H. Lupton. 1923. Muscular exercise, lactic acid, and the supply and utilization of oxygen. *Quarterly Journal of Medicine*, 16:135-171.

Hill, J.O., and E.L. Melanson. 1999. Overview of the determinants of overweight and obesity: current evidence and research issues. *Medicine and Science in Sports and Exercise*, 31 (Suppl): S515-S521.

Hill, R.J., and P.S. Davis. 2002. Energy intake and energy expenditure in elite lightweight female rowers. *Medicine and Science in Sports and Exercise*, 34:1823-1827.

Holloszy, J.O., M. Chen, G.D. Cartee, and J.C. Young. 1991. Skeletal muscle atrophy in old rats: differential changes in the three fiber types. *Mechanisms of Ageing and Development*, 60:199-213.

Holloszy, J.O., and E.F. Coyle. 1984. Adaptations of skeletal muscle to endurance exercise and their metabolic consequences. *Journal of Applied Physiology*, 56:831-838.

Holloszy, J.O., L.B. Oscai, P.A. Mole, and I.J. Don. 1971. Biochemical adaptations to endurance exercise in skeletal muscle. In: *Muscle metabolism during exercise*, ed. B. Pernow and B. Saltin, 51-61. New York: Plenum Press.

Hoppeler, H., H. Howald, K. Concley, et al. 1985. Endurance training in humans: aerobic capacity and structure of skeletal muscle. *Journal of Applied Physiology*, 59:320-327.

Horowitz, J.F., and S. Klein. 2000. Oxidation of nonplasma fatty acids during exercise is increased in women with abdominal obesity. *Journal of Applied Physiology*, 89:2276-2282.

Horton, T.J., M.J. Pagliassotti, K. Hobbs, and J.O. Hill. 1998. Fuel metabolism in men and women during and after long-duration exercise. *Journal of Applied Physiology*, 85:1823-1832.

Houde-Nadeau, M., L. de Jonge, and D.R. Garrel. 1993. Thermogenic response to food: intraindividual variability and measurement reliability. *Journal of the American College of Nutrition*, 12:511-516.

Houmard, J.A., M.L. Weidner, K.E. Gavigan, G.L. Tyndall, M.S. Hickey, and A. Alshami. 1998. Fiber type and citrate synthase activity in human gastrocnemius and vastus lateralis with aging. *Journal of Applied Physiology*, 85:1337-1341.

Hukshorn, C.J., and W.H.M. Saris. 2004. Leptin and energy expenditure. *Current Opinion in Clinical Nutrition and Metabolic Care*, 7:629-633.

Hunter, G., L. Blackman, L. Dunnam, and G. Flemming. 1988. Bench press metabolic rate as a function of exercise intensity. *Journal of Applied Sports Science Research*, 2:1-6.

Hunter, G.R., T. Kekes-Szabo, and A. Schnitzler. 1992. Metabolic cost: vertical work ratio during knee extension and knee flexion weight-training exercise. *Journal of Applied Sports Science Research*, 6:42-48.

Hurley, B.F., J.M. Hagberg, W.K. Allen, D.R. Seals, J.C. Young, R.W. Cuddihee, and J.O. Holloszy. 1984. Effect of training on blood lactate levels during submaximal exercise. *Journal of Applied Physiology*, 56:1260-1264.

Hurley, B.F., P.M. Nemeth, W.H. Martin, 3rd, J.M. Hagberg, G.P. Dalsky, and J.O. Holloszy. 1986. Muscle triglyceride utilization during exercise: effect of training. *Journal of Applied Physiology*, 60:562-567.

Ingjer, F. 1979. Capillary supply and mitochondrial content of different skeletal muscle fiber types in untrained and endurance trained men: a histochemical and ultra structural study. *European Journal of Applied Physiology*, 40:197-209.

Issekutz, B., and P. Paul. 1968. Intramuscular energy sources in exercising normal and pancreatecotomized dogs. *American Journal of Physiology*, 215:197-204.

Ivy, J.L., B.A. Frishberg, S.W. Farrell, W.J. Miller, and W.M. Sherman. 1985. Effect of elevated and exercise-induced muscle glycogen levels on insulin sensitivity. *Journal of Applied Physiology*, 59:154-159.

Izawa, T., T. Komabayashi, T. Mochizuki, K. Suda, and M. Tsuboi. 1991. Enhanced coupling of adenylate cyclase to lipolysis in permeabilized adipocytes from trained rats. *Journal of Applied Physiology*, 71:23-29.

Jackson, A., S. Blair, M. Mahar, L. Weir, R. Ross, and J. Stuteville. 1990. Prediction of functional aerobic capacity without exercise testing. *Medicine and Science in Sports and Exercise*, 22:863-870.

Jakicic, J.M., M. Marcus, K.I. Gallagher, C. Randall, E. Thomas, F.L. Goss, and R.J. Robertson. 2004. Evaluation of SenseWear Pro Armband™ to assess energy expenditure during exercise. *Medicine and Science in Sports and Exercise*, 36:897-904.

Jakicic, J.M., R.R. Wing, B.A. Butler, and R.J. Robertson. 1995. Prescribing exercise in multiple short bouts versus one continuous bout: effect on adherence, cardiorespiratory fitness, and weight loss in overweight women. *International Journal of Obesity,* 19:893-901.

Jakicic, J.M., C. Winters, W. Lang, and R.R. Wing. 1999. Effects of intermittent exercise and use of home exercise equipment on adherence, weight loss, and fitness in overweight women: a randomized trial. *Journal of the American Medical Association,* 282:1554-1560.

James, R.C., T.W. Burns, and G.R. Chase. 1971. Lipolysis of human adipose tissue cells: influence of donor factors. *Journal of Laboratory and Clinical Medicine,* 77:254-266.

James, W.P.T. 1992. From SDA to DIT to TEF. In: *Energy metabolism: tissue determinants and cellular corollaries,* ed. J.M. Kinney and H.N. Tucker, 163-186. New York: Raven Press.

Jansson, E., and L. Kaijser. 1977. Muscle adaptation to extreme endurance training in men. *Acta Physiologica Scandinavica,* 100:315-324.

Jansson, E., and L. Kaijser. 1987. Substrate utilization and enzymes in skeletal muscle of extremely endurance-trained men. *Journal of Applied Physiology,* 62:999-1005.

Jansson, P.A., U. Smith, and P. Lonnroth. 1990. Interstitial glycerol concentration measured by microdialysis in two subcutaneous regions in humans. *American Journal of Physiology,* 258:E918-E922.

Janz, K.F., J. Witt, and L.T. Mahoney. 1995. The stability of children's physical activity as measured by accelerometry and self-report. *Medicine and Science in Sports and Exercise,* 27:1326-1332.

Jenkins, A.B., D.J. Chisholm, D.E. James, K.Y. Ho, and E.W. Kraegen. 1985. Exercise induced hepatic glucose output is precisely sensitive to the rate of systemic glucose supply. *Metabolism,* 34:431-441.

Jenkins, A.B., S.M. Furler, D.J. Chisholm, and E.W. Kraegen. 1986. Regulation of hepatic glucose output during exercise by circulating glucose and insulin in humans. *American Journal of Physiology,* 250:R411-R417.

Jeukendrup, A., and M. Gleeson. 2004. *Sports nutrition: an introduction to energy production and performance.* Champaign, IL: Human Kinetics.

Johnson, L.N. 1992. Glycogen phosphorylase: control by phosphorylation and allosteric effectors. *FASEB Journal,* 6:2274-2282.

Jones, A.M. 1998. A 5-year physiological case study of an Olympic runner. *British Journal of Sports Medicine,* 32:39-43.

Jones, A.M., and H. Carter. 2000. The effect of endurance training on parameters of aerobic fitness. *Sports Medicine,* 29:373-386.

Kalkhoff, R. 1982. Metabolic effects of progesterone. *American Journal of Obstetrics and Gynecology,* 142:735-738.

Kang, J., E.C. Edward, M.A. Mastrangelo, J.R. Hoffman, N.A. Ratamess, and E. O'Connor. 2005a. Metabolic and perceptual responses during Spinning® cycle exercise. *Medicine and Science in Sports and Exercise,* 37:853-859.

Kang, J., J.R. Hoffman, J. Im, B.A. Spiering, N.A. Ratamess, K.W. Rundell, S. Nioka, J. Cooper, and B. Chance. 2005b. Evaluation of physiological responses during recovery following three resistance exercise programs. *Journal of Strength and Conditioning Research,* 19:305-309.

Kang, J., J.R. Hoffman, M. Wendell, H. Walker, and M. Hebert. 2004. Effect of contraction frequency on energy expenditure and substrate utilization during upper and lower body exercise. *British Journal of Sports Medicine,* 38:31-35.

Kang, J., D.E. Kelley, R.J. Robertson, F.L. Goss, R.R. Suminski, A.C. Utter, and S.G. Dasilva. 1999. Substrate utilization and glucose turnover during exercise of varying intensities in individuals with NIDDM. *Medicine and Science in Sports and Exercise,* 31:82-89.

Kang, J., R.J. Robertson, F.L. Goss, S.G. DaSilva, R.R. Suminski, A.C. Utter, R.F. Zoeller, and K.F. Metz. 1997. Metabolic efficiency during arm and leg exercise at the same relative intensities. *Medicine and Science in Sports and Exercise,* 29:377-382.

Kang, J., R.J. Robertson, J.M. Hagberg, D.E. Kelley, F.L. Goss, S.G. DaSilva, R.R. Suminski, and A.C. Utter. 1996. Effect of exercise intensity on glucose and insulin metabolism in obese individuals and obese NIDDM patients. *Diabetes Care,* 19:341-349.

Kaplan, G.B., D.J. Greenblatt, B.L. Ehrenberg, J.E. Goddard, M.M. Cotreau, J.S. Harmatz, and R.I. Shader. 1997. Dose-dependent pharmacokinetics and psychomotor effects of caffeine in humans. *Journal of Clinical Pharmacology,* 37:693-703.

Karlsson, J., L.-O. Nordesjo, L. Jorfeldt, and B. Saltin. 1972. Muscle lactate, ATP, and CP levels during exercise after physical training in man. *Journal of Applied Physiology,* 33:199-203.

Karvonen, J., J. Chwalbinska-Moneta, and S. Saynajakangas. 1984. Comparison of heart rate measured by ECG and by microcomputer. *Physician and Sports Medicine,* 12:65-69.

Kashiwazaki, H., Y. Dejima, and T. Suzuki. 1990. Influence of upper and lower thermoneutral room temperatures (20 °C and 25 °C) on fasting and postprandial resting metabolism under different outdoor temperatures. *European Journal of Clinical Nutrition,* 44:405-413.

Kasper, H., H. Thiel, and M. Ehl. 1973. Response of body weight to a low carbohydrate high fat diet in normal and obese subjects. *American Journal of Clinical Nutrition,* 26:197-204.

Kelley, D.E. 2005. Skeletal muscle fat oxidation: timing and flexibility are everything. *Journal of Clinical Investigation,* 115:1699-1702.

Kelley, D.E., M. Mokan, and L.J. Mandarino. 1992. Intracellular defects in glucose metabolism in obese patients with NIDDM. *Diabetes,* 41:698-706.

Kendrick, Z.V., and G.S. Ellis. 1991. Effect of estradiol on tissue glycogen metabolism and lipid availability in exercised male rats. *Journal of Applied Physiology,* 71:1694-1699.

Kendrick, Z.V., C. Steffen, W. Rumsey, and D. Goldberg. 1987. Effect of estradiol on tissue glycogen metabolism in exercised oophorectomized rats. *Journal of Applied Physiology,* 63:492-496.

Kennedy, C., and L. Sokoloff. 1957. An adaptation of the nitrous oxide methods to the study of the cerebral circulation of children: normal values for cerebral blood flow and cerebral metabolic rate in childhood. *Journal of Clinical Investigation,* 36:1130-1137.

Kiens, B., B. Essen-Gustavsson, N.J. Christensen, and B. Saltin. 1993. Skeletal muscle substrate utilization during submaximal exercise in man: effect of endurance training. *Journal of Physiology,* 469:459-478.

Kiens, B., and H. Lithell. 1989. Lipoprotein metabolism influenced by training-induced changes in human skeletal muscle. *Journal of Clinical Investigation,* 83:558-564.

King, D.S., G.P. Dalsky, M.A. Staten, W.E. Clutter, D.R. van Houten, and J.O. Holloszy. 1987. Insulin action and secretion in endurance-trained and untrained humans. *Journal of Applied Physiology,* 63:2247-2252.

Kjaer, M. 1989. Epinephrine and some other hormonal responses to exercise in man: with special reference to physical training. *International Journal of Sports Medicine,* 10:2-15.

Kjaer, M. 1995. Hepatic fuel metabolism during exercise. In: *Exercise metabolism,* ed. M. Hargreaves, 73-97. Champaign, IL: Human Kinetics.

Kjaer, M., P.A. Farrell, N.J. Christensen, and H. Galbo. 1986. Increased epinephrine response and inaccurate glucoregulation in exercising athletes. *Journal of Applied Physiology,* 61:1693-1700.

Kjaer, M., B. Kiens, M. Hargreaves, and E.A. Richter. 1991. Influence of active muscle mass on glucose homeostasis during exercise in humans. *Journal of Applied Physiology,* 71:552-557.

Kjaer, M., N.H. Secher, F.W. Bach, and H. Galbo. 1987. Role of motor center activity for hormonal changes and substrate mobilization in humans. *American Journal of Physiology,* 253:R687-R695.

Klesges, R.C., L.M. Klesges, A.M. Swenson, and A.M. Phely. 1985. A validation of two motion sensors in the prediction of child and adult physical activity levels. *American Journal of Epidemiology,* 122:400-410.

Knuttgen, H.G., and K. Emerson, Jr. 1974. Physiological responses to pregnancy at rest and during exercise. *Journal of Applied Physiology,* 36:549-553.

Kokkinos, P.F., and B.F. Hurley. 1990. Strength training and lipoprotein-lipid profiles: A critical analysis and recommendations for further study. *Sports Medicine,* 9:266-272.

Koranyi, L.I., R.E. Bourey, C.A. Slentz, and J.O. Holloszy. 1991. Coordinate reduction of rat pancreatic islet glucokinase and proinsulin mRNA by exercise training. *Diabetes,* 40:401-404.

Kraemer, W.J., J.S. Volek, K.L. Clark, S.E. Gordon, T. Incledon, S. Puhl, N.T. Triplett-McBride, J.M. McBride, M. Putukian, and W.J. Sebastianelli. 1997. Physiological adaptations to a weight loss dietary regimen and exercise program in women. *Journal of Applied Physiology,* 83:270-279.

Kremer, R.R., H. Chu, and V.D. Castracane. 2002. Leptin and exercise. *Experimental Biology and Medicine,* 227:701-708.

Kyle, C.R. 1989. The aerodynamics of helmets and handlebars. *Cycling Science,* 1:22-25.

Kyle, C.R. 1991. The effect of crosswinds upon time trials. *Cycling Science,* 3:51-56.

Kyrolainen, H., T. Pullinen, R. Candau, J. Avela, P. Huttunen, and P.V. Komi. 2000. Effects of marathon running on running economy and kinematics. *European Journal of Applied Physiology,* 82:297-304.

Lafontan, M., L. Dang-Tran, and M. Berlan. 1979. Alpha-adrenergic antilipolytic effect of adrenaline in human fat cells of the thigh: comparison with adrenaline responsiveness of different fat deposits. *European Journal of Clinical Investigation,* 9:261-266.

Lake, M., and P. Cavanagh. 1996. Six weeks of training does not change running mechanics or improve running economy. *Medicine and Science in Sports and Exercise,* 28:860-869.

Lamont, L.S., A.J. McCullough, and S.C. Kalhan. 1999. Comparison of leucine kinetics in endurance-trained and sedentary humans. *Journal of Applied Physiology,* 86:320-325.

Lang, P.B., R.W. Latin, K.E. Berg, et al. 1992. The accuracy of the ACSM cycle ergometry equation. *Medicine and Science in Sports and Exercise,* 24:272-276.

LaPorte, R.E. 1979. An objective measure of physical activity for epidemiologic research. *American Journal of Epidemiology,* 109:158-168.

Latin, R.W., and K.E. Berg. 1994. The accuracy of the ACSM and a new cycle ergometry equation for young women. *Medicine and Science in Sports Exercise,* 26:642-646.

Laville, M., C. Cornu, S. Normand, G. Mithieux, M. Beylot, and J.P. Riou. 1993. Decreased glucose-induced thermogenesis at the onset of obesity. *American Journal of Clinical Nutrition,* 57:851-856.

Lawrence, R.D. 1926. The effect of exercise on insulin action in diabetes. *British Medical Journal,* 1:648-650.

Lawrie, R.A. 1953. Effect of enforced exercise on myoglobin in muscle. *Nature,* 171:1069-1070.

Lawson, S., J.D. Webster, P.J. Pacy, and J.S. Garrow. 1987. Effect of a 10-week aerobic exercise programme on metabolic rate, body composition and fitness in lean sedentary females. *British Journal of Clinical Practice,* 41:684-688.

LeBlanc, J., P. Diamond, J. Cote, and A. Labrie. 1984a. Hormonal factors in reduced postprandial heat production of exercise-trained subjects. *Journal of Applied Physiology,* 56:772-776.

LeBlanc, J., P. Mercier, and P. Samson. 1984b. Diet-induced thermogenesis with relation to training state in female subjects. *Canadian Journal of Physiology and Pharmacology,* 62:334-337.

Lee, J.S., C.R. Bruce, R.J. Tunstall, D. Cameron-Smith, H. Hugel, and J.A. Hawley. 2002. Interaction of exercise and diet on GLUT-4 protein and gene expression in type I and type II rat skeletal muscle. *Acta Physiologica Scandinavica,* 175:37-44.

Leger, L., and M. Thivierge. 1988. Heart rate monitor: validity, stability, and functionality. *Physician and Sports Medicine,* 16:143-151.

Leibel, R.L., M. Rosenbaum, and J. Hirsch. 1995. Changes in energy expenditure resulting from altered body weight. *New England Journal of Medicine,* 332:621-628.

Lejeune, T.M., P.A. Willems, and N.C. Heglund. 1998. Mechanics and energetics of human locomotion on sand. *Journal of Experimental Biology,* 201:2071-2080.

Lemon, P., and J. Mullin. 1980. Effect of initial muscle glycogen levels on protein catabolism during exercise. *Journal of Applied Physiology,* 48:624-629.

Lemon, P.W.R., M.A. Tarnopolsky, J.D. MacDougall, and S.A. Atkinson. 1992. Protein requirements and muscle mass/strength changes during intensive training in novice bodybuilders. *Journal of Applied Physiology,* 73:767-775.

Lennon, D., F. Nagel, F. Stratman, E. Shrago, and S. Dennis. 1984. Diet and exercise training effects on resting metabolic rate. *International Journal of Obesity*, 9:39-47.

Lieberman, H.R., R.J. Wurtman, G.G. Emde, and I.L. Coviella. 1987a. The effects of caffeine and aspirin on mood and performance. *Journal of Clinical Psychopharmacology*, 7:315-320.

Lieberman, H.R., R.J. Wurtman, G.G. Emde, C. Roberts, and I.L. Coviella. 1987b. The effects of low doses of caffeine on human performance and mood. *Psychopharmacology* (Berl), 92:308-312.

Londeree, B.R., J. Moffitt-Gerstenberger, J.A. Padfield, and D. Lottmann. 1997. Oxygen consumption of cycle ergometry is nonlinearly related to work rate and pedal rate. *Medicine and Science in Sports and Exercise*, 29:775-780.

Lonnqvist, F., B. Nyberg, H. Wahrenberg, and P. Arner. 1990. Catecholamine-induced lipolysis in adipose tissue of the elderly. *Journal of Clinical Investigation*, 85:1614-1621.

Louard, R.J., E.J. Barrett, and R.A. Gelfand. 1990. Effect of infused branched-chain amino acids on muscle and whole-body amino acid metabolism in man. *Clinical Science*, 79:457-466.

Lowell, B.B., and B.M. Spiegelman. 2000. Towards a molecular understanding of adaptive thermogenesis. *Nature*, 404:652-660.

Luke, A., K.C. Maki, N. Barkey, R. Cooper, and D. McGee. 1997. Simultaneous monitoring of heart rate and motion to assess energy expenditure. *Medicine and Science in Sports and Exercise*, 29:144-148.

Lusk, G. 1924. Animal calorimetry: analysis of the oxidation of mixtures of carbohydrate and fat. A correction. *Journal of Biological Chemistry*, 59:41-42.

Lutz, P.L. 2002. *The rise of experimental biology: an illustrated history*. Totowa, NJ: Plenum Press.

MacDonald, I. 1984. Differences in dietary-induced thermogenesis following the ingestion of various carbohydrates. *Annals of Nutrition and Metabolism*, 28:226-230.

MacDougall, J.D. 1986. Morphological changes in human skeletal muscle following strength training and immobilization. In: *Human muscle power*, ed. N.L. Jones, 269-288. Champaign, IL: Human Kinetics.

MacDougall, J., A.L. Hicks, J.R. MacDonald, R.S. McKelvie, H.J. Green, and K.M. Smith. 1998. Muscle performance and enzymatic adaptations to sprint interval training. *Journal of Applied Physiology*, 84:2138-2142.

MacDougall, J.D., G.R. Ward, D.G. Sale, and J.R. Sutton. 1977. Biochemical adaptation of human skeletal muscle to heavy resistance training and immobilization. *Journal of Applied Physiology*, 43:700-703.

Macek, M., J. Vavra, and J. Novosadova. 1976. Prolonged exercise in pre-pubertal boys II: Changes in plasma volume and in some blood constituents. *European Journal of Applied Physiology*, 35:299-303.

Mackintosh, R.M., and J. Hirsch. 2001. The effects of leptin administration in non-obese human subjects. *Obesity Research*, 9:462-469.

Maehlum, S., A.T. Hostmark, and L. Hermansen. 1977. Synthesis of muscle glycogen during recovery after prolonged, severe exercise in diabetic and nondiabetic subjects. *Scandinavian Journal of Clinical and Laboratory Investigation*, 37:309-316.

Maffei, M., M. Stoffel, M. Barone, B. Moon, M. Dammerman, E. Ravussin, C. Bogardus, D.S. Ludwig, J.S. Flier, and M. Talley. 1996. Absence of mutations in the human OB gene in obese/diabetic subjects. *Diabetes*, 45:679-682.

Malchow-Moller, A., S. Larsen, H. Hey, K.H. Stokholm, E. Juhl, and F. Quaade. 1981. Ephedrine as an anorectic: the story of the "Elsinore pill." *International Journal of Obesity*, 5:183-187.

Margaret-Mary, G.W., and J.E. Morley. 2003. Physiology of aging invited review: aging and energy balance. *Journal of Applied Physiology*, 95:1728-1736.

Margaria, R., P. Cerretelli, P. Aghemo, and G. Sassi. 1963. Energy cost of running. *Journal of Applied Physiology*, 18:367-370.

Margaria, R., H.T. Edward, and O.B. Dill. 1933. The possible mechanisms of contracting and paying the oxygen debt and the role of lactic acid in muscular contraction. *American Journal of Physiology*, 106:689-715.

Marker, J.C., I.B. Hirsch, L.J. Smith, C.A. Parvin, J.O. Holloszy, and P.E. Cryer. 1991. Catecholamines in prevention of hyperglycemia during exercise in humans. *American Journal of Physiology*, 260:E705-E712.

Marliss, E.B., S.H. Kreisman, A. Manzon, J.B. Halter, M. Vranic, and S.J. Nessim. 2000. Gender differences in glucoregulatory responses to intense exercise. *Journal of Applied Physiology*, 88:457-466.

Martin, I.K., A. Katz, and J. Wahren. 1995. Splanchnic and muscle metabolism during exercise in NIDDM patients. *American Journal of Physiology*, 269:E583-E590.

Martin, P.E., and D.W. Morgan. 1992. Biomechanical considerations for economical walking and running. *Medicine and Science in Sports and Exercise*, 24:467-474.

Martinez, L.R., and E.M. Haymes. 1992. Substrate utilization during treadmill running in pre-pubertal girls and women. *Medicine and Science in Sports and Exercise*, 24:975-983.

Matute, M.L., and R. Kalkhoff. 1973. Sex steroid influence on hepatic gluconeogenesis and glycogen formation. *Endocrinology*, 92:762-768.

McArdle, W.D., F.I. Katch, and V.L. Katch. 2001. *Exercise physiology: energy, nutrition, and human performance*. 5th ed. Baltimore: Lippincott Williams & Wilkins.

McArdle, W.D., F.I. Katch, and V.L. Katch. 2005. *Sports and exercise nutrition*. 2nd ed. Baltimore: Lippincott Williams & Wilkins.

McCartney, N., L.L. Spriet, G.I.F. Heigenhauser, J.M. Kowalchuk, J.R. Sutton, and N.L. Jones. 1986. Muscle power and metabolism in maximal intermittent exercise. *Journal of Applied Physiology*, 60:1164-1169.

McCormack, J., and R. Denton. 1994. Signal transduction by intra-mitochondrial calcium in mammalian energy metabolism. *News in Physiological Sciences*, 9:71-76.

McGarry, J.D., and N.F. Brown. 1997. The mitochondrial carnitine palmitoyltransferase system from concept to molecule analysis. *European Journal of Biochemistry*, 244:1-14.

McKenzie, S., S.M. Phillips, S.L. Carter, S. Lowther, M.J. Gibala, and M.A. Tarnopolsky. 2000. Endurance exercise training attenuates leucine oxidation and BCOAD activation during exercise in humans. *American Journal of Physiology*, 278: E580-587.

Melanson, E.L., T.A. Sharp, H.M. Seagle, W.T. Donahoo, G.K. Grunwald, J.C. Peters, J.T. Hamilton, and J.O. Hill. 2002. Resistance and aerobic exercise have similar effects on 24-h nutrient oxidation. *Medicine and Science in Sports and Exercise*, 34:1793-1800.

Melby, C., C. Scholl, G. Edwards, and R. Bullough. 1993. Effect of acute resistance exercise on postexercise energy expenditure and resting metabolic rate. *Journal of Applied Physiology*, 75:1847-1853.

Menier, D.R., and L.G.C.E. Pugh. 1968. The relation of oxygen intake and velocity of walking and running in competition walkers. *Journal of Physiology*, 197:717-721.

Merrill, A.L., and B.K. Watt. 1973. *Energy value of foods: basis and derivation*. Agriculture Handbook No. 74. Washington, DC: U.S. Department of Agriculture. www.nal.usda.gov/fnic/foodcomp/Data/Classics/ah74.pdf

Mikines, K.J., B. Sonne, P.A. Farrell, B. Tronier, and H. Galbo. 1988. Effect of physical exercise on sensitivity and responsiveness to insulin in humans. *American Journal of Physiology*, 254: E248-E259.

Mikines, K.J., B. Sonne, P.A. Farrell, B. Tronier, and H. Galbo. 1989a. Effect of training on the dose-response relationship for insulin action in men. *Journal of Applied Physiology*, 66:695-703.

Mikines, K.J., B. Sonne, P.A. Farrell, B. Tronier, and H. Galbo. 1989b. Effects of training and detraining on dose-response relationship between glucose and insulin secretion. *American Journal of Physiology*, 256:E588-E596.

Miller, D.J., P.S. Freedson, and G.M. Kline. 1994. Comparison of activity levels using the Caltrac accelerometer and five questionnaires. *Medicine and Science in Sports and Exercise*, 26:376-382.

Miller, J.F., and B.A. Stamford. 1987. Intensity and energy cost of weighted walking vs. running for men and women. *Journal of Applied Physiology*, 62:1497-1501.

Miller, W.C., D.M. Koceja, and E.J. Hamilton. 1997. A meta-analysis of the past 25 years of weight loss research using diet, exercise or diet plus exercise intervention. *International Journal of Obesity Related Metabolic Disorders*, 21:941-947.

Millet, L., H. Vidal, F. Andreelli, D. Larrouy, J.P. Riou, D. Ricquier, M. Laville, and D. Langin. 1997. Increased uncoupling protein-2 and -3 mRNA expression during fasting in obese and lean humans. *Journal of Clinical Investigation*, 100:2665-2670.

Montoye, H.J., H.C.G. Kemper, W.H.M. Saris, and R.A. Washburn. 1996. Movement assessment device. In: *Measuring physical activity and energy expenditure*, 72-96. Champaign, IL: Human Kinetics.

Montoye, H.J., R. Washburn, S. Servais, A. Ertl, J.G. Webster, and F.J. Nagle. 1983. Estimation of energy expenditure by a portable accelerometer. *Medicine and Science in Sports and Exercise*, 15:403-407.

Morgan, D.W., P.E. Martin, F.D. Baldini, and G.S. Krahenbuhl. 1990. Effects of a prolonged maximal run on running economy and running mechanics. *Medicine and Science in Sports and Exercise*, 22:834-840.

Morse, M., F.W. Schlutz, and E. Cassels. 1949. Relation of age to physiological response of the older boy (10-17 years) to exercise. *Journal of Applied Physiology*, 1:638-709.

Mudambo, K.S.M.T., C. Mc Scrimgeour, and M.J. Rennie. 1997. Adequacy of food ratings in soldiers during exercise in hot, day-time, conditions assessed by doubly labeled water and energy balance methods. *European Journal of Applied Physiology*, 76:346-351.

Nagle, F.J., B. Balke, G. Baptista, J. Alleyia, and E. Howley. 1971. Compatibility of progressive treadmill, bicycle, and step tests based on oxygen uptake responses. *Medicine and Science in Sports*, 3:149-154.

Nagle, F.J., B. Balke, and J.P. Naughton. 1965. Gradational step tests for assessing work capacity. *Journal of Applied Physiology*, 20:745-748.

Nagy, T.R., M.I. Goran, R.L. Weinsier, M.J. Toth, Y. Schutz, and E.T. Poehlman. 1996. Determinations of basal fat oxidation in healthy Caucasians. *Journal of Applied Physiology*, 80:1743-1748.

Nair, K.S., D.E. Matthews, S.L. Welle, and T. Braiman. 1992. Effect of leucine on amino acid and glucose metabolism in humans. *Metabolism*, 41:643-648.

National Institutes of Health, National Heart, Lung, and Blood Institute. 1998. Clinical guidelines on the identification, evaluation, and treatment of overweight and obesity in adults: the evidence report. *Obesity Research*, 6 (Suppl):S51-S210.

Nelson, K., R. Weinsier, L. James, B. Darnell, G. Hunter, and C.L. Long. 1992. Effect of weight reduction on resting energy expenditure, substrate utilization, and the thermic effect of food in moderately obese women. *American Journal of Clinical Nutrition*, 55:924-933.

Nichol, C., P.V. Komi, and P. Marconnet. 1991. Effects of marathon fatigue on running kinematics and economy. *Scandinavian Journal of Medical Science in Sports*, 1:195-204.

Nieman, D.C. 1999. *Exercise testing and prescription: a health-related approach*. 4th ed. Mountain View, CA: Mayfield.

Ohtake, P.J., and L.A. Wolfe. 1998. Physical conditioning attenuates respiratory responses to exercise in late gestation. *Medicine and Science in Sports and Exercise*, 30:17-27.

Olds, T.S., and P.J. Abernethy. 1993. Post-exercise oxygen consumption following heavy and light resistance exercise. *Journal of Strength and Conditioning Research*, 7:147-152.

Overend, T.J., D.H. Peterson, and D.A. Cunningham. 1992. The effect of interval and continuous training on the aerobic parameters. *Canadian Journal of Applied Sports Science*, 17:129-134.

Palmer, G.S., T.D. Noakes, and J.A. Hawley. 1999. Metabolic and performance responses to constant load vs. variable intensity exercise. *Journal of Applied Physiology*, 87:1186-1196.

Papa, S. 1996. Mitochondrial oxidative phosphorylation changes in the life span: molecular aspects and pathophysiological implications. *Biochimica et Biophysica Acta*, 1276:87-105.

Pasman, W.J., M.S. Westerterp-Plantenga, and W.H. Saris. 1998. The effect of exercise training on leptin levels in obese males. *American Journal of Physiology,* 274:E280-E286.

Pasquali, R., M.P. Cesari, L. Besteghi, N. Melchionda, and V. Balestra. 1987a. Thermogenic agents in the treatment of human obesity: preliminary results. *International Journal of Obesity,* 11 (Suppl 3):23-26.

Pasquali, R., M.P. Cesari, N. Melchionda, C. Stefanini, A. Raitano, and G. Labo. 1987b. Does ephedrine promote weight loss in low-energy-adapted obese women? *International Journal of Obesity,* 11:163-168.

Pattengale, P.K., and J.O. Holloszy. 1967. Augmentation of skeletal muscle myoglobin by a program of treadmill running. *American Journal of Physiology,* 213:783-787.

Paul, G.L. 1989. Dietary protein requirements of physically active individuals. *Sports Medicine,* 8:154-176.

Payne, P.R., E.F. Wheeler, and C.B. Salvosa. 1971. Prediction of daily energy expenditure from average pulse rate. *American Journal of Clinical Nutrition,* 24:1164-1170.

Pelleymounter, M.A., M.J. Cullen, M.B. Baker, R. Hecht, D. Winters, T. Boone, and F. Collins. 1995. Effects of the obese gene production on body weight regulation in ob/ob mice. *Science,* 269:540-543.

Pencek, R.R., F.D. James, D.B. Lacy, K. Jabbour, P.E. Williams, P.T. Fueger, and D.H. Wasserman. 2003. Interaction of insulin and prior exercise in control of hepatic metabolism of a glucose load. *Diabetes,* 52:1897-1903.

Penetar, D., U. McCann, D. Thorne, G. Kamimori, C. Galinski, H. Sing, M. Thomas, and G. Belenky. 1993. Caffeine reversal of sleep deprivation effects on alertness and mood. *Psychopharmacology* (Berl), 112:359-365.

Pereira, M.A., S.J. FitzGerald, E.W. Gregg, M.L. Joswiak, W.J. Ryan, R.R. Suminski, A.C. Utter, and J.M. Zmuda. 1997. A collection of physical activity questionnaires for health-related research. In: *Medicine and science in sports and exercise,* ed. A.M. Kriska and C.J. Caspersen, (Suppl) 29:S1-S204.

Peterson, M.R., M. Rothschildt, C.R. Weinberg, R.D. Fell, K.R. McLeish, and M.A. Pfeiffer. 1986. Body fat and the activity of autonomic nervous system. *New England Journal of Medicine,* 318:1077-1083.

Phillips, S.M., S.A. Atkinson, M.A. Tarnopolsky, and J.D. MacDougall. 1993. Gender differences in leucine kinetics and nitrogen balance in endurance athletes. *Journal of Applied Physiology,* 75:2134-2141.

Phillips, S.M., H.J. Green, M.A. Tarnopolsky, and S.M. Grant. 1995. Decreased glucose turnover following short-term training is unaccompanied by changes in muscle oxidative potential. *American Journal of Physiology,* 269:E222-E230.

Phillips, W.T., and J.R. Ziuraitis. 2003. Energy cost of the ACSM single-set resistance training protocol. *Journal of Strength and Conditioning Research,* 17:350-355.

Piers, L.S., M.J. Soares, T. Makran, and P.S. Shetty. 1992. Thermic effect of a meal. I. Methodology and variations in normal young adults. *British Journal of Nutrition,* 67:165-175.

Poehlman, E.T., and C. Melby. 1998. Resistance training and energy balance. *International Journal of Sports Nutrition,* 8:143-159.

Poehlman, E.T., C.L. Melby, and S.F. Badylak. 1988. Resting metabolic rate and postprandial thermogenesis in highly trained and untrained males. *American Journal of Clinical Nutrition,* 47:793-798.

Poehlman, E.T., C.L. Melby, S.F. Badylak, and J. Calles. 1989. Aerobic fitness and resting energy expenditure in young adult males. *Metabolism,* 38:85-90.

Poehlman, E.T., A. Tremblay, A. Nadeau, J. Dussault, G. Theriault, and C. Bouchard. 1986. Heredity and changes in hormones and metabolic rates with short term training. *American Journal Physiology,* 260:E711-717.

Pollock, M.L., H.S. Miller, A.C. Linnerud, and K.H. Cooper. 1975. Frequency of training as a determinant for improvement in cardiovascular function and body composition of middle-aged men. *Archives of Physical Medicine and Rehabilitation,* 56:141-145.

Poole, D.C., and R.S. Richardson. 1997. Determinants of oxygen uptake. *Sports Medicine,* 24:308-320.

Poole, D.C., W. Schaffartzik, D.R. Knight, T. Derion, B. Kennedy, H.J. Guy, R. Prediletto, and P.D. Wagner. 1991. Contribution of exercising legs to the slow component of oxygen uptake in humans. *Journal of Applied Physiology,* 71:1245-1260.

Poole, D.C., S.A. Ward, and B.J. Whipp. 1990. The effects of training on the metabolic and respiratory profile of high-intensity cycle ergometer exercise. *European Journal of Applied Physiology,* 59:421-429.

Powers, S.K., and E.T. Howley. 2007. *Exercise physiology: theory and application to fitness and performance.* 6th ed. New York: McGraw-Hill.

Powers, S.K., and E.T. Howley. 2001. Nutrition, body composition, and performance. In: *Exercise physiology: theory and application to fitness and performance.* 4th ed., 437-456. New York: McGraw-Hill.

Pratley, R., B. Nichlas, M. Rubin, J. Miler, A. Smith, M. Smith, B. Hurley, and A. Goldberg. 1994. Strength increases resting metabolic rate and norepinephrine levels in healthy 50- to 65-year old men. *Journal of Applied Physiology,* 76:133-137.

Pugh, L.G. 1970. Oxygen uptake in track and treadmill running with observations on the air resistance. *Journal of Physiology,* 207:823-835.

Pullinen, T., A. Mero, E. MacDonald, A. Pakarinen, and P.V. Komi. 1998. Plasma catecholamine and serum testosterone responses to four units of resistance exercise in young and adult male athletes. *European Journal of Applied Physiology,* 77:413-420.

Raguso, C.A., A.R. Coggan, L.S. Sidossis, A. Gastaldelli, and R.R. Wolfe. 1996. Effect of theophylline on substrate metabolism during exercise. *Metabolism,* 45:1153-1160.

Rakowski, W., and V. Mor. 1992. The association of physical activity with mortality among older adults in the longitudinal study of aging (1984-1988). *Journal of Gerontology,* 47:M122-129.

Randle, P.J., P.B. Garland, C.N. Hales, and E.A. Newsholme. 1963. The glucose-fatty acid cycle: Its role in insulin sensitivity and metabolic disturbances of diabetes mellitus. *Lancet,* 1:785-789.

Randle, P.J., E.A. Newsholme, and P.B. Garland. 1964. Effects of fatty acids, ketone bodies, and pyruvate and of alloxan-diabetes and starvation on the uptake and metabolic fate of glucose in rat heart and diaphragm muscle. *Biochemistry Journal*, 93:652-665.

Ravussin, E., K.J. Acheson, O. Vernet, E. Danforth, and E. Jequier. 1985. Evidence that insulin resistance is responsible for the decreased thermic effect of glucose in human obesity. *Journal of Clinical Investigation*, 76:1268-1273.

Ravussin, E., C. Bogardus, R. Schwartz, D.C. Robbins, R.R. Wolfe, E.S. Horton, E. Danforth, Jr., and E.A. Sims. 1983. Thermic effect of infused glucose and insulin in man: decreased response with increased insulin resistant diabetes mellitus. *Journal of Clinical Investigation*, 72:893-902.

Ravussin, E., S. Lillioja, W.C. Knowler, L. Christin, D. Freymond, W. Abbott, V. Boyce, B.V. Howard, and C. Bogardus. 1988. Reduced rate of energy expenditure as a risk factor for body weight gain. *New England Journal of Medicine*, 318:467-472.

Ravussin, E., and R. Rising. 1992. Daily energy expenditure in humans: measurement in a respiratory chamber and by doubly labeled water. In: *Energy metabolism: tissue determinants and cellular corollaries*, ed. J.M. Kinney and H.N. Tucker, 81-96. New York: Raven Press.

Reaven, G., and R. Miller. 1968. Study of the relationship between glucose and insulin responses to an oral glucose load in man. *Diabetes*, 17:560-569.

Reichard, G.A., B. Issekutz, Jr., P. Kimbel, R.C. Putnam, N.J. Hochella, and S. Weinhouse. 1961. Blood glucose metabolism in man during muscular work. *Journal of Applied Physiology*, 16:1001-1005.

Ren, J.M., C.F. Semenkovich, E.A. Gulve, J. Gao, and J.O. Holloszy. 1994. Exercise induces rapid increases in GLUT4 expression, glucose transport capacity, and insulin-stimulated glycogen storage in muscle. *Journal of Biological Chemistry*, 269:14396-14401.

Rennie, K., T. Rowsell, S.A. Jebb, D. Holburn, and N.J. Wareham. 2000. A combined heart rate and movement sensor: proof of concept and preliminary testing study. *European Journal of Clinical Nutrition*, 54:409-414.

Richelsen, B., S.B. Pedersen, T. Moller-Pedersen, and J.F. Bak. 1991. Regional differences in triglyceride breakdown in human adipose tissue: effects of catecholamines, insulin, and prostaglandin E2. *Metabolism*, 40:990-996.

Richter, E.A. 1996. Glucose utilization. In: *Handbook of physiology*, ed. L.B. Rowell and J.T. Shepherd, 912-951. New York: Oxford University Press.

Richter, E.A., and H. Galbo. 1986. High glycogen levels enhance glycogen breakdown in isolated contracting skeletal muscle. *Journal of Applied Physiology*, 61:827-831.

Richter, E.A., H. Galbo, and N.J. Christensen. 1981. Control of exercise-induced muscular glycogenolysis by adrenal medullary hormones in rats. *Journal of Applied Physiology*, 50:21-26.

Richter, E.A., L.P. Garetto, M. Goodman, and N.B. Ruderman. 1982. Muscle glucose metabolism following exercise in the rat: Increase sensitivity to insulin. *Journal of Clinical Investigation*, 69:785-793.

Richter, E.A., L.P. Garetto, M. Goodman, and N.B. Ruderman. 1984. Enhanced glucose metabolism after exercise: modulation by local factors. *American Journal of Physiology*, 246:E476-E482.

Richter, E.A., K.J. Mikines, H. Galbo, and B. Kiens. 1989. Effect of exercise on insulin action in human skeletal muscle. *Journal of Applied Physiology*, 66:876-885.

Richter, E.A., T. Ploug, and H. Galbo. 1985. Increased muscle glucose uptake after exercise: No need for insulin during exercise. *Diabetes*, 34:1041-1048.

Rico-Sanz, J., T. Rankinen, D.R. Joanisse, A.S. Leon, J.S. Skinner, J.H. Wilmore, D.C. Rao, and C. Bouchard. 2003. Familial resemblance for muscle phenotypes in The Heritage Family Study. *Medicine and Science in Sports and Exercise*, 35:1360-1366.

Riumallo, J.A., D. Schoeller, G. Barrera, V. Gattas, and R. Vauy. 1989. Energy expenditure in underweight free-living adults: impact of energy supplementation as determined by doubly labeled water and indirect calorimetry. *American Journal of Clinical Nutrition*, 49:239-246.

Roberts, S.B., P. Fuss, G.E. Dallal, A. Atkinson, W.J. Evans, L. Joseph, M.A. Fiatarone, A.S. Greenberg, and V.R. Young. 1996. Effect of age on energy expenditure and substrate oxidation during experimental overfeeding in healthy men. *Journal of Gerontology*, 51:B148-157.

Robinson, S. 1938. Experimental studies of physical fitness in relation to age. *Arbeitsphysiologie*, 10:251-323.

Roch-Norlund, A.E. 1972. Muscle glycogen and glycogen synthase in diabetic man. *Scandinavian Journal of Clinical and Laboratory Investigation*, 125:1-27.

Romijn, J.A., E.F. Coyle, L.S. Sidossis, A. Gastaldelli, J.F. Horowitz, E. Endert, and R.R. Wolfe. 1993. Regulation of endogenous fat and carbohydrate metabolism in relation to exercise intensity and duration. *American Journal of Physiology*, 265: E380-391.

Romijn, J.A., E.F. Coyle, L.S. Sidossis, J. Rosenblatt, and R.R. Wolfe. 2000. Substrate metabolism during different exercise intensities in endurance-trained women. *Journal of Applied Physiology*, 88:1707-1714.

Rooney, T.P., Z.V. Kendrick, J. Carlson, G.S. Ellis, B. Matakevich, S.M. Lorusso, and J.A. McCall. 1993. Effect of estradiol on the temporal pattern of exercise-induced tissue glycogen depletion in male rats. *Journal of Applied Physiology*, 75:1502-1506.

Rooyackers, O.E., D.B. Adey, P.A. Ades, and K.S. Nair. 1996. Effect of age on in vivo rates of mitochondrial protein synthesis in human skeletal muscle. *Proceedings of the National Academy of Science, USA*, 93:15364-15369.

Rooyackers, O.E., and K.S. Nair. 1997. Hormonal regulation of human muscle protein metabolism. *Annual Review of Nutrition*, 17:457-485.

Ross, R., D. Dagnone, P.J. Jones et al. 2000. Reduction in obesity and related comorbid conditions after diet-induced weight loss or exercise-induced weight loss in men: a randomized, controlled trial. *Annual Internal Medicine*, 133:92-103.

Ross, R., and I. Janssen. 2001. Physical activity, total and regional obesity: dose-response considerations. *Medicine and Science in Sports and Exercise*, 6 (Suppl):S521-S527.

Rothwell, N.J., and M.J. Stock. 1980. Similarities between cold- and diet-induced thermogenesis in the rat. *Canadian Journal of Physiology and Pharmacology*, 58:842-848.

Rothwell, N.J., and M.J. Stock. 1983. Diet-induced thermogenesis. *Advances in Nutritional Research*, 5:201-220.

Rowell, L.B., E.J. Masoro, and M.J. Spencer. 1965. Splanchnic metabolism in exercising man. *Journal of Applied Physiology,* 20:1032-1037.

Rowland, T.W., J.A. Auchinachie, T.J. Keenan, and G.M. Green. 1987. Physiological responses to treadmill running in adult and pre-pubertal males. *International Journal of Sports Medicine,* 8:292-297.

Rowland, T.W., and T.A. Rimany. 1995. Physiological responses to prolonged exercise in pre-menarcheal and adult females. *Pediatric Exercise Science,* 7:183-191.

Rowlands, A.V., R.G. Eston, and D.K. Ingledew. 1997. Measurement of physical activity in children with particular reference to the use of heart rate and pedometry. *Sports Medicine,* 258-272.

Ruby, B.C., R.A. Robergs, D.L. Waters, M. Burge, C. Mermier, and L. Stolarczyk. 1997. Effects of estradiol on substrate turnover during exercise in amenorrheic females. *Medicine and Science in Sports and Exercise,* 29:1160-1169.

Rudderman, N.B., A.K. Saha, D. Vavvas, and L.A. Witters. 1999. Malonyl-CoA, fuel sensing, and insulin resistance. *American Journal of Physiology,* 276:E1-E18.

Ryan, A.S., R.E. Pratley, D. Elahi, and A.P. Goldberg. 1995. Resistance training increases fat free mass and maintains RMR despite weight loss in postmenopausal women. *Journal of Applied Physiology,* 79:818-823.

Sady, S. 1981. Transient oxygen uptake and heart rate responses at the onset of relative endurance exercise in pre-pubertal boys and adult men. *International Journal of Sports Medicine,* 2:240-244.

Sahlin, B., A. Katz, and S. Broberg. 1990. Tricarboxylic acid cycle intermediates in human muscle during prolonged exercise. *American Journal of Physiology,* 259:C834-C841.

Sahlin, K. 1992. Metabolic factors in fatigue. *Sports Medicine,* 13:99-107.

Sahlin, K., M. Tonkonogi, and K. Söderlund. 1998. Energy supply and muscle fatigue in humans. *Acta Physiologica Scandinavica,* 162:261-266.

Saibene, F., and A.E. Minetti. 2003. Biomechanical and physiological aspects of legged locomotion in humans. *European Journal of Applied Physiology,* 88:297-316.

Sallis, J.F., M.J. Buono, and P.S. Freedson. 1991. Bias in estimating caloric expenditure from physical activity in children: implications for epidemiological studies. *Sports Medicine,* 11:203-209.

Sallis, J.F., M.J. Buono, J.J. Roby, D. Carlson, and J.A. Nelson. 1990. The Caltrac accelerometer as a physical activity monitor for school-age children. *Medicine and Science in Sports and Exercise,* 22:698-703.

Sallis, J.F., and B.E. Saelens. 2000. Assessment of physical activity by self-report: status, limitations, and future directions. *Research Quarterly for Exercise and Sport,* 71:1-14.

Saltin, B., and P.O. Åstrand. 1993. Free fatty acids and exercise. *American Journal of Clinical Nutrition,* 57 (Suppl):S752-S758.

Saltin, B., M. Houston, E. Nygaard, T. Graham, and J. Wahren. 1979. Muscle fiber characteristics in healthy men and patients with juvenile diabetes. *Diabetes,* 28 (Suppl 1):93-99.

Saltin, B., K. Nazar, D.L. Costill, E. Stein, E. Jansson, B. Essén, and P.D. Gollnick. 1976. The nature of the training response: peripheral and central adaptation to one-legged exercise. *Acta Physiologica Scandinavica,* 96:289-305.

Saris, W.H.M. 1993. The role of exercise in the dietary treatment of obesity. *International Journal of Obesity,* 17 (Suppl 1): S17-S21.

Saris, W.H.M., and R.A. Binkhorst. 1977. The use of pedometer and actometer in studying daily physical activity in man. Part I: Reliability of pedometer and actometer. *European Journal of Applied Physiology,* 37:219-228.

Sawka, M.N. 1986. Physiology of upper body exercise. *Exercise and Sport Sciences Reviews,* 14:175-211.

Scalfi, L., A. Coltorti, E. D'Arrigo, V. Carandente, C. Mazzacano, M. DiPalo, and F. Contaldo. 1987. The effect of dietary fibers on postprandial thermogenesis. *International Journal of Obesity,* 11(Suppl 1):95-99.

Schoeller, D.A., L.K. Cella, M.K. Sinha, and J.F. Caro. 1997. Entrainment of the diurnal rhythm of plasma leptin to meal timing. *Journal of Clinical Investigation,* 100:1882-1887.

Schoeller, D.A., and E. van Santen. 1982. Measurement of energy expenditure in human by doubly labeled water method. *Journal of Applied Physiology,* 53:955-959.

Schrauwen, P., F.J. Troost, J. Xia, E. Ravussin, and W.H.M. Saris. 1999. Skeletal muscle UCP2 and UCP3 expression in trained and untrained male subjects. *International Journal of Obesity,* 23:966-972.

Schuenke, M.D., R.P. Mikat, and J.M. McBride. 2002. Effect of an acute period of resistance exercise on excess post-exercise oxygen consumption: implication for body mass management. *European Journal of Applied Physiology,* 86:411-417.

Schutz, Y., T. Bessard, and E. Jequier. 1987. Exercise and postprandial thermogenesis in obese women before and after weight loss. *American Journal of Clinical Nutrition,* 45:1424-1432.

Schwartz, R., J. Halter, and E. Bierman. 1983. Reduced thermic effect of feeding in obesity: role of norepinephrine. *Metabolism,* 32:114-117.

Seagle, H., D.H. Bessesen, and J.O. Hill. 1998. Effects of sibutramine on resting metabolic rate and weight loss in overweight women. *Obesity Research,* 6:115-121.

Seals, D.R., J.M. Hagberg, W.K. Allen, B.F. Hurley, G.P. Dalsky, A.A. Ehsani, and J.O. Holloszy. 1984. Glucose tolerance in young and older athletes and sedentary men. *Journal of Applied Physiology,* 56:1521-1525.

Segal, K.R., and B. Gutin. 1983a. Exercise efficiency in lean and obese women. *Medicine and Science in Sports and Exercise,* 15:106-107.

Segal, K.R., and B. Gutin. 1983b. Thermic effects of food and exercise in lean and obese women. *Metabolism,* 32:531-589.

Segal, K.R., A. Jeanine, A. Chun, A. Edano, B. Legaspi, and F.X. Pi-Sunyer. 1992. Independent effects of obesity and insulin resistance on postprandial thermogenesis in men. *Journal of Clinical Investigation,* 89:824-833.

Segal, R.S., E. Presta, and B. Gutin. 1984. Thermic effect of food during graded exercise in normal weight and obese men. *American Journal of Clinical Nutrition,* 40:995-1000.

Seidell, J.C., D.C. Muller, J.D. Sorkin, and R. Andres. 1992. Fasting respiratory exchange ratio and resting metabolic rate as predictors of weight gain: the Baltimore Longitudinal Study on Aging. *International Journal of Obesity Related Metabolic Disorders*, 16:667-674.

Semih, S.Y., and T. Feluni. 1998. A comparison of the endurance training responses to road and sand running in high school and college students. *Journal of Strength and Conditioning Research*, 12:79-83.

Shaw, S.M. 1985. Gender and leisure: inequality in the distribution of leisure time. *Journal of Leisure Research*, 17:266-282.

Shephard, R.J. 2000. Exercise and training in women. Part II: Influence of menstrual cycle and pregnancy on exercise responses. *Canadian Journal of Applied Physiology*, 25:35-54.

Shephard, R.J. 2003. Limits to the measurement of habitual physical activity by questionnaires. *British Journal of Sports Medicine*, 37:197-206.

Sherman, W.M., D.L. Costill, W.J. Fink, and J.M. Miller. 1981. Effect of exercise-diet manipulation on muscle glycogen and its subsequent utilization during performance. *International Journal of Sports Medicine*, 2:114-118.

Sherman, W.M., D.M. Morris, T.E. Kirby, R.A. Petosa, B.A. Smith, and D.J. Frid. 1998. Evaluation of a commercial accelerometer (Tritrac-R3D) to measure energy expenditure during ambulation. *International Journal of Sports Medicine*, 19:43-47.

Shier, D., B. Jackie, and R. Lewis. 1999. Chemical basis of life. In: *Human anatomy and physiology*, 36-58. New York: WCB/McGraw-Hill.

Short, K.R., and D.A. Sedlock. 1997. Excess postexercise oxygen consumption and recovery rate in trained and untrained subjects. *Journal of Applied Physiology*, 83:153-159.

Sial, S., A.R. Coggan, R.C. Hickner, and S. Klein. 1998. Training-induced alterations in fat and carbohydrate metabolism during exercise in elderly subjects. *American Journal of Physiology*, 274:E785-E790.

Simoneau, J-A., D.E. Kelley, M. Neverova, and C.H. Warden. 1998. Over-expression of muscle uncoupling protein-2 content in human obesity associated with reduced skeletal muscle lipid oxidation. *FASEB Journal*, 12:1739-1745.

Sinha, M.K., J.P. Ohannesian, M.L. Heiman, A. Kriauciunas, T.W. Stephens, S. Magosin, C. Marco, and J.F. Caro. 1996. Nocturnal rise of leptin in lean, obese, and non-insulin-dependent diabetes mellitus subjects. *Journal of Clinical Investigation*, 97:1344-1347.

Sirard, J.R., and R.R. Pate. 2001. Physical activity assessment in children and adolescents. *Sports Medicine*, 31:439-454.

Sivan, E., X. Chen, C.J. Homko, E.A. Reece, and G. Boden. 1997. Longitudinal study of carbohydrate metabolism in healthy obese pregnant women. *Diabetes Care*, 20:1470-1475.

Skinner, J.S., and T.H. McLellan. 1980. The transition from aerobic to anaerobic metabolism. *Research Quarterly for Exercise and Sport*, 51:234-248.

Smith, K., and M.J. Rennie. 1990. Protein turnover and amino acid metabolism in human skeletal muscle. *Clinical Endocrinology and Metabolism*, 4:461-498.

Smolander, J., V. Louhevaara, T. Hakola, E. Ahonen, and T. Klen. 1989. Cardiorespiratory strain during walking in snow with boots of differing weights. *Ergonomics*, 32:3-13.

Smolin, L.A., and M.B. Grosvenor. 2003. *Nutrition: science and application*. 4th ed., 176-214. Hoboken, NJ: Wiley.

Snitker, S., I. Macdonald, E. Ravussin, and A. Astrup. 2000. The sympathetic nervous system and obesity: role in aetiology and treatment. *Obesity Reviews*, 1:5-15.

Sonne, B., K.J. Mikines, E.A. Richter, N.J. Christensen, and H. Galbo. 1985. Role of liver nerves and adrenal medulla in glucose turnover of running rats. *Journal of Applied Physiology*, 59:1640-1646.

Spraul, M., E. Ravussin, A.M. Fontvieille, R. Rising, D.E. Larson, and E.A. Anderson. 1993. Reduced sympathetic nervous activity. A potential mechanism predisposing to body weight gain. *Journal of Clinical Investigation*, 92:1730-1735.

Spurr, G.B., A.M. Prentice, P.R. Murgatroyd, G.R. Goldberg, J.C. Reina, and N.T. Christman. 1988. Energy expenditure from minute-by-minute heart rate recording: comparison with indirect calorimetry. *American Journal of Clinical Nutrition*, 48:552-559.

Standl, E., N. Lotz, T. Dexel, H-U. Janka, and J.H. Kolb. 1980. Muscle triglycerides in diabetic subjects. *Diabetologia*, 18:463-469.

Stanley, W., and R. Connett. 1991. Regulation of muscle carbohydrate metabolism during exercise. *FASEB Journal*, 5:2155-2159.

Starritt, E.C., R.A. Howlett, G.J. Heigennauser, and L.L. Spriet. 2000. Sensitivity of CPT I to malonyl-CoA in trained and untrained human skeletal muscle. *American Journal of Physiology*, 278:E462-E468.

Stensrud, T., F. Ingjer, H. Holm, and S.B. Stromme. 1992. L-tryptophan supplementation does not improve endurance performance. *International Journal of Sports Medicine*, 13:481-485.

Stock, M.J., and J.A. Stuart. 1974. Thermic effects of ethanol in the rat and man. *Annals of Nutrition and Metabolism*, 17:297-305.

Strath, S.J., D.R. Bassett, Jr., A.M. Swartz, and D.L. Thompson. 2001a. Simultaneous heart rate-motion sensor technique to estimate energy expenditure. *Medicine and Science in Sports and Exercise*, 33:2118-2123.

Strath, S.J., D.R. Bassett, Jr., D.L. Thompson, and A.M. Swartz. 2001b. Validity of the simultaneous heart rate-motion sensor technique for measuring energy expenditure. *Medicine and Science in Sports and Exercise*, 34:888-894.

Strath, S.J., A.M. Swartz, D.R. Bassett, W.L. O'Brien, G.A. King, and B.E. Ainsworth. 2000. Evaluation of heart rate as a method for assessing moderate intensity physical activity. *Medicine and Science in Sports and Exercise*, 32 (Suppl):S465-S470.

Stroud, M.A., P. Ritz, W.A. Coward, M.B. Sawyer, D. Constantin-Teodosiu, P.L. Greenhaff, and I.A. Macdonald. 1997. Energy expenditure using isotope-labeled water (2H18O), exercise performance, skeletal muscle enzyme activities and plasma biochemical parameters in humans during 95 days of endurance exercise with inadequate energy intake. *European Journal of Applied Physiology*, 76:243-252.

Stryer, L. 1988. *Biochemistry*. 3rd ed. New York: Freeman.

Stunkard, A.J., and D. Kaplan. 1977. Eating in public places: a review of reports of the direct observation of eating behavior. *International Journal of Obesity*, 1:89-101.

Stunkard, A.J., and M. Waxman. 1981. Accuracy of self-reports of food intake. *Journal of the American Dietetic Association*, 79:547-551.

Swain, D.P. 2000. Energy cost calculation for exercise prescription. *Sports Medicine*, 30:17-22.

Swain, D.P., and B.C. Leutholtz. 1997. Heart rate reserve is equivalent to %VO2 reserve, not to %VO2max. *Medicine and Science in Sports and Exercise*, 29:410-414.

Swain, D.P., B.C. Leutholtz, M.E. King, L.A. Haas, and J.D. Branch. 1998. Relationship between % heart rate reserve and % VO2 reserve in treadmill exercise. *Medicine and Science in Sports and Exercise*, 30:318-321.

Swain, R.A., D.M. Harsha, and J. Baenziger. 1997. Do pseudoephedrine or phenylpropanolamine improve maximum oxygen uptake and time to exhaustion? *Clinical Journal of Sports Medicine*, 7:168-173.

Takaishi, T., Y. Yasuda, and T. Moritani. 1994. Neuromuscular fatigue during prolonged pedaling rates. *European Journal of Applied Physiology*, 69:154-158.

Takaishi, T., Y. Yasuda, T. Ono, and T. Moritani. 1996. Optimal pedaling rate estimated from neuromuscular fatigue for cyclists. *Medicine and Science in Sports and Exercise*, 28:1492-1497.

Tappy, L., and E. Jequier. 1993. Fructose and dietary thermogenesis. *American Journal of Clinical Nutrition*, 58 (Suppl):766S-770S.

Tappy, L., J.P. Randin, J.P. Felber, R. Chiolero, D.C. Simonson, E. Jequier, and R.A. DeFronzo. 1986. Comparison of thermogenic effect of fructose and glucose in normal humans. *American Journal of Physiology*, 250:E718-E724.

Tarnopolsky, M.A., S.A. Atkinson, S.M. Phillips, and J.D. MacDougall. 1995. Carbohydrate loading and metabolism during exercise in men and women. *Journal of Applied Physiology*, 78:1360-1368.

Tarnopolsky, M.A., M. Bosman, J.R. MacDonald, D. Vandeputte, J. Martin, and B.D. Roy. 1997. Post-exercise protein-carbohydrate and carbohydrate supplements increase muscle glycogen in men and women. *Journal of Applied Physiology*, 83:1877-1883.

Tarnopolsky, L.J., J.D. MacDougall, S.A. Atkinson, M.A. Tarnopolsky, and J.R. Sutton. 1990. Gender differences in substrate for endurance exercise. *Journal of Applied Physiology*, 68:302-308.

Tepperman, J., and H.M. Tepperman. 1987. *Metabolic and Endocrine Physiology*, 5th ed. Chicago: Year Book Medical Publisher.

Terada, S., T. Yokozeki, K. Kawanaka, K. Ogawa, M. Higuchi, O. Ezaki, and I. Tabata. 2001. Effects of high-intensity swimming training on GLUT-4 and glucose transport activity in rat skeletal muscle. *Journal of Applied Physiology*, 90:2019-2024.

Thiebaud, D., K. Acheson, Y. Schutz, J.P. Felber, A. Golay, R.A. Defronzo, and E. Jequier. 1983a. Stimulation of thermogenesis in men after combined glucose long-chain triglyceride infusion. *American Journal of Clinical Nutrition*, 37:603-611.

Thiebaud, D., Y. Schutz, K.J. Acheson, et al. 1983b. Energy cost of glucose storage in human subjects during glucose-insulin infusions. *American Journal of Physiology*, 244:E216-E221.

Thorne, A., and J. Wahren. 1989. Beta-adrenergic blockade does not influence the thermogenic response to a meal in man. *Clinical Physiology*, 9:321-332.

Thorne, A., and J. Wahren. 1990. Diminished meal-induced thermogenesis in elderly man. *Clinical Physiology*, 10:427-437.

Thornton, M.K., and J.A. Potteiger. 2002. Effect of resistance exercise bouts of different intensities but equal work on EPOC. *Medicine and Science in Sports and Exercise*, 34:715-722.

Thorstensson, A., B. Hultén, W. van Döbeln, and J. Karlsson. 1976. Effect of strength training on enzyme activities, and fiber characteristics of human skeletal muscle. *Acta Physiologica Scandinavica*, 96:392-398.

Toner, M.M., and W.D. McArdle. 1996. Human thermoregulatory responses to acute cold stress with special reference to water immersion. In: *Handbook of physiology*, Section 4: *Environmental physiology*, ed. M.J. Fregly and C.M. Blatteis. New York: Oxford University Press.

Tonkonogi, M., and K. Sahlin. 2002. Physical exercise and mitochondrial function in human skeletal muscle. *Exercise and Sport Sciences Reviews*, 30:129-137.

Toth, M.J., P.J. Arciero, A.W. Gardner, J. Calles-Escandon, and E.T. Poehlman. 1996. Rates of free fatty acid appearance and fat oxidation in healthy younger and older men. *Journal of Applied Physiology*, 80:506-511.

Toubro, S., A. Astrup, L. Breum, and F. Quaade. 1993. The acute and chronic effects of ephedrine/caffeine mixtures on energy expenditure and glucose metabolism in humans. *International Journal of Obesity and Related Metabolic Disorders*, 17 (Suppl 3):S73-S77.

Treiber, F.A., L. Musante, S. Hartdagan, H. Davis, M. Levy, and W.B. Strong. 1989. Validation of a heart rate monitor with children in laboratory and field settings. *Medicine and Science in Sports and Exercise*, 21:338-342.

Tremblay, A., J. Cote, and J. LeBlanc. 1983. Diminished dietary thermogenesis in exercise-trained human subjects. *European Journal of Applied Physiology*, 52:1-4.

Tremblay, A., J.P. Despres, C. Leblanc et al. 1990. Effect of intensities of physical activity on body fatness and fat distribution. *American Journal of Clinical Nutrition*, 51:153-157.

Tremblay, A., E. Fontaine, and A. Nadeau. 1985. Contribution of postexercise increment in glucose storage to variations in glucose-induced thermogenesis in endurance athletes. *Canadian Journal of Physiology and Pharmacology*, 63:1165-1169.

Tremblay, A., E. Fontaine, E.T. Poehlman, D. Mitchell, L. Perron, and C. Bouchard. 1986. The effect of exercise-training on resting metabolic rate in lean and moderately obese individuals. *International Journal of Obesity*, 10:511-517.

Tremblay, A., J.A. Simoneau, and C. Bouchard. 1994. Impact of exercise intensity on body fatness and skeletal muscle metabolism. *Metabolism*, 43:814-818.

Treuth, M.S., G.R. Hunter, T. Kekes-Szabo, W.L. Weinsier, M.I. Goran, and L. Berland. 1995a. Reduction in intra-abdominal adipose tissue after strength training in older women. *Journal of Applied Physiology*, 78:1425-1431.

Treuth, M.S., G.R. Hunter, R. Weinsier, and S. Kell. 1995b. Energy expenditure and substrate utilization in older women after strength training: 24 hour calorimeter results. *Journal of Applied Physiology*, 78:2140-2146.

Troisi, R.J., S.T. Weiss, D.R. Parker, D. Parrow, J.B. Young, and L. Landsberg. 1991. Relation of obesity and diet to the sympathetic nervous system. *Hypertension*, 17:669-677.

Tudor-Locke, C.E., and A.M. Myers. 2001. Challenges and opportunities for measuring physical activity in sedentary adults. *Sports Medicine*, 31:91-100.

Turcotte, L.P., E.A. Richter, and B. Kiens. 1992. Increased plasma FFA uptake and oxidation during prolonged exercise in trained vs. untrained humans. *American Journal of Physiology*, 262:E791-E799.

Tzankoff, S.P., and A.H. Norris. 1978. Longitudinal changes in basal metabolism in man. *Journal of Applied Physiology*, 45:536-539.

Ukropcova, B., M. McNeil, O. Sereda, L. de Jonge, H. Xie, G.A. Bray, and S.R. Smith. 2005. Dynamic changes in fat oxidation in human primary myocytes mirror metabolic characteristics of the donor. *Journal of Clinical Investigation*, 115:1934-1941.

United States Department of Health and Human Services. 1996. *Physical activity and health: a report of the Surgeon General.* Atlanta: U.S. Department of Health and Human Services, Centers for Disease Control and Prevention, National Center for Chronic Disease Prevention and Health Promotion.

Vander, A.J., J.H. Sherman, and D.S. Luciano. 2001. *Human physiology: mechanisms of body function.* 7th ed. New York: McGraw-Hill.

Van Etten, L.M.L.A., K.R. Westerterp, and F.T.J. Verstappen. 1995. Effect of weight-training on energy expenditure and substrate utilization during sleep. *Medicine and Science in Sports and Exercise*, 27:188-193.

Visser, M., P. Deurenberg, W.A. van Stavern, and J. Hautvast. 1995. Resting metabolic rate and diet-induced thermogenesis in young and elderly subjects: relationship with body composition, fat distribution, and physical activity level. *American Journal of Clinical Nutrition*, 61:772-778.

Visser, M., L.J. Launer, P. Deurenberg, and D.J.H. Deeg. 1997. Total and sports activity in older men and women: relation with body fat distribution. *American Journal of Epidemiology*, 145:752-761.

Volek, J.S., J.L. VanHeest, and C.E. Forsythe. 2005. Diet and exercise for weight loss: A review of current issues. *Sports Medicine*, 35:1-9.

Vukovich, M.D., D.L. Costill, M.S. Hickey, S.W. Trappe, K.J. Cole, and W.J. Fink. 1993. Effect of fat emulsion infusion and fat feeding on muscle glycogen utilization during cycle exercise. *Journal of Applied Physiology*, 75:1513-1518.

Wagenmakers, A.J.M., E.J. Beckers, F. Brouns, H. Kuipers, P.B. Soeters, G.L. van der Vusse, and V.H.M. Saris. 1991. Carbohydrate supplementation, glycogen depletion, and amino acid metabolism during exercise. *American Journal of Physiology*, 260:E883-E890.

Wagenmakers, A.J.M., J.H. Brooks, J.H. Coakley, T. Reilly, and R.H.T. Edwards. 1989. Exercise induced activation of branched-chain 2-oxo acid dehydrogenase in human muscle. *European Journal of Applied Physiology*, 59:159-167.

Wahren, J., P. Felig, and L. Hagenfeldt. 1978. Physical exercise and fuel homeostasis in diabetes mellitus. *Diabetologia*, 14:213-222.

Wahren, J., L. Hagenfeldt, and P. Felig. 1975. Splanchnic and leg exchange of glucose, amino acids, and free fatty acids during exercise in diabetes mellitus. *Journal of Clinical Investigation*, 55:1303-1314.

Wahren, J., Y. Sato, J. Ostman, L. Hagenfeldt, and P. Felig. 1984. Turnover and splanchnic metabolism of free fatty acids and ketones in insulin-dependent diabetics at rest and in response to exercise. *Journal of Clinical Investigation*, 73:1367-1376.

Wahrenberg, H., J. Bolinder, and P. Arner. 1991. Adrenergic regulation of lipolysis in human fat cells during exercise. *European Journal of Clinical Investigation*, 21:534-541.

Walker, M. 1995. Obesity, insulin resistance, and its link to non-insulin-dependent diabetes mellitus. *Metabolism*, 44 (Suppl 3):18-20.

Wallace, J.P. 1997. Obesity. In: *ACSM's exercise management for persons with chronic diseases and disabilities*, 106-111. Champaign, IL: Human Kinetics.

Wallberg-Henriksson, H., R. Gunnarsson, J. Henriksson, R. DeFronzo, P. Felig, J. Östman, and J. Wahren. 1982. Increased peripheral insulin sensitivity and muscle mitochondrial enzymes but unchanged blood glucose control in type 1 diabetics after physical training. *Diabetes*, 31:1044-1050.

Waluga, M., M. Janusz, E. Karpel, M. Hartleb, and A. Nowak. 1998. Cardiovascular effects of ephedrine, caffeine and yohimbine measured by thoracic electrical bioimpedance in obese women. *Clinical Physiology*, 18:69-76.

Wang, C., S. Strouse, and A. Saunders. 1924. Studies on the metabolism of obesity: III. The specific dynamic action of food. *Archives of Internal Medicine*, 34:573-583.

Washburn, R., M.K. Chin, and H.J. Montoye. 1980. Accuracy of pedometer in walking and running. *Research Quarterly for Exercise and Sport*, 51:695-702.

Washburn, R.A., and H.J. Montoye. 1986. Validity of heart rate as a measure of mean daily energy expenditure. *Exercise Physiology*, 2:161-172.

Washburn, R.C., T.C. Cook, and R.E. LaPorte. 1989. The objective assessment of physical activity in an occupationally active group. *Journal of Sports Medicine and Physical Fitness*, 29:279-284.

Wasserman, D.H., H.L.A. Lickley, and M. Vranic. 1984. Interactions between glucagon and other counter-regulatory hormones during normoglycemic and hypoglycemic exercise in dogs. *Journal of Clinical Investigation*, 74:1404-1413.

Wasserman, D.H., J.A. Spalding, D.B. Lacy, C.A. Colburn, R.E. Goldstein, and A.D. Cherrington. 1989. Glucagon is a primary controller of hepatic glycogenolysis and gluconeogenesis during muscular work. *American Journal of Physiology*, 257:E108-117.

Wasserman, D.H., P.E. Williams, D.B. Lacy, D.R. Green, and A.D. Cherrington. 1988. Importance of intrahepatic metabolisms to gluconeogenesis from alanine during exercise and recovery. *American Journal of Physiology*, 254:E518-E525.

Webb, K.A., L.A. Wolfe, and M.J. McGrath. 1994. Effects of acute and chronic maternal exercise on fetal heart rate. *Journal of Applied Physiology*, 97:2207-2213.

Weinsier, R.L., G.R. Hunter, P.A. Zuckerman, and B.E. Darnell. 2003. Low resting and sleeping energy expenditure and fat use do not contribute to obesity in women. *Obesity Research*, 11:937-944.

Weinsier, R.L., K.M. Nelson, D.D. Hendsrud, B.E. Darnell, G.R. Hunter, and Y. Schutz. 1995. Metabolic predictors of obesity: contribution of resting energy expenditure, thermic effect of food, and fuel utilization to four year weight gain of post-obese and never-obese women. *Journal of Clinical Investigation*, 95:980-985.

Welk, G.J., and C.B. Corbin. 1995. The validity of the Tritrac-R3D activity monitor for assessment of physical activity in children. *Research Quarterly for Exercise and Sport*, 66:202-209.

Westerterp, K.R., G.A. Meijer, E.M. Janssen, W.H. Saris, and F. Hoor Ten. 1992. Long-term effect of physical activity on energy balance and body composition. *British Journal of Nutrition*, 68:21-30.

Westman, E.C., W.S. Yancy, Jr., M.K. Olsen, T. Dudley, and J.R. Guyton. 2006. Effect of a low-carbohydrate, ketogenic diet program compared to a low-fat diet on fasting lipoprotein subclasses. *International Journal of Cardiology*, 110:212-216.

Whitaker, R.C., J.A. Wright, M.S. Pepe, K.D. Seidel, and W.H. Dietz. 1997. Predicting obesity in young adulthood from childhood and parental obesity. *New England Journal of Medicine*, 337:869-873.

White, C.R., and R.S. Seymour. 2005. Allometric scaling of mammalian metabolism. *The Journal of Experimental Biology*, 208:1611-1619.

Wilcox, A., and R. Bulbulian. 1984. Changes in running economy relative to VO2max during a cross-country season. *Journal of Sports Medicine and Physical Fitness*, 24:321-326.

Wildman, R., and B. Miller. 2004. Amino acids, protein, and exercise. In: *Sports and fitness nutrition*, 119-156. Belmont, CA: Wadsworth.

Williams, M.H. 2005. *Nutrition for health, fitness, and sport*. 7th ed. New York: McGraw-Hill.

Williamson, D.F., J. Madans, R.F. Anda, J.C. Kleinman, H. Kahn, and T. Byers. 1993. Recreational physical activity and ten-year weight change in a US national cohort. *International Journal of Obesity*, 17:279-286.

Willoughby, D.S., D.R. Chilek, D.A. Schiller, and J.R. Coast. 1991. The metabolic effects of three different free weight parallel squatting intensities. *Journal of Human Movement Studies*, 21:51-67.

Wilmore, J.H. 1995. Variations in physical activity habits and body composition. *International Journal of Obesity*, 19 (Suppl 4): S107-S112.

Wilmore, J.H., and D.L. Costill. 2004. *Physiology of sports and exercise*. 3rd ed. Champaign, IL: Human Kinetics.

Wilmore, J.H., R.B. Parr, P. Ward, P.A. Vodak, T.J. Barstow, T.V. Pipes, G. Grimditch, and P. Leslie. 1978. Energy cost of circuit weight training. *Medicine and Science in Sports*, 10:75-78.

Winder, W.W. 2001. Energy-sensing and signaling by AMP-activated protein kinase in skeletal muscle. *Journal of Applied Physiology*, 91:1017-1028.

Winder, W.W., K.M. Baldwin, and J.O. Holloszy. 1974. Enzymes involved in ketone utilization in different types of muscle: adaptation to exercise. *European Journal of Biochemistry*, 47:461-467.

Wolfe, L.A., I.K.M. Brenner, and M.F. Mottola. 1994. Maternal exercise, fetal well-being, and pregnancy outcome. *Exercise and Sport Sciences Reviews*, 22:145-194.

Wolfe, R.R., E.J. Peters, S. Klein, O.B. Holland, J. Rosenblatt, and H. Gary, Jr. 1987. Effect of short-term fasting on lipolytic responsiveness in normal and obese human subjects. *American Journal of Physiology*, 252:E189-E196.

Womack, C.J., S.E. Davis, J.L. Blumer, E. Barrett, A.L. Weltman, and G.A. Gaesser. 1995. Slow component of O_2 uptake during heavy exercise: adaptations to endurance training. *Journal of Applied Physiology*, 79:838-845.

Wong, T.C., J.G. Webster, H.J. Montoye, and R. Washburn. 1981. Portable accelerometer device for measuring human energy expenditure. *IEEE Transactions on Biomedical Engineering*, 28:467-471.

World Health Organization. 1998. *Obesity: preventing and managing the global epidemic*. Geneva: World Health Organization.

Yeomans, M.R., S. Caton, and M.M. Hetherington. 2003. Alcohol and food intake. *Current Opinion in Clinical Nutrition and Metabolic Care*, 6:639-644.

Yki-Jarvinen, H., V.A. Koivisto, M.R. Tashinen, and E.A. Nikkila. 1984. Glucose tolerance, plasma lipoproteins and tissue lipoprotein lipase activities in body builders. *European Journal of Applied Physiology*, 53:253-259.

Zanconato, S., S. Buchthal, T.J. Barstow, and D.M. Cooper. 1993. ^{31}P-magnetic resonance spectroscopy of leg muscle metabolism during exercise in children and adults. *Journal of Applied Physiology*, 74:2214-2218.

Zed, C., and W.P.T. James. 1986. Dietary thermogenesis in obesity: response to carbohydrate and protein meals: the effect of β-adrenergic blockade and semi-starvation. *International Journal of Obesity*, 10:391-405.

Zhang, Y., R. Proenca, M. Maffei, M. Barone, M. Lepold, and J.M. Friedman. 1994. Positional cloning of the mouse obese gene and its human homologue. *Nature*, 373:425-432.

Zinman, B., S. Zuniga-Guajardo, and D. Kelly. 1984. Comparison of the acute and long term effects of physical training on glucose control in type 1 diabetes. *Diabetes Care*, 7:515-519.

Zwillich, C., B. Martin, F. Hofeldt, A. Charles, V. Subryen, and K. Burman. 1981. Lack of effects of β-sympathetic blockade on the metabolic and respiratory responses to carbohydrate feeding. *Metabolism*, 30:451-456.

Index

Note: The italicized *f* and *t* following page numbers refer to figures and tables, respectively.

A

acceleration 55
α-cells 36
acetylcholine (ACh) 35
acetyl-CoA (acetylcoenzyme A) 11
adaptations to anaerobic and resistance training
  resistance training 102-103
  short-term exercise 101-102
adenosine triphosphate (ATP)
  adenosine diphosphate (ADP) 4, 4*f*
  adenosine monophosphate (AMP) 4, 4*f*
  adenosine triphosphatase (ATPase) 4, 4*f*
  description of 4, 5
  as energy currency 4
  energy in 3
  energy systems for replenishing 9-14
  hydrolysis 4
  stored 5
  three-part structure of 4, 4*f*
adenylate cyclase 108
adenylate kinase 9
adipocytes 8
adipose tissue lipolysis. *See* lipolysis regulation
adolescence 113
adrenal hormones
  adrenal cortex 36, 37
  adrenal gland 36
  adrenal medulla 36, 37
  cortisol and aldosterone 37
  epinephrine and norepinephrine 37
  regulation of cortisol secretion by negative feedback mechanism 37, 37*f*
  selected hormones, catabolic role in energy homeostasis 37, 37*t*
adrenergic agonist 156
aerobic capacity, children and adolescents 114, 114*f*
aerobic training and cellular adaptations
  autonomic nervous system 93
  mitochondrial and capillary density 93-94, 94*t*
  muscle fiber types 95, 96*t*
  myoglobin 94, 95
  specificity 93
aerobic training effect on RMR 137-138

age and gender on metabolism. *See also* gender differences in substrate metabolism
  energy metabolism in children and adolescents 113-115
  exercise metabolism in elderly people 111-113
  pregnancy 109-111, 109*t*
age and thermic effect of food (TEF) 147
airway resistance 51
alcohol consumption and thermic effect of food (TEF) 148
aldosterone 37
algorithm 59
amino acids 5-6
amino acids, energy metabolism of
  alanine cycle 23*f*, 29
  catabolism of 28
  endogenous tissue protein 28
  major metabolic pathways for 29, 29*f*
  principal sources of 28
  transamination 23*f*, 28-29
anaerobic capacity in children 115
arm cranking, metabolic equations for 75, 76, 76*f*
assimilation 143
asthenia 155
ATP-PCr system (phosphagen system). *See also* energy transformation
  adenylate kinase 9
  ADP 9
  ATP 9
  components of 9
  creatine kinase 9
  hydrolysis 9
  mitochondria 9
  PCr 9
Atwater, Wibur Olin 6
Atwater general factors
  for calculating caloric values of food 6, 7*t*
  description of 6
  uses for 6
autonomic nervous and endocrine system 93

B

basal metabolic rate (BMR) 52
β-cells 36
bench stepping, metabolic equations for 76, 76*f*
Bernard, Claude 33

bioenergetic pathway, enzyme changes in elderly 112
biological control system
  control system and its operation 34, 34*f*, 35
  homeostasis and steady state 33-34, 34*f*
  homeostatic control systems and exercise 33
biosynthesis 4
blood-brain barrier 156
blood glucose 123-124
blood lipid profile 86
bodily energy stores
  adipocytes 8
  availability of energy substrates in body 6, 8*t*
  blood glucose, maintenance of 6, 8, 8*t*
  description of 6
  energy from fat 8, 8*t*
  fatty acids 8
  glucose 6
  ketones 8
body fat distribution and insulin resistance
  fat distribution, role of different patterns of 121
  subcutaneous adipose tissue in legs 121-122
  visceral adiposity 121
body mass effects and energy cost during walking and running 71
body mass index (BMI) 140
body size and resting metabolic rate (RMR) 134, 135*f*
bomb calorimeter 5, 5*f*
β-oxidation 25, 26*f*
bradycardia 159
branched-chain amino acids (BCAA)
  central fatigue hypothesis 30
  description of 30
  ergogenic effect of ingesting carbohydrate 30
  protein and 27
  supplementation of 30

C

caffeine
  cognitive function 159
  description of 159
  fat utilization 160
  sport performance 159-160
  working mechanism of 159

caloric equivalent of oxygen 52, 53*t*
caloric values of food, calculating. *See* Atwater general factors
calories (cal) 3
calorimetry 49
Cannon, Walter 33
capillary density 80
carbohydrate
    energy provision, carbohydrates *vs.* fat 19
    and fat metabolism, alterations in 112-113
    glycogen 19, 20*f*
    hypoglycemia 21
    liver sources of 20*f*, 21-24, 23*f*
    macronutrients for replenishing ATP 19
    PCr (phosphocreatine) for replenishing ATP 19, 21
    storage and utilization in children 115
    and thermic effect of food 145, 145*f*
    utilization at onset of exercise 19, 21
carbohydrate utilization
    endurance training and muscle glycogen utilization, studies 96-97
    muscle biopsy 96
    plasma glucose 97
carbohydrate utilization and fatty acid availability
    carbohydrate utilization 40, 41
    glucose-fatty acid cycle (Randle cycle) 40, 40*f*
    hexokinase 40
    pyruvate dehydrogenase complex 40
    reduced enzymatic activities 40, 41
cardiac output 98
carnitine 25
carnitine acyltransferase I 25
catabolism 4
catecholamines
    adrenal hormones and 37
    lipolysis and 25
    working mechanisms and 37, 38, 38*t*
cellular adaptations to aerobic training
    autonomic nervous system 93
    mitochondrial and capillary density 93-94, 94*t*
    muscle fiber types 95, 96*t*
    myoglobin 94, 95
    specificity 93
central fatigue hypothesis 30
chemiosmotic hypothesis
    cellular energy production 12, 13
    description of 11, 12
    electron transfer through respiratory chain leads to pumping of protons 12, 13*f*
    uncoupling 13
    uncoupling proteins (UCP) 13
childhood 113
children and adolescents and energy metabolism
    aerobic capacity 114, 114*f*
    age-related functional deficiency 113

anaerobic capacity 115
carbohydrate storage and utilization 115
    metabolic efficiency 114-115, 115*t*
    oxygen deficit and respiratory exchange ratio 114
    phases to adulthood 113
circuit training 86, 87*f*
citrate synthase 102
climate conditions and RMR 134, 136
closed-circuit spirometry
    airway resistance 51
    description of 51
    inspiration and expiration 51
coefficient of digestibility. *See* digestive efficiency
comorbidities 117
concentric contraction 76
conformational change 119
control system and its operation
    afferent and efferent pathways 34
    extracellular fluid 35
    general components of 34, 34*f*
    insulin 35
    negative feedback 35
    parathyroid hormone 35
    thermoregulation 35
corpus luteum 107
cortex 36
cortisol 37
creatine kinase 9
cycling, metabolic calculations of 75, 76*f*
cycling and energy costs
    brake resistance 72
    cycling outdoors 73
    energy costs during 70*t*, 72
    energy expenditure and work rate 72-73
    gait 72
    servomechanism 72
    stationary cycle ergometer 72
cytochrome c oxidase 112

D
deamination 14
diabetes mellitus
    alterations in metabolism during exercise 123-125
    description of 119
    types of 118-119
dietary fiber content and thermic effect of food (TEF) 147
digestive efficiency
    coefficient of digestibility 6
    definition of 6
    digestibility, heat of combustion, and net energy values of nutrients 6, 7*t*
direct calorimetry
    description of 49
    direct calorimetry chamber 49, 50*f*
    energy metabolism 49
    insulated calorimeter 49, 50
disaccharides. *See* energy consumption
diseases, obesity, and diabetes

diabetes mellitus 118-119
    obesity 117-118
diurnal rhythm of leptin 156
dizygotic twins 139
dopamine 153
doubly labeled water technique
    advantages and disadvantages of 55
    mass spectrometer 54-55
    stable isotopes 54
"Douglas bag" 49
duration of exercise 81

E
eccentric contraction 76
ectothermic 134
efferent signals of central nervous system 22
efficiency of human movement 66, 67
elderly and exercise metabolism
    aging 111, 111*t*
    carbohydrate and fat metabolism, alterations in 112-113
    enzymes in bioenergetic pathway, changes in 112
    metabolic changes and their physiological changes 111, 111*t*
    reduced $\dot{V}O_2$max and energy expenditure 111, 112
electronic transport chain 11
electrocardiography (ECG) 57
endogenous tissue protein 28
endothelium 156
endothermic 134
endurance performance and oxygen uptake. *See also* oxygen deficit
    excess postexercise oxygen consumption 99-100
    exercise efficiency 101
    lactate threshold 100, 100*f*, 101
    metabolic responses during submaximal exercise 98-99
    $\dot{V}O_2$max 98
energy
    in ATP 3, 4
    biologically usable 4-5, 4*f*
    definition of 3
    forms of 3
    hydrolysis 4
    law of the conservation of energy 3
    potential and kinetic 3-4
    units of 3
energy balance and physical activity
    capillary density of muscle tissue 80
    effect of body weight on energy expenditure of walking 80, 80*f*
    fat utilization 80
    level of physical activity and overweight/obesity risk, studies 79-80
    obesity 79
    physical activity and energy expenditure, defining 80, 80*f*
energy consumption. *See also* energy content of foods, measuring
    Atwater general factors 6, 7*f*
    bodily energy stores 6, 8, 8*t*

carbohydrate, fat, and protein 5
digestive efficiency 6, 7t
disaccharides 5
glycogen 5
hepatic-portal vein 5
hydrolytic reactions of macronutrients 5
monosaccharides 5
triglycerides 5
energy content of foods, measuring
bomb calorimeter 5, 5f
drawbacks 5-6
energy cost during various exercises
cycling 72-73
resistance exercise 70t, 73, 74, 74t
walking and running 69, 70f, 70t, 71-72
energy cost of physical activities and sports
energy cost during various exercises 69-75
energy utilization during exercise, principles of 65-69
metabolic calculation 75-77
energy expenditure
description of 129
gross and net energy expenditure 65-66
and reduced $\dot{V}O_2$max in elderly 111, 112
energy expenditure, subjective measures for. See subjective measures for assessing energy expenditure
energy expenditure through exercise and physical activity
caloric threshold 83, 83t
duration 81
exercise and physical activity 81
frequency 82-83
guidelines for energy expenditure and long-term weight control 83, 83t
heart rate reserve (HRR) 81
intensity 81, 82
leisure-time activities 81, 82t
physical activity, categories of 80-81
weight management program 81
energy metabolism, measurement of
laboratory approaches 49-54
subjective measures 60-62
energy metabolism, obesity, and diabetes
alterations during exercise 123-125
diseases, description of 117-119
improving insulin sensitivity, role of exercise 125-127
insulin resistance 120-122, 123f
energy metabolism, regulation of
biological control system overview 33-35
neural and hormonal control systems 35-38
regulation of substrate metabolism during exercise 38, 39-44
energy metabolism in children and adolescents
aerobic capacity 114, 114f

age-related functional deficiency 113
anaerobic capacity 115
carbohydrate storage and utilization 115
metabolic efficiency 114-115, 115t
oxygen deficit and respiratory exchange ratio 114
phases to adulthood 113
energy metabolism of amino acids
alanine cycle 23f, 29
catabolism of 28
endogenous tissue protein 28
major metabolic pathways for 29, 29f
principal sources of 28
transamination 23f, 28-29
energy transformation
ATP-PCr system (phosphagen system) 9
ATP stores, energy systems for replenishing 9
description of 1, 8-9
energy-yielding systems 9
glycolytic system (glycolysis) 9-11, 10f
law of the conservation of energy 8-9
metabolic stress and 91
oxidative pathway 11-14, 11f-13f, 14t
energy transformation in sport and physical activity
classification of three energy systems 15
energy sources of muscular work for various athletic activities 14, 15, 15t
incremental exercise, phases and transition of energy sources 16
power and capacity 14
response during exercise of three energy systems 15
three energy systems 14
energy utilization, exercise strategies for
enhancing energy expenditure with exercise and physical activity 80-83, 82t, 83t
exercise intensity and fat utilization 83, 84, 84t
limitations of exercise alone in weight management 87-88, 88f
other exercise strategies 85-87, 87f
physical activity and energy balance 79-80, 80f
energy utilization during exercise
adenosine triphosphate (ATP) 65
exercise efficiency 66, 67
gross and net energy expenditure 65-66
postexercise oxygen consumption 68, 68f, 69
slow component of oxygen uptake 66, 67f
environmental temperature and thermic effect of food (TEF) 147-148
enzyme changes in bioenergetic pathway, age-related 112
ephedrine
and caffeine, interactive effect 157, 158, 158f
description of 156

effect on weight loss 157
function as ergogenic aid 157
potential side effects 158
working mechanism of 156-157
epinephrine 37
ergogenic 30
estrogen 107
euglycemic 120
excess postexercise oxygen consumption (EPOC)
athletes performing two or more exercise bouts on same day 69
oxygen debt 68
quantifying energy cost of activity 68-69
trained vs. untrained individuals 100
exercise and energy utilization principles
adenosine triphosphate (ATP) 65
exercise efficiency 66, 67
gross and net energy expenditure 65-66
postexercise oxygen consumption 68, 68f, 69
slow component of oxygen uptake 66, 67f
exercise and improving insulin sensitivity 125-127
exercise and leptin 156
exercise during pregnancy. See pregnancy
exercise efficiency
assessing athletic performance 67
definition of 101
description of 66
efficiency of human movement 66
mathematical expression of 66
motor unit recruitment patterns 101
studies 101
exercise influence on RMR
aerobic training 137-138
experimental approaches 136-137
resistance training 138
exercise in obesity and diabetes, alterations in metabolism
blood glucose 123-124
fatty acids and triglycerides 125, 126t
muscle glycogen 124-125
exercise intensity and duration and carbohydrate
glycogen utilization and reestablishment of glycogen stores 21
hypoglycemia 21
percentage of energy derived from sources of fuel during exercise 21, 22f
exercise intensity and fat utilization
fat vs. total calories expended during stationary cycling 84, 84t
intensity and total energy expenditure 27f, 84
low-intensity exercise 53t, 83, 84, 84t
studies 84
exercise limitations in weight management
average weight loss during 15-week diet and exercise program 88, 88f
leptin and 155-156

exercise limitations in weight management (*continued*)
  limitations of exercise alone 87-88, 88*f*
  negative caloric balance 87
  resting metabolic rate 88
exercise metabolism in elderly people
  aging 111, 111*t*
  carbohydrate and fat metabolism, alterations in 112-113
  enzymes in bioenergetic pathway, changes in 112
  metabolic changes and their physiological changes 111, 111*t*
  reduced $\dot{V}O_2$max and energy expenditure 111, 112
exercise strategies for energy utilization
  enhancing energy expenditure with exercise and physical activity 80-83, 82*t*, 83*t*
  exercise intensity and fat utilization 83, 84, 84*t*
  limitations of exercise alone in weight management 87-88, 88*f*
  other exercise strategies 85-87, 87*f*
  physical activity and energy balance 79-80, 80*f*
exogenous estrogen 108
expiration 51
extracellular 35. *See also* control system and its operation

F
facultative component of TEF 143
fat
  carbohydrate metabolism alterations in elderly and 112-113
  energy and 8
  glucose utilization in obesity and diabetes and 122, 123*f*
  thermic effect of food and 146
fat-free mass (FFM) 86
fat oxidation, gender comparisons 105, 106-107
fat oxidation and exercise intensity and duration
  glucose-fatty acid cycle theory 27
  studies 22*f*, 25, 26, 27*f*
fatty acid availability on carbohydrate utilization
  carbohydrate utilization 40, 41
  glucose-fatty acid cycle (Randle cycle) 40, 40*f*
  hexokinase 40
  pyruvate dehydrogenase complex 40
  reduced enzymatic activities 40, 41
fatty acids
  bodily energy stores and 8
  triglycerides and 125, 126*t*
fat utilization
  β-oxidation 25, 26*f*
  carnitine 25
  carnitine acyltransferase I 25
  effect of caffeine on 160
  during exercise 97

glucose and 122, 123*t*
  oxidation 25
fat utilization and exercise intensity
  fat *vs.* total calories expended during stationary cycling 84, 84*t*
  intensity and total energy expenditure 27*f*, 84
  low-intensity exercise 53*t*, 83, 84, 84*t*
  studies 84
field-based techniques for measuring energy metabolism
  accurate measurement, importance of 54
  advantages of 54
  doubly labeled water technique 54-55
  heart rate and motion monitoring combined 57*t*, 59-60
  heart rate monitoring 57-59, 57*t*, 58*f*
  motion sensors 55-56, 55*f*, 56*f*, 57*t*
  multisensor monitoring system 60
flavin adenine dinucleotide (FAD) 11
follicular phase 108
frequency of exercise 82
fuel utilization changes
  carbohydrate utilization 96-97
  depletion of muscle glycogen during strenuous exercise 95, 96
  energy from fat oxidation 96, 96*f*
  fat utilization 97
  fuels used during exercise, modification of rate 96, 96*f*
  protein utilization 97-98
  respiratory exchange ratio (RER) 96

G
gait 72
gender and age on metabolism. *See also* gender differences in substrate metabolism
  energy metabolism in children and adolescents 113-115
  exercise metabolism in elderly people 111-113
  pregnancy 109-111, 109*t*
gender differences in substrate metabolism
  fat oxidation 105, 106-107
  performance, men *vs.* women 105, 106*t*
  sex hormones, effects of 107-109, 108*t*
gestational diabetes 110
glucagon 36
glucokinase 127
glucomannan-supplemented meal 147
gluconeogenesis 20*f*, 22-24, 23*f*
glucose 6
glucose and fat utilization in obesity and diabetes 122, 123*f*
glucose-fatty acid cycle (Randle cycle) 27, 40, 40*f*
glucose tolerance 112
glucose transporters (GLUT) 119
glycerol 13. *See also* oxidative pathway
glycogen. *See also* energy consumption
  degradation of 19, 20*f*

particle, structure of 19, 20*f*
  stores 6, 8, 8*t*
glycogenolysis 10. *See also* glycolytic system (glycolysis)
glycogen phosphorylase 39
glycolysis and Krebs cycle regulation
  isocitrate dehydrogenase (IDH) 39, 40
  phosphofructokinase (PFK) 39, 40
  rate-limiting enzymes 39-40
glycolysis (Meyerhof pathway) 9-10, 10*f*
glycolytic system (glycolysis). *See also* energy transformation
  advantages of 11
  description of 9
  disadvantages of 10-11
  enzymatic reactions for glycogen breakdown 10
  glycogenolysis 10
  glycolysis 9
  glycolytic pathway, summary of 9, 10*f*
  Meyerhof pathway 9, 10
  phosphorylation 10
  pyruvic acid 9
  substrate-level phosphorylation 10
  summary of 10
gross and net energy expenditure
  gross energy expenditure 66
  net energy expenditure 66
  pattern of substrate utilization 65-66
growth and development and RMR 132*t*, 134, 135*t*
growth hormone 42

H
heart rate and motion monitoring combined
  accelerometry and HR monitoring 59
  algorithm 59
  studies 57*t*, 59-60
heart rate monitoring
  average pulse rate 58-59
  electrocardiography (ECG) monitoring 57
  methods 58-59, 58*f*
  R-R waves 57-58
heart rate reserve (HRR) 81
hepatic glucose output 21, 22
hepatic glucose output, regulation of
  adrenergetic activities 41
  description of 41
  epinephrine and norepinephrine 41
  exercise-induced hyperglycemia 41
  hyperglycemia 41
  insulin and glucose 41
hepatic-portal vein. *See* energy consumption
hexokinase 40. *See also* fatty acid availability on carbohydrate utilization
homeostasis and steady state
  homeostasis, description of 33, 34
  internal environment 33, 34
  physiological environment 33
  steady state 33-34, 34*f*
hormonal and neural control systems

acetylcholine (ACh) 35
adrenal hormones 36, 37, 37f, 37t
autonomic nervous system 35-36
  description of 35
  hormonal secretion from endocrine
    glands 36
  membrane permeability 36
  neurotransmitters 35-36
  norepinephrine 35-36
  pancreatic hormones 36, 36f
  parasympathetic division 35
  sympathetic division 35
  working mechanisms 37, 38, 38t
hormonal concentration and RMR 136
hydrolysis 4, 9
hyperglycemia 41
hyperinsulinemic 120
hyperlipidemia 113
hyperthyroidism 136
hypoglycemia 21

I
inclusion of resistance exercise 86-87, 87f
indirect calorimetry
  airway resistance 51
  closed-circuit spirometry 51
  indirect vs. direct 50
  inspiration and expiration 51
  measurements 50-51
  open-circuit spirometry 51, 51f, 52
  principle of 50
  respiratory chamber 52
infancy 113
insomnia 154
inspiration 51
insulin
  control system operations and 35
  responsiveness 121
  sensitivity 121, 125-127
insulin-dependent diabetes mellitus
    (IDDM) 119. See also diabetes
    mellitus
insulin-like growth factors 43
insulin resistance
  body fat distribution and 121-122
  description of 120
  euglycemic, hyperinsulinemic glucose
    clamp 120
  glucose and fat utilization 122, 123f
  insulin responsiveness 121
  insulin sensitivity 121
  during pregnancy 110
  testing for 120-121, 120f, 121f
intensity 81, 82
intermittent exercise for weight manage-
    ment, studies 85
intramuscular triglycerides 97
islets of Langerhans 36
isocitrate dehydrogenase (IDH) 39, 40
isoforms 119
isotopes 54

J
Joule, Sir Prescott 3
joules (J) 3

K
ketones 8
kilocalories (kcal or Cal) 3
kilojoules (kJ) 3
kinetic energy (energy of motion) 3-4
Krebs, Hans 11
Krebs cycle 11

L
laboratory techniques for measuring
    energy metabolism
  direct calorimetry 49, 50, 50f
  indirect calorimetry 50-52
  measuring substrate oxidation 52, 53t,
    54
lactate 22
lactate dehydrogenase 102
lactate threshold
  aerobic conditioning and 100, 100f, 101
  description of 22
  trained vs. untrained state 100, 100f
Langerhans, Paul 36
Lavoisier, Antoine 131
law of the conservation of energy 3, 8-9
leg cycling, metabolic equations for 75,
    76, 76f
leptin
  description of 155
  effect of exercise on 156
  weight management and 155-156
  working mechanism of 155, 155f
lipase 24, 25
lipid(s). See also oxidative pathway
  β-oxidation 25
  carbohydrate and fat utilization 26, 27
  carnitine 25
  carnitine acyltransferase I 25
  catecholamines 25
  energy sources from 24, 25
  exercise intensity and duration on fat
    utilization 22f, 25, 26, 27f
  fat catabolism 24
  fat oxidation 25, 26, 27f
  fat utilization, preparatory stages for
    25, 26f
  glucose-fatty acid cycle theory 27
  lipase 24, 25
  lipolysis 24, 25, 25f
  triglyceride molecule 24, 24f
  utilization 24
lipolysis 13. See also lipids; regulation
    of lipolysis
lipoprotein lipase (LPL) 107
lipoproteins 86
liver sources of carbohydrate
  efferent signals of central nervous
    system 22
  gluconeogenesis, processes of 20f, 22-
    24, 23f
  hepatic glucose output 21, 22
  hypoglycemia 21
  lactate 22
  lactate threshold 22
  source and quantity of glucose, factors
    determining 22

splanchnic glucose output 22
logarithmic plot 134
luteal phase 108

M
macronutrient composition and thermic
    effect of food (TEF) 146, 147
macronutrients and metabolism during
    exercise
  carbohydrate 19-24, 20f, 22f, 23f
  lipid 24-27, 24f-27f
  protein and amino acids 27-30, 29f
malonyl-CoA 42
mass spectrometer 54-55
maximal aerobic capacity (VO$_2$max) 62
measurement of energy content of foods
  amino acids 5-6
  bomb calorimeter 5, 5f
  drawbacks of 5-6
  urea 6
measurement of energy metabolism
  field-based techniques 54-60
  laboratory approaches 49-54
  subjective measures 60-62
measurement of resting metabolic rate
  contribution of organs and tissues to
    body weight and BMR 132, 132t,
    133
  description of 132
  estimating 133, 133t
  thermogenesis 132
measurement of substrate oxidation
  caloric equivalent of oxygen 52, 53t
  estimation composition of fuels oxi-
    dized 52
  respiratory exchange ratio (RER) 54
  respiratory quotient (RQ) 52, 53t, 54
measurement of thermic effect of food
    143-144
mechanical energy
  description of 3-4
  as potential or kinetic energy 4
medulla 36
membrane permeability 36
metabolic adaptations to exercise train-
    ing
  adaptations to anaerobic and resistance
    training 101-103
  cellular adaptations to aerobic training
    93-95
  fuel utilization, changes in 95-98
  oxygen uptake and endurance perfor-
    mance, responses of 98-101
metabolic calculation
  cycling 75, 76f
  energy cost and intensity indices, linear
    relation between 75
  metabolic equations for other activities
    75-77, 76f, 77t
  walking and running 75, 76f
metabolic efficiency, children and adoles-
    cents 114-115, 115t
metabolic equations for
  arm cranking and leg cycling 75, 76,
    76f

metabolic equations for *(continued)*
    bench stepping 76, 76*f*
  concentric contraction 76
  eccentric contraction 76
  metabolic equations, accuracy of 76-78, 77*t*
metabolic equivalent (MET) 61-62
metabolic inflexibility 122
metabolic responses during submaximal exercise 98-99
metabolism of macronutrients in exercise
  carbohydrate 19-24, 20*f*, 22*f*, 23*f*
  lipid 24-27, 24*f*-27*f*
  protein and amino acids 27-30, 29*f*
Meyerhof, Otto Fritz 9
Meyerhof pathway. *See* glycolysis
micro-electromechanical system (MEMS) 55
Mitchell, Peter 11, 12
mitochondrial and capillary density and aerobic training
  capillary density 94
  role of mitochondria in bioenergetics 93-94, 94*t*
mitochondria (powerhouses of the cell). *See also* ATP-PCr system; oxidative pathway
  acetyl-CoA (acetylcoenzyme A) 11
  structure of 11, 11*f*
monosaccharides. *See* energy consumption
monozygotic twins 139
motion sensors
  accelerometers 56-57, 56*f*, 57*t*
  description of 55
  micro-electromechanical system (MEMS) 55
  pedometers 55, 55*f*, 57*t*
  piezoceramic material 56
  transducer 56
motor unit recruitment 101
multisensor monitoring system 60
muscle biopsy 96
muscle fiber types 95, 96*t*
muscle glycogen 124-125
muscle glycogen degradation, regulation of
  epinephrine and glycogen phosphorylase 39
  glycogen phosphorylase 39
  regulation of 39
  substrate availability and muscle glycogenolysis 39
muscle hypertrophy 102
mutations 155
myocardial contraction 98
myoglobin
  description of 94, 95
  fast-twitch muscle fibers 94, 95
  slow-twitch muscle fibers 94

N
negative caloric balance 87
negative feedback 35. *See also* control system and its operation

net energy expenditure 66
net energy values 6
neural and hormonal control systems
  acetylcholine (ACh) 35
  adrenal hormones 36, 37, 37*f*, 37*t*
  autonomic nervous system 35-36
  description of 35
  hormonal secretion from endocrine glands 36
  membrane permeability 36
  neurotransmitters 35-36
  norepinephrine 35-36
  pancreatic hormones 36, 36*f*
  parasympathetic division 35
  sympathetic division 35
  working mechanisms 37, 38, 38*t*
neurotransmitters 35-36
Newsholme, Eric 30
nicotinamide adenine dinucleotide (NAD) 11
nitrogen balance 28
non-insulin-dependent diabetes mellitus (NIDDM) 119. *See also* diabetes mellitus
norepinephrine 35-36

O
obesity
  alterations in metabolism during exercise 123-125
  childhood obesity 117
  energy output 118
  food intake 117-118
  role of RMR in pathogenesis of 139-141
  thermic effect of food and 149-150
  total energy expenditure 118
obligatory component of TEF 143
open-circuit spirometry
  description of 51
  open-circuit indirect calorimetry system 51, 51*f*, 52
oxidative pathway. *See also* energy transformation
  acetyl-CoA (acetylcoenzyme A) 11
  aerobic pathway 13
  amino acids 14
  chemiosmotic hypothesis 11-13, 13*f*
  deamination 14
  electronic transport chain 11
  fat oxidation 13
  flavin adenine dinucleotide (FAD) 11
  glycerol 13
  hydrogen removal 11
  Krebs cycle 11
  lipolysis 13
  mitochondria (powerhouses of the cell) 11, 11*f*
  nicotinamide adenine dinucleotide (NAD) 11
  oxidative phosphorylation 11
  protein 13-14
  stages of 11, 12*f*
  three energy systems, comparisons of 14, 14*t*
  transamination 14

uncoupling 13
uncoupling proteins (UCP) 13
oxidative phosphorylation 11
oxygen debt 68
oxygen deficit
  description of 99
  endurance training and 99, 99*f*
  O_2 deficit at onset of exercise 99, 99*f*
  postexercise oxygen consumption and 68, 68*f*
  and respiratory exchange ratio, children, and adolescents 114
  training-induced increase in mitochondrial content 99
oxygen uptake, slow component of 66, 67*f*
oxygen uptake and endurance performance. *See also* oxygen deficit
  excess postexercise oxygen consumption 99-100
  exercise efficiency 101
  lactate threshold 100, 100*f*, 101
  metabolic responses during submaximal exercise 98-99
  $\dot{V}O_2$max 98

P
pancreatic hormones
  α-cells and β-cells 36
  glucagon 36
  insulin and glucagon 36, 36*f*
  islets of Langerhans 36
parasympathetic divisions 35
parathyroid hormone 35. *See also* control system and its operation
peptides 38
pharmacologic and nutritional substances
  caffeine 159-160
  ephedrine 156-158
  leptin 155-156
  sibutramine (Meridia; Reductil) 153-155, 154*f*
phenobarbital 157
phosphagen system. *See* ATP-PCr system
phosphocreatine (PCr)
  in ATP-PCr system 9
  for replenishing ATP at onset of exercise 19, 21
phosphofructokinase (PFK) 39, 40
phosphorylase 102
phosphorylation 10. *See also* glycolytic system (glycolysis)
physical activities and sports and energy cost
  energy cost during various exercises 69-75
  energy utilization during exercise, principles of 65-69
  metabolic calculation 75-77
physical activity, exercise, and energy expenditure
  caloric threshold 83, 83*t*
  duration 81
  exercise and physical activity 81

frequency 82-83
guidelines for energy expenditure and long-term weight control 83, 83t
heart rate reserve (HRR) 81
intensity 81, 82
leisure-time activities 81, 82t
physical activity, categories of 80-81
weight management program 81
physical activity and energy balance
capillary density of muscle tissue 80
effect of body weight on energy expenditure of walking 80, 80f
fat utilization 80
level of physical activity and overweight/obesity risk, studies 79-80
obesity 79
physical activity and energy expenditure, defining 80, 80f
physical activity and thermic effect of food (TEF) 148-149
piezoceramic material 56
pituitary gland 42
placenta 107
placental lactogen 110
portal circulation 121
postabsorptive state 125
postexercise oxygen consumption
excess postexercise oxygen consumption (EPOC) 68, 69
oxygen debt 68
oxygen deficit 68
response of oxygen uptake during steady exercise 68, 68f
postobese model 139
postprandial thermogenesis 147
potential and kinetic energy
biosynthesis 4
catabolism 4
kinetic energy 4
mechanical energy 3-4
potential energy, role of 3
pregnancy
demands on metabolism 109
energy cost of household activities in women, pregnant vs. nonpregnant 109, 109t
exercise during 110-111
substrate metabolism during 110
progesterone 107
proinsulin 127
prolactin 110
prospective study 141
protein and amino acids
branched-chain amino acids (BCAA) issues 27, 30
contractile- and noncontractile-related proteins 27-28
efflux of 3-MH (3-methylhistidine) 28
energy metabolism of amino acids 23f, 28-29, 29f
during fasting and starvation 27
nitrogen balance 28
protein degradation and synthesis 27-28

protein and thermic effect of food 144-145
protein degradation and synthesis
amino acids 28
assessment of 28
contractile- and noncontractile-related proteins 27-28
during exercise, findings 28
3-methyhistidine (3-MH), measurement of 28
nitrogen balance 28
proteins in skeletal muscle, classes of 27
protein synthesis and degradation, regulation of
cortisol 43
growth hormone 43, 44
insulin 43
insulin-like growth factors 43
protein degradation 43
protein synthesis 43-44
somatomedins 43
testosterone and testosterone analogs 43, 44
protein utilization
exercise training and 97-98
leucine oxidation 97-98
puberty 113
pulmonary ventilation 52
pyruvate dehydrogenase complex 40. See also fatty acid availability on carbohydrate utilization

R
Randle, Philip 27, 40
Randle cycle 27, 40, 40f
recombinant leptin 155
regulation of energy metabolism
biological control system, overview of 33-35
neural and hormonal control systems 35-38
substrate metabolism regulation during exercise 38, 39-44
regulation of glycolysis and Krebs cycle. See glycolysis and Krebs cycle regulation
regulation of hepatic glucose output. See hepatic glucose output, regulation of
regulation of lipolysis
adipose tissue lipolysis 42
catecholamines 42
growth hormone 42
insulin 42
lipolysis, description of 42
pituitary gland 42
plasma glucose concentration 42
regulation of muscle glycogen degradation. See muscle glycogen degradation, regulation of
regulation of protein synthesis and degradation. See protein synthesis and degradation, regulation of
regulation of substrate metabolism during exercise

fatty acid availability on carbohydrate utilization 40, 40f, 41
glycolysis and Krebs cycle 39-40
hepatic glucose output 41
lipolysis 42
muscle glycogen degradation 39
protein synthesis and degradation 43-44
substrate utilization 38
triglyceride utilization 42
regulation of triglyceride utilization. See triglyceride utilization, regulation of
resistance exercise and energy costs
benefits of 73
circuit weight training 73, 74t
energy expenditure during various physical activities 70t, 73
EPOC 74
resistance exercise in comprehensive exercise program
advantages of 86
blood lipid profile 86
circuit training program 86, 87, 87f
fat-free mass (FFM) 86
lipoproteins 86
strength and power 86
resistance training
adaptations to 102-103
effect on RMR 138
respiratory chamber 52. See also indirect calorimetry
respiratory exchange ratio (RER) 54. See also oxygen deficit
respiratory quotient (RQ) 52, 53t
resting metabolic rate (RMR). See also weight management
anabolic reactions 131
basal metabolic ratio (BMR) 132
concept founder 131
description of 131-132
factors influencing 132t, 134-136, 135f, 135t
influence of exercise on 136-138
measurement of 132-133, 132t, 133t
role in pathogenesis of obesity 139-141
total daily energy expenditure 131
R-R waves of ECG 57-58
Rubner, Max 134, 144
running and walking, energy costs during
body mass, effects of 71
energy cost during walking 69, 70f
energy costs of walking vs. running 69, 70f, 71
energy expenditure during running/walking 69, 70t
running economy 71-72
terrain, effects of 71
running and walking, metabolic calculation of
metabolic equation for various activities 76, 76f
oxygen demands 75

running economy 71-72

S

second messengers 38
SenseWare armband (SWA) 60
serotonin 153
servomechanism 72
sex hormones, effects on substrate metabolism
   adenylate cyclase 108
   corpus luteum 107
   estrogen and progesterone studies 107, 108t
   exogenous estrogen 108
   follicular phase of menstrual cycle 108
   gender differences 107
   lipoprotein lipase (LPL) 107
   luteal phase of menstrual cycle 108
   observations with human subjects 108-109
   placenta 107
short-term exercise adaptations
   citrate synthase 102
   lactate dehydrogenase 102
   muscle hypertrophy 102
   phosphorylase 102
   short-term 102
   studies 101-102
   succinate dehydrogenase 102
   ultra-short-term 101
sibutramine (Meridia; Reductil)
   description of 153, 154
   potential side effects 154-155
   working mechanism of 153, 154, 154f
slow component of oxygen uptake 66, 67f
somatomedins 43
specificity 93
spirometry 51
splanchnic glucose output 22
sport performance, effect of caffeine on 159-160
steady state and homeostasis
   homeostasis, description of 33, 34
   oxygen consumption during exercise 34, 34f
   stable internal environment 33, 34
   steady state, description of 33-34
stroke volume 98
subcutaneous 121
subjective measures for assessing energy expenditure
   activity logs and diaries 60-61
   Compendium of Physical Activity 61-62
   maximal aerobic capacity ($\dot{V}O_2$max) 62
   metabolic equivalent (MET) 61-62

questionnaires 60, 61
submaximal exercise and metabolic responses 98-99
substrate metabolism and gender differences
   fat oxidation 105, 106-107
   performance, men *vs.* women 105, 106t
   sex hormones, effects of 107-109, 108t
substrate metabolism during pregnancy 110
substrate metabolism regulation during exercise
   fatty acid availability on carbohydrate utilization 40, 40f, 41
   glycolysis and Krebs cycle 39-40
   hepatic glucose output 41
   lipolysis 42
   muscle glycogen degradation 39
   protein synthesis and degradation 43-44
   substrate utilization 38
   triglyceride utilization 42
substrate oxidation, measuring. *See* measurement of substrate oxidation
succinate dehydrogenase 102
sympathetic divisions 35
sympathomimetic 154
synthase 126

T

tachycardia 158
terrain effects and energy cost during walking and running 71
testosterone 43, 44
thermic effect of food (TEF)
   carbohydrate, influence of 145, 145f
   fat, influence of 146
   measuring 143-144
   obesity and 149-150
   other factors influencing 146-148, 146t
   and physical activity, interaction between 148-149
   protein, influence of 144-145
   respiratory chamber 52
thermogenesis 132
transamination 14
transducer 56
triglycerides
   energy consumption and 5
   fatty acids and 125, 126t
   oxidative pathway and 13
triglyceride utilization, regulation of
   carnitine palmitoyl transferase I (CPT-1) 42
   malonyl-CoA 42
   regulation of fat utilization 42
type I muscle fibers 95, 96t
type II muscle fibers 95, 96t

U

uncoupling proteins (UCP) 13, 140
units of energy
   calories (cal) 3
   converting calories to joules or kilocalories to kilojoules 3
   joules (J) 3
   kilocalories (kcal or Cal) 3
   kilojoules (kJ) 3
urea 6

V

variable-intensity protocol 85-86
vasodilation 136
vastus lateralis 106
$\dot{V}O_2$max
   cardiac output 98
   description of 98
   enhanced $\dot{V}O_2$max 98
   myocardial contraction 98
   stroke volume 98
$\dot{V}O_2$max reduction and energy expenditure and aging 111, 112
visceral 121

W

walking and running, energy costs during
   body mass, effects of 71
   energy cost during walking 69, 70f
   energy costs of walking *vs.* running 69, 70f, 71
   energy expenditure during running/walking 69, 70t
   running economy 71-72
   terrain, effects of 71
walking and running, metabolic calculation of
   metabolic equation for various activities 75, 76f
   oxygen demands 75
weightlifting. *See* resistance exercise and energy costs
weight management
   average weight loss during 15-week diet and exercise program 88, 88f
   leptin and 155-156
   limitations of exercise alone 87-88, 88f
   negative caloric balance 87
   resting metabolic rate 88
working mechanisms
   α and β receptors 38, 38t
   α and β receptors (adrenergic) 37, 38, 38t
   cortisol and catecholamines 38
   norepinephrine and epinephrine 38
   peptides 38
   second messengers 38

About the Author

Jie Kang, PhD, is a professor in the department of health and exercise science at the College of New Jersey in Ewing, where he directs graduate and undergraduate student research and teaches a variety of exercise science courses. Kang's interest in exercise metabolism originated with his doctoral work at the University of Pittsburgh, where he conducted clinical research dealing with exercise and diabetes. He received his PhD in exercise physiology in 1994.

Kang has published over 50 peer-reviewed articles and has been an invited speaker in various workshops and symposiums discussing topics related to bioenergetics and exercise metabolism, specifically the alterations in energy metabolism, substrate utilization, cardiorespiratory activity, and perceived exertion in response to acute and chronic exercise.

For the past several years, Kang has served as director of the ACSM Health/Fitness Instructor workshop and certification at the College of New Jersey. He is also an ACSM-certified exercise specialist. His work as both a clinician in several clinical, health, and fitness facilities and as a researcher provide opportunities for Kang to transfer bioenergetics theory to practical application.

Kang is also a member and fellow of the American College of Sports Medicine (ACSM) and a past member of the executive committee for the Mid-Atlantic Regional Chapter of the ACSM.

During his free time, Kang enjoys reading, photography, and traveling in addition to physical pursuits of jogging, swimming, and tennis. He and his wife, Julie Ye, live in Medford, New Jersey.

*You'll find
other outstanding
exercise science resources at*

www.HumanKinetics.com

In the U.S. call

1-800-747-4457

Australia..08 8372 0999
Canada 1-800-465-7301
Europe...+44 (0) 113 255 5665
New Zealand.......................................0064 9 448 1207

HUMAN KINETICS
The Information Leader in Physical Activity
P.O. Box 5076 • Champaign, IL 61825-5076 USA